前环衬图片：袁隆平与杂交水稻研究协作组成员一起攻关

Volume
7
Yuan Longping Collection

袁隆平全集

第七卷
学术论文
1966—2010年

Volume 7
Academic Papers
1966—2010

主　编　——柏连阳

执行主编——袁定阳

辛业芸

「十四五」国家重点图书出版规划

湖南科学技术出版社·长沙

本卷编著人员

主　编　辛业芸

出版说明

　　袁隆平先生是我国研究与发展杂交水稻的开创者，也是世界上第一个成功利用水稻杂种优势的科学家，被誉为"杂交水稻之父"。他一生致力于杂交水稻技术的研究、应用与推广，发明"三系法"籼型杂交水稻，成功研究出"两系法"杂交水稻，创建了超级杂交稻技术体系，为我国粮食安全、农业科学发展和世界粮食供给做出杰出贡献。2019年，袁隆平荣获"共和国勋章"荣誉称号。中共中央总书记、国家主席、中央军委主席习近平高度肯定袁隆平同志为我国粮食安全、农业科技创新、世界粮食发展做出的重大贡献，并要求广大党员、干部和科技工作者向袁隆平同志学习。

　　为了弘扬袁隆平先生的科学思想、崇高品德和高尚情操，为了传播袁隆平的科学家精神、积累我国现代科学史的珍贵史料，我社策划、组织出版《袁隆平全集》(以下简称《全集》)。《全集》是袁隆平先生留给我们的巨大科学成果和宝贵精神财富，是他为祖国和世界人民的粮食安全不懈奋斗的历史见证。《全集》出版，有助于读者学习、传承一代科学家胸怀人民、献身科学的精神，具有重要的科学价值和史料价值。

　　《全集》收录了20世纪60年代初期至2021年5月逝世前袁隆平院士出版或发表的学术著作、学术论文，以及许多首次公开整理出版的教案、书信、科研日记等，共分12卷。第一卷至第六卷为学术著作，第七卷、第八卷为学术论文，第九卷、第十卷为教案手稿，第十一卷为书信手稿，第十二卷为科研日记手稿（附大事年表）。学术著作按出版时间的先后为序分卷，学术论文在分类编入各卷之后均按发表时间先后编排；教案手稿按照内容分育种讲稿和作物栽培学讲稿两卷，书信手稿和科研日记手稿分别按写信日期和记录日期先后编排（日记手稿中没有注明记录日期的统一排在末尾）。教案手稿、书信手稿、科研日记手稿三部分，实行原件扫描与电脑录入图文对照并列排版，逐一对应，方便阅读。因时间紧迫、任务繁重，《全集》收入的资料可能不完全，如有遗漏，我们将在机会成熟之时出版续集。

《全集》时间跨度大，各时期的文章在写作形式、编辑出版规范、行政事业机构名称、社会流行语言、学术名词术语以及外文译法等方面都存在差异和变迁，这些都真实反映了不同时代的文化背景和变化轨迹，具有重要史料价值。我们编辑时以保持文稿原貌为基本原则，对作者文章中的观点、表达方式一般都不做改动，只在必要时加注说明。

《全集》第九卷至第十二卷为袁隆平先生珍贵手稿，其中绝大部分是首次与读者见面。第七卷至第八卷为袁隆平先生发表于各期刊的学术论文。第一卷至第六卷收录的学术著作在编入前均已公开出版，第一卷收入的《杂交水稻简明教程（中英对照）》《杂交水稻育种栽培学》由湖南科学技术出版社分别于1985年、1988年出版，第二卷收入的《杂交水稻学》由中国农业出版社于2002年出版，第三卷收入的《耐盐碱水稻育种技术》《盐碱地稻作改良》、第四卷收入的《第三代杂交水稻育种技术》《稻米食味品质研究》由山东科学技术出版社于2019年出版，第五卷收入的《中国杂交水稻发展简史》由天津科学技术出版社于2020年出版，第六卷收入的《超级杂交水稻育种栽培学》由湖南科学技术出版社于2020年出版。谨对兄弟单位在《全集》编写、出版过程中给予的大力支持表示衷心的感谢。湖南杂交水稻研究中心和袁隆平先生的家属，出版前辈熊穆葛、彭少富等对《全集》的编写给予了指导和帮助，在此一并向他们表示诚挚的谢意。

湖南科学技术出版社

总　序

一粒种子，改变世界

一粒种子让"世无饥馑、岁晏余粮"。这是世人对杂交水稻最朴素也是最崇高的褒奖，袁隆平先生领衔培育的杂交水稻不仅填补了中国水稻产量的巨大缺口，也为世界各国提供了重要的粮食支持，使数以亿计的人摆脱了饥饿的威胁，由此，袁隆平被授予"共和国勋章"，他在国际上还被誉为"杂交水稻之父"。

从杂交水稻三系配套成功，到两系法杂交水稻，再到第三代杂交水稻、耐盐碱水稻，袁隆平先生及其团队不断改良"这粒种子"，直至改变世界。走过91年光辉岁月的袁隆平先生虽然已经离开了我们，但他留下的学术著作、学术论文、科研日记和教案、书信都是宝贵的财富。1988年4月，袁隆平先生第一本学术著作《杂交水稻育种栽培学》由湖南科学技术出版社出版，近几十年来，先生在湖南科学技术出版社陆续出版了多部学术专著。这次该社将袁隆平先生的毕生累累硕果分门别类，结集出版十二卷本《袁隆平全集》，完整归纳与总结袁隆平先生的科研成果，为我们展现出一位院士立体的、丰富的科研人生，同时，这套书也能为杂交水稻科研道路上的后来者们提供不竭动力源泉，激励青年一代奋发有为，为实现中华民族伟大复兴的中国梦不懈奋斗。

袁隆平先生的人生故事见证时代沧桑巨变。先生出生于20世纪30年代。青少年时期，历经战乱，颠沛流离。在很长一段时期，饥饿像乌云一样笼罩在这片土地上，他胸怀"国之大者"，毅然投身农业，立志与饥饿做斗争，通过农业科技创新，提高粮食产量，让人们吃饱饭。

在改革开放刚刚开始的1978年，我国粮食总产量为3.04亿吨，到1990年就达4.46亿吨，增长率高达46.7%。如此惊人的增长率，杂交水稻功莫大焉。袁隆平先生曾说："我是搞育种的，我觉得人就像一粒种子。要做一粒好的种子，身体、精神、情感都要健康。种子健康了，事业才能够根深叶茂，枝粗果硕。"每一粒种子的成长，都承载着时代的力量，也见证着时代的变迁。袁隆平先生凭借卓越的智慧和毅力，带领团队成功培育出世界上第一代杂交水稻，并将杂交水稻科研水平推向一个又一个不可逾越的高度。1950年我国水稻平均亩产只有141千克，2000年我国超级杂交稻攻关第一期亩产达到700千克，2018年突破1100千克，大幅增长的数据是我们国家年复一年粮食丰收的产量，让中国人的"饭碗"牢牢端在自己手中，"神农"袁隆平也在人们心中矗立成新时代的中国脊梁。

袁隆平先生的科研精神激励我们勇攀高峰。马克思有句名言："在科学的道路上没有平坦的大道，只有不畏劳苦沿着陡峭山路攀登的人，才有希望达到光辉的顶点。"袁隆平先生的杂交水稻研究同样历经波折、千难万难。我国种植水稻的历史已经持续了六千多年，水稻的育种和种植都已经相对成熟和固化，想要突破谈何容易。在经历了无数的失败与挫折、争议与不解、彷徨与等待之后，终于一步一步育种成功，一次一次突破新的记录，面对排山倒海的赞誉和掌声，他却把成功看得云淡风轻。"有人问我，你成功的秘诀是什么？我想我没有什么秘诀，我的体会是在禾田道路上，我有八个字：知识、汗水、灵感、机遇。"

"书本上种不出水稻，电脑上面也种不出水稻"，实践出真知，将论文写在大地上，袁隆平先生的杰出成就不仅仅是科技领域的突破，更是一种精神的象征。他的坚持和毅力，以及对科学事业的无私奉献，都激励着我们每个人追求卓越、追求梦想。他的精神也激励我们每个人继续努力奋斗，为实现中国梦、实现中华民族伟大复兴贡献自己的力量。

袁隆平先生的伟大贡献解决世界粮食危机。世界粮食基金会曾于2004年授予袁隆平先生年度"世界粮食奖"，这是他所获得的众多国际荣誉中的一项。2021年5月

22 日，先生去世的消息牵动着全世界无数人的心，许多国际机构和外国媒体纷纷赞颂袁隆平先生对世界粮食安全的卓越贡献，赞扬他的壮举"成功养活了世界近五分之一人口"。这也是他生前两大梦想"禾下乘凉梦""杂交水稻覆盖全球梦"其中的一个。

一粒种子，改变世界。袁隆平先生和他的科研团队自 1979 年起，在亚洲、非洲、美洲、大洋洲近 70 个国家研究和推广杂交水稻技术，种子出口 50 多个国家和地区，累计为 80 多个发展中国家培训 1.4 万多名专业人才，帮助贫困国家提高粮食产量，改善当地人民的生活条件。目前，杂交水稻已在印度、越南、菲律宾、孟加拉国、巴基斯坦、美国、印度尼西亚、缅甸、巴西、马达加斯加等国家大面积推广，种植超 800 万公顷，年增产粮食 1600 万吨，可以多养活 4000 万至 5000 万人，杂交水稻为世界农业科学发展、为全球粮食供给、为人类解决粮食安全问题做出了杰出贡献，袁隆平先生的壮举，让世界各国看到了中国人的智慧与担当。

喜看稻菽千重浪，遍地英雄下夕烟。2023 年是中国攻克杂交水稻难关五十周年。五十年来，以袁隆平先生为代表的中国科学家群体用他们的集体智慧、个人才华为中国也为世界科技发展做出了卓越贡献。在这一年，我们出版《袁隆平全集》，这套书呈现了中国杂交水稻的求索与发展之路，记录了中国杂交水稻的成长与进步之途，是中国科学家探索创新的一座丰碑，也是中国科研成果的巨大收获，更是中国科学家精神的伟大结晶，总结了中国经验，回顾了中国道路，彰显了中国力量。我们相信，这套书必将给中国读者带来心灵震撼和精神洗礼，也能够给世界读者带去中国文化和情感共鸣。

预祝《袁隆平全集》在全球一纸风行。

刘旭，著名作物种质资源学家，主要从事作物种质资源研究。2009 年当选中国工程院院士，十三届全国政协常务委员，曾任中国工程院党组成员、副院长，中国农业科学院党组成员、副院长。

凡　例

1.《袁隆平全集》收录袁隆平 20 世纪 60 年代初到 2021 年 5 月出版或发表的学术著作、学术论文,以及首次公开整理出版的教案、书信、科研日记等,共分 12 卷。本书具有文献价值,文字内容尽量照原样录入。

2. 学术著作按出版时间先后顺序分卷;学术论文按发表时间先后编排;书信按落款时间先后编排;科研日记按记录日期先后编排,不能确定记录日期的 4 篇日记排在末尾。

3. 第七卷、第八卷收录的论文,发表时间跨度大,发表的期刊不同,当时编辑处理体例也不统一,编入本《全集》时体例、层次、图表及参考文献等均遵照论文发表的原刊排录,不作改动。

4. 第十一卷目录,由编者按照"×年×月×日写给××的信"的格式编写;第十二卷目录,由编者根据日记内容概括其要点编写。

5. 文稿中原有注释均照旧排印。编者对文稿某处作说明,一般采用页下注形式。作者原有页下注以"※"形式标注,编者所加页下注以带圈数字形式标注。

7. 第七卷、第八卷收录的学术论文,作者名上标有"#"者表示该作者对该论文有同等贡献,标有"*"者表示该作者为该论文的通讯作者。对于已经废止的非法定计量单位如亩、平方寸、寸、厘、斤等,在每卷第一次出现时以页下注的形式标注。

8. 第一卷至第八卷中的数字用法一般按中华人民共和国国家标准《出版物上数字

用法的规定》执行，第九卷至第十二卷为手稿，数字用法按手稿原样照录。第九卷至第十二卷手稿中个别标题序号的错误，按手稿原样照录，不做修改。日期统一修改为"××××年××月××日"格式，如"85—88年"改为"1985—1988年""12.26"改为"12月26日"。

9.第九卷至第十二卷的教案、书信、科研日记均有手稿，编者将手稿扫描处理为图片排入，并对应录入文字，对手稿中一些不规范的文字和符号，酌情修改或保留。如"弗"在表示费用时直接修改为"费"；如"∴"表示"所以"，予以保留。

10.原稿错别字用〔〕在相应文字后标出正解，如"付信件"改为"付〔附〕信件"；同一错别字多次出现，第一次之后直接修改，不一一注明，避免影响阅读。

11.有的教案或日记有残缺，编者加注说明。有缺字漏字，在相应位置使用〔〕补充，如"无融生殖"修改为"无融〔合〕生殖"；无法识别的文字以"□"代替。

12.某些病句，某些不规范的文字使用，只要不影响阅读，均照原稿排录。如"其它""机率""2百90""三～四年内""过P酸Ca"及"做""作"的使用，等等。

13.第十一卷中，英文书信翻译成中文，以便阅读。部分书信手稿为袁隆平所拟初稿，并非最终寄出的书信。

14.第十二卷中，手稿上有许多下划线。标题下划线在录入时删除，其余下划线均照录，有利于版式悦目。

目录

02

04

水稻的雄性不孕性

　　水稻具有杂种优势现象，尤以籼粳杂种更为突出[1]，但因人工杂交制种困难，到现在为止尚未能利用。显然，要想利用水稻的杂种优势，首先必须解决大量生产杂种的制种技术，从农作物杂种优势育种的研究趋势和实际成果来看，解决这个问题的有效途径，首推利用雄性不孕性。

　　为了获得水稻的雄性不孕材料，我们最近两年在水稻大田里进行了逐穗检查工作，观察到一些雄性不孕植株，现将初步观察结果报道如下。

方法和经过

　　水稻雄性不孕植株，是 1964—1965 年在湖南省安江农校实习农场及附近生产队的水稻大田中检查出来的。已知花药不开裂是许多作物的雄性不孕性在外表上的共同特征之一，因此就根据这个特征按图索骥，于抽穗期间晴日中午前后，在田间进行逐穴逐穗检查，将注意力集中到正在开花和刚开过花的稻穗花药上。正常植株的颖花，刚开花时，花药膨松，颜色鲜黄，用手轻振便有大量花粉散出；开花后不久，花药即已裂开，药囊变空，呈白色薄膜状挂在花丝上。在检查时，发现有开花后花药不开裂、振动亦不散粉的稻穗，再用 5 倍放大镜进一步检视，确证为花药不开裂的，就视作雄性不孕植株，加以标记，2~3 日内复查几次，并采集花药进行显微镜检验，用碘化钾液染色法观察花粉反应。前后总共检查了 14 000 余穗，在 4 个品种中共找出 6 株雄性不孕植株，成熟时分株采收自然传粉种子（个别的作了人工杂交）。

　　为了加速鉴定和选育过程，凡成熟早的，在当年就将其部分种子进行"翻秋"播种，其余的则在次年春播，均采用盆钵育苗，分系单本移栽，每个系统种植一小区，其旁种一行同品种的正常植株作对照。于抽穗期间进行逐株观察记载，用花粉染色法和套袋自交的结实率鉴定孕性程度。

观察结果

　　检查出的 6 株雄性不孕植株，按表现不同，可分以下三种类型。

（1）无花粉型　花药较小而瘦瘪（图1），白色，全部不开裂，其内不含花粉或仅有很少量极细的颗粒，为完全雄性不孕（2株，1965年7月自品种胜利籼中找出）。

正常的稻穗（左）；雄性不孕的稻穗（无花粉型）（右）。
图1　开花后2小时花药的状态

（2）花粉败育型　花药细小，黄白色，全部不开裂；花粉数量少且发育不完全，较正常的显著为小，大多数形状不规则，表面有皱纹，对碘化钾液无蓝色反应，为完全雄性不孕（2株，分别在1964年和1965年6月自南特号中找出）。

（3）部分雄性不孕　其中又有两种情况：①大多数颖花（80%以上）的花药不开裂，但花药和花粉的其他性状皆与正常植株的相同（1株，1964年7月自早粳4号中找出）；②大多数颖花（85%以上）的花药不开裂，这些不开裂花药中的花粉数量较少且大部分发育不完全，仅个别花粉对碘化钾液有蓝色反应（1株，1964年自洞庭早籼中找出）。

此外，还检查到一些非遗传的部分雄性不孕植株。

表1是对前两种完全雄性不孕类型的花药和花粉大小测定的结果。

表1　水稻完全雄性不孕植株的花药和花粉大小与正常植株的比较

单位：mm

株号	孕性类型	花药		花粉直径 *
		长 $\bar{x}$+s	宽 $\bar{x}$+s	$\bar{x}$+s
1	花粉败育型	1.0±0.32	0.32±0.031	50±5.0
对照	正常	1.9±0.11	0.42±0.025	52±4.8
5	无花粉型	1.8±0.19	0.27±0.021	—
对照	正常	2.1±0.11	0.40±0.015	—

注：* 败育花粉测算的是其长度。

　　所有上述雄性不孕植株，其他方面的形态特征，包括雌蕊在内，与同品种的正常植株看不出差异，唯在自然传粉情况下的结实率皆偏低，最高的一株为 70%，其余为 40%～50%，其原因可能与抽穗扬花期间均遇到过连续几日阴雨或大风而不利异花传粉有关。

　　截至目前，已有 4 株繁殖了 1～2 代，兹将该 4 株自然传粉（大田中同品种的花粉）F_1 代的孕性分离情况列于表 2。

表 2　水稻雄性不孕植株自然传粉种子后代的孕性分离情况

株号	品种	不孕类型	播种年月	F_1 总株数	完全雄性不孕株数	部分雄性不孕株数	正常孕株数
1	南特号	花粉败育	1964 年 7 月	21	2	16	3
			1965 年 4 月	62	5	39	18
			合计	83	7	55	21
2	洞庭早籼	部分不孕	1964 年 7 月	12	0	3	9
			1965 年 4 月	14	0	2	12
			合计	26	0	5	21
3	早粳 4 号	部分不孕	1965 年 3 月	39	0	3	36
4	南特号	花粉败育	1965 年 7 月	16	4	7	5

　　不同类型雄性不孕原始植株后代分离的情况有所不同。两棵花粉败育型完全不孕植株，都分离出完全雄性不孕、部分雄性不孕和正常孕性后代。这些完全不孕的 F_1，表现与母本相同，套袋隔离穗的结实率为 0，但人工授粉杂交的结实率很高（表 3），15 个杂交组合中有 3 个组合达 90% 以上，只是亲缘关系较远的 2 个籼粳杂交组合在 50% 以下，这表明它们不具有雌性不孕性，分离出的部分雄性不孕植株表现为：部分颖花（不同植株的百分率不同）的花药开裂散粉，花粉发育正常；不开裂花药中的花粉绝大多数是败育的。

表 3　败育型完全雄性不孕植株人工授粉杂交的结实情况

父本	授粉花数	结实粒数	结实率 /%
南选 1 号	56	54	81.8
南选 2 号	16	12	75.0
南选 3 号	40	36	90.0
南选 5 号	38	31	81.5

续表

父本	授粉花数	结实粒数	结实率 /%
南选 6 号	31	29	93.5
南 R 1 号	124	101	81.4
矮脚南特号	82	59	71.9
南陆矮	92	67	72.8
陆才号	24	17	70.8
洞庭早籼	83	60	72.2
西湖早	65	59	90.7
南京 1 号	20	17	85.0
矮南选系	48	43	83.3
元子 2 号（粳）	70	35	50.0
早粳 16 号	68	26	38.2
合计	867	643	74.1

第 2、第 3 号部分雄性不孕原始植株都只出现部分不孕和正常孕的后代，这些部分不孕的 F_1 在花药、花粉特性上各自与其母本相同，但孕性程度个体间有差异。

两棵无花粉型的雄性不孕植株，目前虽未获得种子后代，但刈割后约一月稻椿上所生出的再生穗，仍保持着不能形成花粉的完全雄性不孕特性，人工授粉杂交的结实率亦很高，5 个组合均在 80% 以上。

讨 论

雄性不孕在许多作物中常有发现，但水稻的雄性不孕性在国内尚未见有报道。从这次初步检查结果来看，我们接连两年都能在一般大田中找到雄性不孕植株，既包括籼稻和粳稻品种，又有几种不同的不孕类型，这表明水稻的雄性不孕在自然情况下出现的频率较高，按这次调查估计，约 0.13%。从外部表现看，水稻与其他作物的雄性不孕现象是大同小异的，主要表现在花药不开裂、花粉败育或不能形成花粉等方面，但我们却未找到那种花药不外露及雄蕊完全退化的类型。值得指出的是，正因这些雄性不孕水稻植株的花药伸出颖外，开花后数日内又大多不脱落，由于萎缩而显得更加细瘦，所以用肉眼观察也很易识别出来，只要根据这个特点耐心仔细逐穗检查，就不难在现有水稻品种中找到雄性不孕材料。

雄性不孕性在遗传上一般分核质型和胞质型两类，以后者在杂交优势育种中最有利用价值。

Roades 的研究[2] 指出的是，由他所发现的第一个胞质型的玉米雄性不孕植株，当用某一父本类型与之多次重复杂交，后代仍能保持雄性不孕；而在自由传粉的某些情况下，则产生孕性分离的后代。我们所获得的花粉败育型水稻雄性不孕材料（其余 2 种类型现在还不能作推论）与后一情况类似，同时按雄性不孕的一般遗传规律，核质型的多属隐性，通常只有由母本细胞质决定的雄性不孕性才能在 F_1 中重复显现，因此初步认为，它们属于胞质型的可能性较大。由于南特号品种不纯，在群体中存在着多种遗传型的个体，因而在自由传粉的情况下，不同父本核因子与不孕母本细胞质相互作用的特异性，致使 F_1 发生不同的孕性分离现象。由此认为，通过进一步选育，可从中获得雄性不孕系、保持系及恢复系，用作水稻杂种优势育种的材料。

致谢：杨运春、尹华奇、潘立生等同志参加杂交工作，谨此致谢。

───────────── References ─────────────

参考文献

[1] 杨守仁等，农业学报，10（4），1959.　　｜　[2] CmHpHoB B.Γ.，农业译丛，（9），1963.

作者：袁隆平

注：本文发表于《科学通报》1966 年第 4 期。

利用野败育成水稻三系的情况汇报

　　水稻雄性不育研究工作，在我国开展虽晚，但进展较快，尤其是近年来实行全国大协作，大搞群众运动，充分利用我国优越的自然条件，南繁北育，一年三代，因此，在短短的几年时间内，就取得了较大的进展。我省以野败（花粉败育型雄性不育野生稻的简称）为主的研究材料，目前基本上实现了三系配套，同时还初步鉴定出几个具有一定杂种优势、比高产亲本和优良品种都有所增产的杂交组合，进入了小量制种、品比试验和配合力研究阶段。这一科研成果的取得，是毛主席革命路线的胜利，又一次显示了"中国人民有志气，有能力，一定要在不远的将来，赶上和超过世界先进水平"。现将利用野败选育三系的情况分述如下。

一、野败及其第一代杂种的特征特性

　　野败是黔阳地区农校水稻不育系研究小组 1970 年 11 月在海南崖县发现的。它的株型匍匐，分蘖力很强，叶片窄，茎秆细，谷粒小，芒长而红，极易落粒，叶鞘和谷尖紫红色，柱头发达外露；对日照反应敏感，为典型的短日照植物。总之，除雄性不育性外，其他性状与海南的普通野生稻（*O. sativaL. f. Spontanea*）相似。野败原始株的花药瘦小，淡黄色，不开裂，内含典型的败育状花粉。但不育性不够稳定，当气温超过 30 ℃时，数天之后就有少部分花药形成少量染色花粉，并开裂散粉。

　　1971 年，用籼、粳栽培稻品种为父本，与野败作了一些杂交组合，同年冬在海南崖县观察了 10 个杂交组合的子一代共 70 株的育性和某些其他性状。大多数组合的子一代出现育性分离，10 个组合的雄性不育株总数超过可育株，完全不育株率达 41%，表明野败的雄性不育性能通过杂交遗传给后代。但杂种不育株的花药形态与原始株不同，变得空瘪细瘦，呈水浸状乳白色或油浸状乳黄色，花药中花粉全部败育，不育性优于原始株，在海南 4—5 月高温强光照条件下未见到育性恢复。部分不育株占总数的 40%，有两种情况：一是高度不育，每株当中只有很少数颖花里面有一两个正常花药；一是半不育，花药大多呈黄色，一般有

20%～30%的正常花粉，但开裂不良；雄性正常株占总数的18%，花药花粉发育正常，但结实率大多在50%以下，结实正常的只有三株，其中一株（野败×6044）结实率达90%以上，这表明具有野败细胞质的某些杂种，其育性可以达到完全正常，预示有选育恢复系的可能性。

在对子一代不育株的杂交过程中，发现它们还存在着雌育性的分离现象。例如野败×6044子一代四株中，一株雌性正常，两株部分不育，一株高度不育。雌性正常株人工杂交获得完全结实，高度不育株人工杂交的结实率低于10%，而且开花不正常，一穗中只有极少数张颖开花，但雌蕊在形态上无异常现象。

在鉴定育性的同时，对子一代杂种的其他性状也作了一般的观察。野败×栽培稻子一代的几个明显性状都有分离，与一般杂种第一代的一致性现象不相符。例如，水稻各部位的紫色，属显性遗传。野败的叶鞘、叶环、谷尖和柱头都呈紫红色，用这些部位都是无色的6044、广矮3784与它杂交，子一代却出现了有色和无色的植株。此外，株型，穗形，谷粒等方面也发生了较大的分离。这种异常现象涉及野败的起源问题。1972年对野败原始株作了细胞学方面的观察，发现其花粉母细胞在减数分裂时染色体行为异常，类似一般的远缘杂种。把这个观察同杂种第一代在雌雄育性上以及其他性状上均有分离现象联系起来分析，可以初步推断，野败原始株是一个杂合体，很可能是普通野生稻与当地晚熟栽培品种的天然杂种。

二、不育系及其保持系的选育

利用野败选育不育系，实质上是一个核置换过程，即把具有保持力的父本的细胞核，通过杂交和连续回交，转到野败的细胞质中，取代原来的细胞核，这个置换过程一完成，则不育系及其相应的保持系就告育成。根据我们的试验，以野败原始株作母本的，到回交四代或五代，可获得与父本性状基本上一致的不育系，若用已建成的不育系来转育新的不育系，只需要回交二至三代。选育程序可分两个步骤：第一，广泛测交，以筛选保持力良好的各类品种；第二，择优回交，以加速核置换过程。

据不完全统计，在籼稻方面，到1974年夏季止，我们前后共测品种731个，其中保持力较好的624个，占测交品种的85.3%；有恢复力的89个，占12.2%；其余为部分保持的，占2.5%，例如6044、广矮3784、意大利B等，连续回交三至四代，育性仍不稳定，各代都有育性分离，以高不育和半不育株居多数。对于保持力良好的组合，择优回交选育不育系，也就是在杂种中选择雄性完全不育、性状倾向父本和开花正常的植株作母本，例如，以二九南一号和71-72为父本，按此标准与野败连续回交三代，群体分别为3 000多株和

7 000 多株基本上能保持完全不育，在形态上已与父本一致，初步育成了这两个品种同型不育系。1974 年夏，这两个不育系已进入回交七代，育性稳定。现以二九南一号不育系为例，图示其选育过程如下（图 1）：

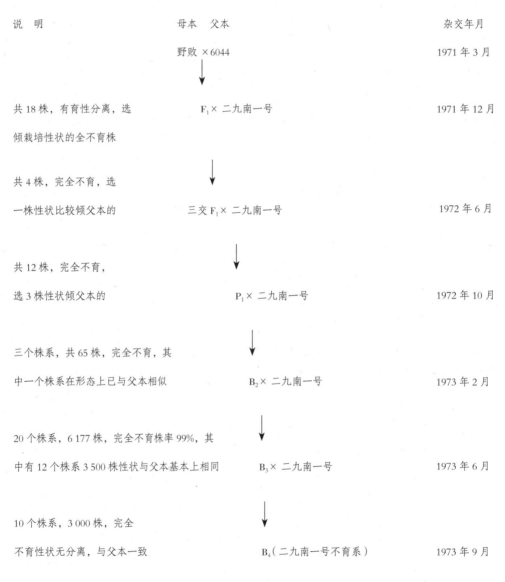

说　明	母本　父本	杂交年月
	野败 ×6044	1971 年 3 月
共 18 株，有育性分离，选倾栽培性状的全不育株	F₁× 二九南一号	1971 年 12 月
共 4 株，完全不育，选一株性状比较倾父本的	三交 F₁× 二九南一号	1972 年 6 月
共 12 株，完全不育，选 3 株性状倾父本的	P₁× 二九南一号	1972 年 10 月
三个株系，共 65 株，完全不育，其中一个株系在形态上已与父本相似	B₂× 二九南一号	1973 年 2 月
20 个株系，6 177 株，完全不育株率 99%，其中有 12 个株系 3 500 株性状与父本基本上相同	B₃× 二九南一号	1973 年 6 月
10 个株系，3 000 株，完全不育性状无分离，与父本一致	B₄（二九南一号不育系）	1973 年 9 月

图 1　二九南一号不育系选育过程

到目前已转育成近 20 个籼稻品种的同型不育系，如 V20、温革 10 号、玻璃占矮、革命一号等。这些不育系的育性稳定，受环境条件的影响很少，雌性正常，其中多数开花正常，柱头发达外露或半外露，有利于制种。1973 年夏，二九南一号 A 与父本以 2∶2 的比例种植，虽然父本早齐穗 3~4 天，花期相遇不好，但在 3 500 株不育株上仍收到 39 斤[①] 种子，折合亩产 110 斤。矮秆野败不育系有不同程度的卡颈现象，一般有五分之一到四分之一长度的稻穗包在剑叶鞘里抽不出来，这是妨碍提高制种产量的一大缺点。不育系的抽穗期比同型保持系要迟 3~5 天，在繁殖和制种时应加注意。此外有少数不育系花时不正，每天的开花时间比正常品种迟，开花数也较少，如 71-72；极个别的全不开花，如 IR880。这在转育过程中尤须注意。

在粳稻方面，几年来共测品种 345 个，除意大利 12 属部分恢复外，其余都是保持的，没有恢复力。大多数组合的不育株开花很少或不开花，有的花时不正。例如，野败 × 京引 66，直到回交五代仍开花不正常，与父本相间种植，自然传粉结实率仅 1.2%。因此，对于选育粳型野败不育系，应特别着重筛选开花正常的组合。到现在我们仅有京引 177 等几个不育系开花接近正常，唯有 "雷卓脱" 不育系开花最好，自然传粉结实率 20% 以上。

三、恢复系的选育

我们主要采用广泛测交的方法来筛选恢复系，在籼稻中测出一些品种对野败不育系具有恢复的能力，根据恢复的程度一般可分三类。

1. 杂种的结实率超亲，高于最好的亲本，或在正常条件下为 90% 以上。这一类品种最少，如黄壳油占等。

2. 杂种的结实率与父本相当，在正常条件下的结实率为 70%~89%。这一类品种有 20 多个，如 IR24 等。

3. 杂种的结实率低于最低的亲本，在 70% 以下。这一类品种为数最多，如莲塘早等。

此外，还有一类可暂称作专一恢复系。例如，二九青一号只能对 71-72A 起正常恢复作用，而对其他品种的不育系则恢复不良。这种现象是否与父母本的亲和力有关或是其他的原因，有待研究。

根据初步观察，恢复系在地理分布和生育期上存在着一定的规律性，以起源于低纬度的晚籼品种及其衍生系居多，次为中迟熟品种，纬度较高的早籼品种最少（表 1）。

1974 年，我们用 IR24、IR661 同 7 个不育系配制了 10 个杂交组合，除三个组合的杂种群体为 600 株外，其余都在 1 000 株以上，其中二九南一号 A×IR24 的种植面积为 1.2 亩[②]。各组合的杂种结实率基本正常，性状比较一致，并且表现了不同程度的优势，最高的杂

---

① 1 斤= 500 g，后同。
② 1 亩≈ 666.7 m²，后同。

种小区产量比高产亲本 IR24、IR661 分别增产 19.5%、33.1%。

表1　籼稻测交品种统计

品种来源	测交品种	保持品种	恢复品种			
			总数	71% 以上	30% ~ 70%	30% 以下
长江流域早、中稻	470	438	32	2	19	11
华南早、中稻	67	59	8	2	5	1
华南晚籼	7	3	4	3	1	
西南籼稻	16	13	3	1	2	
东南亚（包括印巴次大陆[①]）	145	107	38	15	18	5
美洲	15	12	3	1	2	
欧洲	7	6	1		1	
非洲	4	4				
合计	731	642	89	24	48	17

四、关于选育新的恢复系

在已测过的品种中，过硬的野败恢复系很少，而且这些为数很少的恢复系已属迟熟类型，因此在选配优良组合上和在生产应用上都受到限制。解决这个问题的途径，一方面是要多建立一些不同类型和品种的野败不育系，以增加不育系的个数来弥补恢复系少的缺点，扩大选择优良组合的机会；另一方面，是要积极选育新的优良恢复系，其主要途径如下所述。

1. 继续测交筛选。重点放在以 IR24 等恢复系作亲本（无论杂交的或引变的）而育成的优良新品系上（常规育种），初步看来，其效果比一般的筛选较好。当然，外引的特别是来自低纬度的如 IR 系统等材料，亦是主要筛测对象。

2. 转育，即把恢复因子转移到所需要的品种上。有多种方法，我们采用的主要有两种方法。

（1）从杂种后代分离群体中按系选法选育。不育系 × 恢复系的杂种自交后代，往往发生包括育性在内的多种多样的分离，其中常会出现一些符合需要的优良全育株。按照胞质-胞核遗传理论，这种全育株的细胞质既然是不育的，其细胞核里必然会有显性恢复因子。按系选

---

① 印巴次大陆：一般指印度半岛。

法连续选择几代，淘汰那些出现育性分离和性状不良的株系，当各株系整齐一致特别是没有育性分离时，再同不育系进行测交，从中就可能选出新恢复系。例如在朝阳一号 A× 科字 6 号的子二代中，选择了一株结实正常、早熟、性状较好的单株，到自交三代育性和性状仍有分离，我们从中选择两个结实好的单株搞测交，以后每代都在恢复力较好的和没有育性分离的株系中选 1~2 个单株测交，到 1974 年夏（自交五代），有一对测交，共 12 株杂种，其中 4 株结实 80% 以上，其余株为 50%～70%，比上二、三代的恢复力大有提高。虽然我们用此法尚未育成恢复系（原因可能在于父本科字 6 号是一个弱恢复系），但从其他作物已有成功的先例来看，这个方法是可取的。

（2）选育不育系的同型恢复系。用子一代恢复株对原不育系搞反回交，以后再用性状倾母本的杂种全育株与原不育系杂交，反回交三四代后再自交一代，即可通过测交而获得与原不育系同型的恢复系。

例如，我们为了选育粳型野败的恢复系，1972 年用柳州野生稻 × 科字 5 号的杂种作父本同京引 66A 杂交，子一代杂种少量结实，但花粉正常，表明柳州野生稻（或科字 5 号）含有恢复因子，于是便用杂种作父本再同京引 66A 杂交（反回交）。反回交一代 6 株，性状倾粳，均半结实，从中选两株结实较多、性状倾粳的，让其自交两代。各代自交后代都有育性分离，多数为不育和半不育，但也有少数个体结实比较正常。1974 年春在自交二代中选了四个结实较好的单株同几个粳稻不育系作了 4 对测交，根据目前的观察，其中一对，父本无育性分离，并有一定的恢复力。29 株杂种，前期在高温下抽穗的结实不良（结实率 20%～60%），但 8 月下旬晚抽的分蘖穗结实基本正常。这个材料能否育成恢复系，还要看以后的工作，但是，作为一种转育方法，看来是可行的。

作者：湖南省水稻杂种优势研究协作组

注：本文发表于《湖南农业科学》1974 年第 4 期。

杂交水稻制种和高产的关键技术

一、杂交水稻制种技术

杂交水稻的制种，是古人没有搞过，洋人也没有搞过的技术课题。有人曾经很武断地预言："水稻是自花传粉作物，开花时间短，柱头小，异花传粉的结实率很低，制种关肯定过不了，即使水稻的杂种优势再强，也不可能在生产上应用。"

但是，我们依靠群众的智慧和力量，遵照毛主席关于"实践、认识、再实践、再认识"的教导，经过反复的摸索试验，终于突破了制种关，现在已形成了一套比较完整的制种技术，并且在经济上完全达到了实用阶段。亩产由最初的 10 多斤，逐渐提高到现在大面积可过百斤的水平，最高亩产已超过 350 斤。南优 2 号的制种经验概括起来是以下八句话：

（一）育好秧苗是前提

稀播育分蘖秧，适龄插秧，特别是不育系不能插老秧，同时父本秧龄过长会延迟抽穗期，影响花期相遇。

（二）长好苗架是基础

产量是穗数 × 穗粒数 × 结实率，有了较多的穗数和较大穗头，才能有较高的产量，否则，即使花期相遇好、结实率高，产量也是不高的。

行比是 1∶4 的情况下，父本每株要有 15 个以上的穗，母本要有 8～10 个有效穗；1∶6 时父本要有 20 个穗，如果有 25 个穗，可扩大到 1∶8。

（三）安全抽穗是保证

秋季制种要求 9 月上旬抽穗，9 月中旬散完花粉，5 月末至 6 月上旬播种父本；夏季制种要求 7 月 20 日前齐穗或 8 月中旬始穗，以避过我省 7 月下旬至 8 月上旬的高温。前者要求提早播种，保温育秧（而且只能在湘南），后者父本播期要推迟到 5 月中旬。

海南春季制种，3 月中旬抽穗比较安全，父本应在 11 月下旬播种。

（四）花期相遇是关键

1. 父母本播差期应以叶龄而不是以天数为准，春夏制种 9.5～10.1 叶、秋季制种 10.1 叶播母本。

2. 花期预测。以幼穗分化进度来预测较准确，前三期父本早于母本，中三期父母本处在同一时期，后二期母本早于父本，则花期相遇较好。

3. 花期调整措施。前期偏施氮肥，后期旱控水促，效果很好。此外，对父本还可采用断根、割叶的办法（但不宜用于母本，以免产生副作用）。喷"九二〇"[①] 也有作用。

（五）割叶剥苞夺高产

为使穗部裸露在外面接受花粉，分别在始穗期和穗抽出 1/2 时进行割叶、剥苞，提高结实率。

（六）适时喷射"九二〇"

可减轻母本卡颈和促进提早开花，喷一次的，要重点打分蘖穗，15～20 单位，1 克可喷 2 亩左右。

（七）辅助授粉要加强

首先要适时，即母本未开花时不能赶，刮三级以上大风不赶，大量开花时抓紧赶，每日至少三次。

（八）保、杂、劣株要除尽

对于隔离保纯问题，我们的意见是，空间隔离 50 米，时间隔离 20 天。

二、杂交水稻栽培技术特点

（一）培育多蘖壮秧

它的好处是：①穗大而整齐，分蘖穗与主穗的差异小；②返青快，分蘖早，始穗到齐穗的时间短，可提早成熟期；③以蘖代苗，节省秧子。据观察，每亩播 30 斤即 1 平方寸[②] 一粒种子，秧龄 25 天，叶龄 8 叶时，秧苗分蘖便开始死亡，亩播 20 斤即 1.5 平方寸一粒种子，秧龄 32 天，叶龄到 9.5 叶时分蘖秧苗为 100%，没有分蘖死亡现象。其中三权以上的秧苗占 70%（我们的观察因故到 32 天中断了）。估计达 40 天秧龄时，要保证 70% 左右的三权秧，播种量还应减少，并且最好采用宽行窄株条播匀播。

---

① "九二〇"：一般指赤霉素，后同。
② 1 平方寸 ≈ 11.11 平方厘米。

（二）密度和穗数

关键在于要有一个合理的高产穗数指标，有了这个指标就可以因地制宜，根据肥料水平、栽培季节来确定合理的插秧密度和苗数，再通过科学管理来达到这个指标，使之不过多也不过少，绝不能不看条件机械地规定插秧密度。

根据我们去年试种的情况，南优 2 号作中稻，在高肥栽培条件下，每亩过千斤产量的穗数为 11 万~19 万穗，以约 15 万穗的产量最高，因而有些人提出了杂交水稻要以大穗取胜和采取稀植的主张。现在看来，这个主张有片面性，因为在大面积条件下的施肥水平不高，苗架不会长得那么高，穗头也不会那么大，在单位面积内就能容纳较多的穗数，而 15 万穗就不会有较高产量。如桂东县的大水公社，南优 3 号苗架只有 80~90 厘米高，每穗一般只有 140~150 粒，而省农科院的则分别有 110 厘米和近 200 粒。每亩穗数虽差不多，都在 14 万~15 万穗，但产量则相差很大。由此可见，要针对不同的条件，提出不同的合理穗数指标。我们的初步看法是，在条件好肥料足的地方，作中稻每亩仍以 15 万~16 万穗，插 1.0 万~1.5 万穴为宜。在此基础上力争大穗，产量可以达到 1 500 斤左右。如桂东县农科所的丰产田每亩 15.1 万穗，亩产 1 486 斤。在条件较差，施肥水平较低的地方，穗数指标可提高到约 18 万穗，插 1.5 万~2 万穴，穗头虽然小一些，但亩产也可以超千斤。

作晚稻时，穗数指标还应提高，每亩要达到 20 万穗左右，才能获得较高的产量，插秧龄密度 2 万穴左右。

（三）合理施肥

亩产千斤以上的杂交水稻，究竟需要施多少肥料，氮、磷、钾哪种比例较合适？对此，省农科院和上海植物生理所等许多单位正在研究，现只能提一些大概的看法。过去有人认为，杂交水稻的根系发达，吸肥力强，必须重施肥料才能高产。实践证明，这种看法是片面的。杂交水稻虽然需要一定的肥力水平，但决不是施肥越多越好，不适当地施用太多的肥料，特别是氮素化肥，不但经济上不合算，而且增产效果不明显，有时甚至造成减产。例如，桂东县农科所，去年丰产田亩施纯氮达 50 斤，其中一次尿素追肥就有 50 斤，结果纹枯病严重，千粒重、结实率下降，亩产只有 1 100 多斤。今年在同一块试验田里，施氮 29 斤，亩产达 1 486 斤。综合一些地方的试验，施纯氮 30 斤的，亩产可达 1 500 斤，施氮 20 斤的，亩产可达千斤水平。这表明，杂交水稻对肥料的利用率要高于常规品种。

在施肥方法上，也是以有机肥为主，化肥为辅，基肥为主，追肥为辅，并要提高磷、钾肥的比重。追肥要早，前重后轻，后期要防止追肥过多，以免造成贪青、病害和降低结实率。对双季晚稻的追肥要采取"一哄而起"的办法。

（四）怎样提高结实率

现有杂交水稻的空壳率一般较常规品种高，这是它的一个缺点，也是一个潜力。在很多情况下，结实率是高产的决定性因子。"外因通过内因而起作用"，杂交水稻空壳率高的原因，从外因看是不良外界条件的影响，而其内因则是现有恢复系的恢复力还不够强。其机制在于花药中的弹丝（一种促使花药开裂的机械组织）不发达，花药开裂散粉迟缓，以致有一些柱头接收不到花粉。在正常条件下，普通品种开花后五分钟左右花药便开裂散粉，而南优 2 号开裂散粉的时间迟得多，一般在开花后的半小时左右。如果天气不良，遇上高温、低温、阴雨以及因施氮肥过多造成田间阴蔽，则开裂时间更要推迟到开花后一小时甚至两小时以后，而水稻颖花开放的持续时间一般只有 50～60 分钟。因此，当天气不好时，便有很多的柱头接收不到花粉，形成大量空壳。据省气象台观察，开花时最高气温超过 35 ℃或低于 25 ℃结实率便显著下降。

为提高结实率，应采取如下措施：

1. 保证在安全期抽穗扬花，避过高温或低温。中稻早播早插，争取 7 月 20 日以前齐穗，或推迟播种，把抽穗期移到 8 月中下旬（海拔 700 米以上的高寒山区例外）。

2. 增施磷、钾肥，控制氮肥施用量，可使花药的机械组织即弹丝发育较好。

3. 宽行窄株和排水晒田，以降低田间湿度，使田间更通风透光，促使花药开裂。

4. 在有条件的地方，可试行人工辅助授粉。据我们初步观察，辅助授粉有明显效果。

作者：袁隆平

注：本文发表于《遗传与育种》1977 年第 1 期。

杂交水稻培育的实践和理论

　　利用杂种优势以大幅度提高农作物产量，是现代农业科学技术的突出成就之一。由于植物雄性不育性的发现和利用，从而使不少两性花植物，如高粱、向日葵、甜菜等的杂种优势也能广泛应用于生产。近年来，我国的杂交水稻已取得了重大的突破，为大幅度提高水稻产量开创了一条有效的途径。

　　杂交水稻的成功，也为发展遗传育种学的实践和理论提供了新的内容，概括起来有三个方面：一是丰富了雄性不育和"三系"关系的遗传理论；二是否定了稻、麦等自花授粉作物没有杂交优势的旧理论；三是给某些其他自花授粉作物的制种技术提供了良好的借鉴。本着百家争鸣的精神，本文仅就这三方面的问题，谈谈我们的初步体会和看法。

水稻雄性不育的遗传类型及其在"三系"培育中的某些规律性

　　遗传的雄性不育变异在水稻中是经常发生的，按形态可分为花药分裂型、花药或雄蕊退化型、无花粉型和花粉败育型四类（机械性的花药分裂，一般易受环境条件影响）。按来源上可分为自然突变的、人工诱变的和远缘杂交的三类。在遗传学上暂按传统的概念分，则有细胞核遗传、细胞质遗传和核质互作三类。

　　从选育水稻"三系"和利用其杂种优势的目的出发，我们要求的不育材料应该是所谓核质互作型的，因为这种类型比较容易做到"三系"配套。

　　细胞质突变的水稻雄性不育尚未见有报道，已知自然突变和人工诱变的不育材料多属隐性的核遗传，与细胞质遗传直接关系甚少。正反交（不育材料用子一代可育株作杂交亲本）的结果，由于可育性为显性相关，因此，育性正常的父本一般都是这类材料的恢复系，而很难找到它们的保持系；至少在我们的上千个材料的测交中是没有找到的。这主要在于引起这种突变的原因和条件尚不知道。

　　大量的实践证明，用远缘杂交进行的所谓核置换，是创造核质互作型雄性不育的有效方法。例如已经"三系"配套的"野败"型、"野栽"型不育系，是普通野生稻同栽培稻的核置换，"南新"型、"BT"型等不育系是和粳稻的核置换，等等。很明显是质和核的不协调，亦即质核的矛盾，便造成了这一类型的雄性不育系，由于我们已大体知道这种不育性产生的来龙去脉，因此，在找其保持系和恢复系方面就有了较大的自由。就现有的实践，某些籼粳交不育系，多是籼质与粳核矛盾的产物，如加入籼核的成分，来克服或缓和其间的矛盾，便有可能促进其育性恢复。所以说这类籼粳型不育系的恢复因子常存在于籼稻品种中。

然而，并非一切远缘杂交的核置换都能造成雄性不育和实现"三系"配套。综析一些试验结果，可以看出有两点带规律性的现象。

一是亲缘关系的远近（即细胞质分化差异程度）要适度。籼稻和粳稻，属于同一祖先的两个亚种，彼此质核的差异相对来说还是较小，因此大多数籼粳交组合的核置换不能获得不育系，而只有很少数特定的组合，亦即演变上处于较高阶段的现代化粳稻品种的核置换，能获得不育系。这是因为它们细胞质分化的差异程度较大。同样道理，有极少数籼籼交或粳粳交，虽在同一个亚种范围内，但若双亲处在系统发育的两个极端也是可能得到雄性不育的。与此相反，野栽交由于血缘关系较远，彼此的核质差异大，因此很容易获得不育系。但若走向系统发育的两个极端，超出一定范围后，就会因质核矛盾太大，而造就极深刻的雄性不育，亦即转化成了所谓细胞质起主导作用的雄性不育，以致较难找到恢复系。"野败"粳型不育系"京引66"便可能是属于细胞质遗传，至今在原有粳稻品种中还没有它的恢复系。

二是要以在系统发育上处于低阶段的品种作母本。换句话说，就是细胞质要比较原始，而细胞核要比较近代化，核置换才易于获得不育系。通常认为现有栽培稻的演变谱系是从普野到籼再到粳。同一个远缘杂交组合，正反交的核置换结果往往截然相反。例如野 × 栽和某些籼 × 粳，正交会随着回交世代的上升，得到完全的雄性不育的后代，而反交后代的育性会逐渐恢复正常。由此可见，核质矛盾所引起的雄性不育有其特定性，即方向性。不能同意在野生稻或某些籼稻品种的细胞质中，原体就存在着特定的不育因子的假说。我们认为，雄性不育是核中的某些因子与细胞质互作的不协调性，使代谢上的某个环节发生了异常变化而引起的。生物的细胞是一个统一的整体，但从某种意义上看，质是核的环境条件。在进化上处于较高阶段的核，被置换到比较原始的细胞质中，等于从较好的环境转移到较差的环境去，由于核不适应这种环境，因而易于引起雄性不育及其他现象。反之，把比较原始的细胞核置换到比较进化的细胞质中，核对质较易适应，一般不会引起雄性不育等异常现象。

关于水稻的杂种优势

一、植物有无优势，不决定于它们固有的生殖方式

关于自花授粉作物有无优势现象，在学术界存在着几种不同意见。但传统的、迄今仍相当流行的观点，则是自花授粉植物无杂种优势。这个观点对于开展自花授粉作物的杂种优势的育种工作，起着消极的作用。

自花授粉作物无杂种优势的理论基础是显性学说，是根据某些异花授粉作物的自交有衰退而杂种有优势现象推断出来的。显性学说认为，在异体受精生物里存在着许多有害的、呈杂合状态的隐性基因，因而自交表现出衰退现象；而自体受精作物有害的隐性基因，很容易在自交过程中受到自然或人工淘汰，所积累和保存下来的几乎都是有利基因，所以，自交不会衰退，杂交也无优势。国外《遗传学原理》一书中，便是这么写道："已经知道自交对某些生物体（玉蜀黍）是经常有害的，其他则偶尔有害（果蝇），亦有完全无害的（小麦），这种差别是不同物种的生殖生物学所约束的。如小麦，自花受精是正常生殖方法，隐性有害突变在其发生之后立即同质化……所以，小麦自交后，不会使旺势消失，异交一般不表现杂种优势。"

然而这个理论是片面的，我们从中至少可以看到有以下几点难以自圆其说。

（一）与自交系的杂种优势现象相矛盾。大家知道，玉米自交系继续自交不再引起衰退现象，但杂交能产生强大的优势。而天然的自花授粉植物品种（天然自交系），自交也不退化，为什么杂交却不能产生杂种优势呢？

（二）所谓互补余地很少，也是站不住脚的。显性学说认为，形成杂种优势的主要原因是有利显性基因的互补作用，而自花授粉植物品种，本身已经积累了很多有利的、有优势的显性基因，因此显性互补的余地就很少了，杂交自然不会产生较强的优势。

所谓互补，应该包括两个方面。一是指不同性状间的互补，即通常指的互相取长补短，如某些稻麦大穗型品种同多穗型品种杂交，可以获得较多而大穗的杂种；豌豆节间密、节数多的品种同节间长、节数少的品种杂交，就会得到节间既长、节数又多的杂种。从这方面看，毫无疑问，自花、异花授粉作物是受到其共同遗传规律支配的，二者之间绝不会存在着很大的差别。其实，一个经常被显性学说引用的经典例子，就是上述两种节间类型豌豆杂交试验，而豌豆恰好是较严格的自花授粉作物。

互补的第二个方面，就是指控制同一个性状，特别是数量性状的所谓微效多基因间的互补作用（实则为互促作用）。从这方面看，自花授粉植物无优势论也是片面的。例如仅就粒重和其长宽厚这个数量性状而言，现有的高产优良品种，彼此之间千差万别，但我们很难说其中哪

一个品种积累了或每个品种都积累了最多的有利基因。更不能说在这一具体性状上它们之间没有或者很少有互补余地。的确，双亲的互补作用，是构成杂种优势的主要因素之一。但应当指出，互补是建立在有差异的基础上的，这是一条基本原则。无论是不同性状或是同一性状，唯有差异存在，才会有互补作用。

由此可见，决定互补余地大小的不在于植物的生殖方式，而是受该种植物所拥有的类型和品种的丰富度制约。凡是类型和品种越丰富多彩，越千差万别，互补的余地就越大。反之，品种单纯，类型贫乏的品种，互补余地就小。

（三）与显性连锁学说本身相矛盾。显性学说认为，在同一条染色体上隐性不利基因总是与显性有利基因呈紧密连锁关系，所以杂种优势不能在后代固定下来；杂种后代的优势总不如杂种一代。显隐性基因的连锁是生物界常见的现象，自花授粉植物具有同样的连锁现象，但为什么杂交却不能产生优势呢？

（四）显性学说不能解释高超亲现象。在我们水稻杂交试验中，有些组合具有高超亲现象，即杂种某一性状的优势超过双亲之和。例如，二九南 1 号 A × 国际稻 30，杂种从出苗到抽穗的生育期在长沙比双亲之和多 3 天；二九南 1 号 A × 粳 40，杂种每穗粒数多于双亲之和。如按照显性学说，则杂种的性状的数质如果等于双亲之和，就达到了最大限度的互补或累加作用，而不可能出现高超亲现象。

总之，自花授粉植物无优势论者违背了"差异就是矛盾""事物内部的这种矛盾性是事物发展的根本原因"这些辩证唯物主义的基本哲学原理。分析事物不抓事物内部的矛盾本质，因而必然会走向形而上学的推理，把植物有无优势归结于它们的固有生殖方式，显然是错误的。

二、水稻和其他植物的杂种优势受共同的规律所支配

水稻有无杂种优势或优势大小，关键在于选配亲本。这与异花授粉植物具有很大的共同性。

（一）两个杂交亲本在遗传性上要有差异，即它们的远缘关系要远或地理上的分布要远，或生态类型要有所不同，这样的亲本才能形成较强的生活力。我国目前在生产上应用的组合或在优势上比较突出的组合，几乎都是由三方面的亲本所组成的。例如，南优 2 号、南优 6 号是地理上远距离的配偶；二九南 1 号 A × 雪谷早是生态类型上不同的配偶；黎明 A × 培迪，71×72A × 粳 410 是籼粳的配偶。其中又以亲缘关系为第一位，往往可以得到超亲或高超亲的优势。

（二）性状上要有差异，彼此有相互取长补短和相互促进的作用。如南优 2 号父母本在

生育期、株型、抗性以及许多农艺性状上都有较大的差异，才在不同程度上起到互补和互促作用。其粒重偏于较重的亲本，抗稻瘟偏于较抗的父本，株高和每穗粒数都是超亲。生育期、株型等性状则表现为中间型。数量性状多属于中间型遗传，但在杂种优势利用上也有其应用价值。如科卵糯的籽粒过于长大，结实不饱满，与谷粒较小而短圆的珍汕97配制的杂种，谷粒大小表现为中间型，但因结实饱满，故千粒重超亲。

（三）亲本之一应是高产品种，具有优良的综合性状。杂种是在亲本的基础上发挥优势的，即所谓水涨船高，我们需要的是较高的绝对产量。两个差异大的低产品种，虽然可能形成很强的优势，但是绝对产量不见得高。如二九南1号A×粳410的优势强于南优2号，但其绝对产量不如后者。目前生产上应用的杂交组合，无例外地都需要有一个高产亲本。

三、水稻的杂交优势是客观存在且蕴藏着巨大的增产潜力

经过1974年至1976年三年时间的试种，我们对杂交水稻做了比较全面的调查分析，它和常规稻种比较，有如下几个突出的优良特性：

第一，根系发达，功能旺盛，吸收力强。同位素^{32}P示踪观察，表现根系伸展速度快，吸收能力强而持续，养分的积累运转比较协调。

第二，长势苗壮，繁茂，分蘖力特强，光合势强，过氧化物酶的活性和呼吸强度较一般品种低，特别是光呼吸强度低，因而积累多，消耗少，营养增长快。

第三，穗大粒多，一般每穗180~250粒，显著超过双亲。每亩15万~20万穗均可产千斤以上。多数杂交种的千粒重能超过双亲中值，接近大粒亲本。

第四，适应性很强。凡是能种水稻的地方，只要生育期满足它的需要，不论种在什么田里，在大致相同的条件下，杂交水稻比常规稻种易于获得较高的产量。

1976年，湖南省试种杂交水稻约130万亩，绝大多数比常规稻种有明显的增产效果。在湖南现在的生产条件下，一般比当家种每亩可增产100多斤到200多斤。不少地区创造了一季亩产一千三四百斤，甚至一千五六百斤的高产量。证明杂交水稻蕴藏着较大的增产潜力，展示了我国水稻大幅度增产的光辉前景。

杂交水稻虽已应用于生产，但这项工作仅仅才开始，还有广阔的发展前途，还蕴藏着巨大的产量潜力。

（一）水稻是世界上最古老的、分布范围最广的作物之一，资源极其丰富。因此，在选配杂交亲本上比玉米等作物有更多的自由，选择优良组合的概率更大。

（二）水稻由于是自花授粉植物，优良品种的各对基因呈高度纯合状态。因此，杂种第一

代会有高度的一致性，从而有利于发挥高产的作用。

（三）由于基因呈纯合状态，其遗传性状较易控制，也就是说较易把双亲的优点集聚在杂种第一代上。

（四）杂交优势的强弱，是指对亲本相比较而言的相对数值。水稻属高产作物，产量基础本来就高，近年来由于常规育种的不断发展，育成了许多产量相当高的品种，这样，杂交水稻的产量又能在新的水平上更上一层楼。

关于制种问题

"三系"配套以后，由于没有经验，初次制种产量很低，亩产只有10多斤。这时有人说什么"水稻是自花授粉作物，花粉量少，每日开花时间短，柱头很小，且多数品种不外露，这一系列不利于异花传粉的特征特性，注定了杂交水稻过不了制种关"，认为"三系"虽然配了套，又有优势，但不能在生产上应用。事物总是一分为二的。上述的特征特性确有不利异花传粉的一面，但也有有利的一面。例如开颖授粉，花粉轻小而光滑，裂药时几乎全都散出，可借风力传到一定距离（据测定可达40米）。这些保留下来的风媒传粉的特征特性是制种的前提。

"花粉量少，根本不能满足异花授粉的需要"是制种低产论的所谓最有力的根据。这是一种"只见树木，不见森林"的形而上学观点。就单个花药和穗子来看，水稻的花粉量确实比玉米和高粱的少，但就单位面积上累积的花粉量来看，差异并不大。据我们调查，杂交水稻南优2号的父本平均每个花药有600粒，按制种田亩产父本300斤估算，每亩花粉约有300亿粒，以10天散粉计，在散粉均匀的条件下，则每天每平方寸面积上可散落5000粒左右，密度是很大的，完全可以满足授粉的要求。实践证明，影响制种产量的并不是花粉数量不足，而是花粉是否分布均匀并能散落在母本柱头上的问题。水稻的柱头较小，颖壳开张角度小，多数品种柱头不外露或外露不大，这些是不利异花授粉的。但是，我们可以通过人工选择柱头发达外露和开颖角度大的不育系来加以解决。在制种的实践中，群众发现剑叶是阻碍花粉落到柱头上的最大障碍。割去剑叶，不仅可以使母本柱头接触花粉的概率大大增加，而且母本每天开花的时间也随之提早，可使异交率显著提高。这一措施已经在制种中采用，对提高种量效果很显著。

在实践中还发现，每天开花时间拉长，固然可以提高异交率，但由于水稻开花时间一般长达几十分钟，而花粉则在开花后数分钟内就散完。所以，开花时间的长短不是主要因素，关键是父母本开花时间能否相遇。群众采用割叶和喷"九二○"可以提早母本开花时间，对父母本花期相遇有一定作用。

近两年来，湖南省广大群众和科技人员，坚持"实践、认识、再实践、再认识"的辩证唯物论的认识论，充分发挥人的主观能动性，不断认识水稻开花传粉的规律和各种因素，发挥提高异交率的有利一面，克服其不利的因素，终于突破了制种关。全省制种产量一次比一次高。1975 年冬和 1976 年春，全省 97 个县市的 7 000 多名贫下中农、社员群众和科技人员，在海南繁殖、制种 28 400 多亩，平均亩产 75.5 斤，较 1975 年翻了一番。他们不仅创造了较高的制种产量，而且大大发展了杂交水稻的制种技术和理论，这充分说明"群众是真正的英雄"是一个颠扑不破的真理。

作者：袁隆平

注：本文发表于《中国农业科学》1977 年第 1 期。

杂交水稻在国内外的发展近况

我国是世界上第一个在生产上利用水稻杂种优势的国家。近三年来，全国每年种植面积 8 000 万亩左右，其中湖南、四川、江西、江苏、福建、广东等省在 1 000 万亩左右。1982 年，南方十三省（区、市）早、中、晚各季杂交稻种植面积为 8 271 万亩，平均亩产 773 斤，对提高水稻的总产和单产起了重要作用。高产试验田的纪录也不断刷新。1978 年以前，福建、贵州、江苏和湖南省曾出现过亩产 1 500～1 600 斤的高产田块，1979 年江苏邳县山西大队亩产 1 711 斤；1980 年江苏徐州地区农科所亩产为 1 716 斤；1981 年江苏赣榆县朱堵农科站的 1.27 亩赣化 2 号，创造了亩产 1 882 斤的纪录。

在制种方面，由于逐步建立了较健全的繁、制体系，改进了制种技术，产量和质量都有提高，全国从 1979 年的亩产 66 斤提高到 1982 年的 119 斤。其中江苏、四川、浙江、辽宁和湖南省共出现了 48 个亩产 200 斤的县，四川江油县（现为江油市）亩产高达 350 斤，创造了全国以县为单位制种产量的最高水平。

（一）

我国的杂交水稻虽然取得了很大的成绩，从战略上看，目前只是处在发展初期阶段，它还有广阔的发展前途，蕴藏着巨大的增产潜力。从育种理论上看，利用水稻杂种优势具有三大优越性：

1. 是提高水稻生理功能的有效方法。它可以将形态改良与生理功能的提高结合起来，从而能使一些对立的性状（如大穗与多穗、大粒与多粒的矛盾）在较高一级的水平上达到统一。

2. 能较易地把多抗性和高产性结合起来。已知水稻对主要病虫害的抗性，大多数是由显性基因或部分显性基因控制的，只要双亲一方对某一病、虫害有抗性，杂种一代一般也抗该病、虫害。由于现在常规育种的进步，已育成不少抗性强的品种、品系，有了这样的好基础，进而培育能抗四五种甚至更多病、虫害的高产杂交水稻，是完全可以做到的。

3. 杂交稻的遗传基础较丰富，因而具有广泛的适应性。当前在生产上应用的杂交水稻还存在着质源单一，组合很少，生育期偏长，抗性不够等方面的问题。近几年，各地对此进行了大量研究工作，取得了较大的进展，为进一步发展杂交水稻创造了条件。

（1）突破了"早而不优，优而不早"的难关。湖南省科技人员周坤炉同志又育成了生育期与湘矮早9号相似，但可增产15%左右的早稻新组合——威优35，为湖南省和长江流域的双季早稻利用杂种优势展示了美好的前景。衡阳地区农科所从安仁县农科所中峰1号×IR28的后代中育成威优98双季早、晚稻兼用的组合；安江农校育成的威优17，作迟播、迟插的双季晚稻（在湘中、湘南不要专用秧田），表现良好，一般亩产800斤左右。

（2）解决了高产性、多抗性和早熟性三结合的难题。安江农校育成的威优64，生育期比威优6号短15～17天，抗稻瘟、白叶枯、矮缩、飞虱、叶蝉等5种主要病、虫害，产量相当或高于威优6号。

由于这些新组合的出现，在湖南省就能解决双季晚稻早、中、迟熟组合配套的问题，从而有利于全年增产。同时还为湖南省高寒山区发展杂交中稻，湘北发展杂交晚稻创造了有利条件。福建、广东、广西配成的四优30、威优30、汕优30表现高抗白叶枯病、稻飞虱，中抗稻瘟病，已在华南作晚稻大面积推广（这些组合的感光性强，不适宜在湖南省种植）。

（3）大柱头不育系的选育有了苗头。四川省农科院把长药野生稻的硕大柱头及其百分之百的外露性状，转育到栽培稻的不育系和保持系材料上，现在基本趋于稳定，预计不久便可投入应用。湖南省农科院也已育成大柱头不育系大柱1号，其柱头外露率78%，目前正在进行配合力等试验，以确定其生产价值。根据我们今年的初步观察，大柱头不育材料的自然异花授粉结实率高达65.8%，而现有野败不育系不到20%。可见，大柱头不育系的选育成功，将是大幅度提高制种产量的前奏。

（4）新质源不育系的成绩很大，育出一批恢保关系与野败不同和恢复源比野败较广泛的新质源不育系。如云南滇瑞409不育系，安徽的矮败不育系等都已三系配套。湖南省农科院育成的国际2119不育系，抗性强、株型好、米质优，且属于配子体不育类型，恢复谱广，异交率高，是一个有发展前途的新质源不育系。

（5）在粳稻方面出现了一批新组合。北方稻区继黎优57、京优300号等推广以后，又育成秋优57、毫干达歪——黄金A×300号等新组合，近两年的种植面积在100万亩以上。南方稻区有江苏的农虎26A×77302，湖南省农科院的农虎26A×培C等晚稻组合，表现耐寒性强，有的已开始推广，有的正在试种。

（二）

我国杂交水稻的研究成功和在生产上大面积推广，引起世界上许多产稻国家的注目和反响。

国际水稻研究所[1]在 20 世纪 60 年代中期，因育成株型良好的半矮秆良种 IR8 号而一举闻名于世。该所于 1970—1971 年曾进行过选育杂交水稻的研究，后来中断了。我国杂交水稻的成功，使该所受到很大的启发和鼓舞。1979 年 10 月，他们与我国签定了双方合作研究杂交水稻的协议，主要目的是选育适合热带、亚热带地区的高产、多抗杂交稻。两年多来，取得了一定的进展和成绩，但存在的问题也不少。首先是中国的不育系及现有组合不能直接在热带国家利用；第二是基本育成的几个国际稻系统的不育系，配合力太差，用它们配出的组合大多没有优势或优势不强，反之，几个配合力特好的母本保持力又不好，难以转育成不育系；第三是制种技术未过关。因此，估计在 1～2 年内，他们还很难赶上我们。但从长远看，国际水稻所的条件好，资源丰富，研究水平较高，在某些方面尤其是在强优、多抗组合的选育上有可能超过我们。

美国在 20 世纪 70 年代初开始研究杂交水稻，获得了不育系，但不育性不过关。1971—1975 年，加州大学对水稻的杂种优势进行了研究，153 个组合中有 11 个的产量显著超过最好的对照品种，增产幅度平均达 41%，但三系一直未配套，因而在生产上无法利用。

1980 年，杂交水稻作为我国的第一个农业技术转让给美国。迄今，我国的杂交稻在美国试种了三年，每年都表现良好，增产极其显著。如 1981 年在德克萨斯州品比试验中，供试组合、品种 11 个，按产量位次，前 6 名都是中国的杂交稻，第 7、8 名是父本，美国的 3 个对照良种居倒数第 1、2、3 名。由我国专家负责的 1.5 亩的大田对比试验中，威优 6 号亩产 1 515 斤，比当地对照良种增产 61%。1982 年在美国几个农场扩大了对比试种田的面积（每个组合在每个点种 6 亩），完全按美国的栽培方法进行，结果仍然以我国的杂交稻产量最高；如南优 2 号，每英亩[2]产 8 600 多磅[3]，比当地对照良种增产 79%，引起了美国产业界和农业科技界的极大兴趣。在收获后不久，美国即派人来我国同我们联系和洽商，迫切要求与湖南省农科院订立科研合同，为他们培育米质优良的杂交水稻和大柱头不育系。

日本在水稻育种上是世界上最先进的国家之一，也是开展杂交水稻研究最早的国家。1958 年，日本东北大学的胜尾和水岛二人发现中国野生稻有导致藤枚 5 号产生雄性不育的细

---

[1] 国际水稻研究所：简称 IRRI。
[2] 1 英亩 = 4046.86 平方米。
[3] 1 磅 = 453.6 克。

胞质。1968 年日本琉球大学新城长友育成台中 65 号不育系并实现了三系配套。但是，由于这种包台型三系亲缘关系太近，没有优势，又是高秆，即使同它配出的杂种具有很强的优势，也难获得高产，高秆杂交稻过不了倒伏关。因此，日本的杂交水稻始终停留在理论研究上，无实际生产价值。中国农科院访日代表回来对我们说，日本对杂交水稻的认识和态度已有了很大的改变，对我国的杂交稻产生浓厚的兴趣。

作者：袁隆平

注：本文为杂交稻新三系选育列为"六五"国家重点科技攻关项目之际，作者提交给农牧渔业部 1983 年 12 月在长沙召开的南方杂交稻会议的报告。

杂交水稻新组合威优64

威优64是我们新近育成的一个杂交水稻新组合，父本测64系田间代号，原名为IR9761-19-1，1979年从国际水稻所引进，当年用它同V20A、珍汕97A、二九南1号A等进行测交，发现它具有恢复力和较好的配合力，但有的性状包括生育期和抗黄矮病在内还有分离，因而从中再连续二代选择优良单株进行分对提纯测交。通过鉴定，其中以V20A×测64-7表现最好，现定名的威优64就是V20A×测64-7。

该组合从1981年开始，经过三年的试验、试种和大面积示范，都表现良好，是一个很有发展前途的新组合。

特征特性

（一）生育期较短，产量较高。威优64属感温的早熟中稻类型，在湖南省中部作早、中、晚稻栽培的生育期分别为125天、120天、110天左右，比威优6号早熟12~17天。如1981年在湖南省安江农校作中稻栽培，4月16日播种，8月14日成熟，全生育期120天，亩产1 119斤。1982年在同一丘田（1.37亩）用威优64作早、晚两季栽培，早季3月29日播种，8月1日成熟，全生育期125天，亩产1 124斤；晚季7月7日播种，8月4日插秧，10月22日成熟，全生育期108天，亩产897斤，两季合计亩产2 021斤。在安江农校的晚稻组合（品种）比较试验中，1981—1983年威优64的绝对产量和日产量都超过了威优6号，并达到了显著水平。1983年全国南方晚稻生产试验中名列第一，同时在江西、安徽、福建等省的区试中也名列首位。1982年和1983年在湖南省怀化地区杂交晚稻组合配套试验连续两年的结果是，早插（7月20—25日）的威优6号，亩产分别为633斤和980斤，迟插（7月26日至8月1日）的威优64亩产分别为739斤和1 004斤，比威优6号分别增产16.7%和2.4%。可见，威优64作双晚，可迟插夺高产。该组合作中稻栽培也表现高产，1983年在怀化地区无灌溉条件的山区作早熟中稻试验，威优64共种130亩，亩产

944 斤，对照品种为珍珠矮、湘矮早 9 号等，共 13 亩，亩产 672 斤，威优 64 比对照增产
40.4%。在江苏省南京、盐城试种，亩产 1 100 斤，比南京 11 号增产 16%，且早熟 3~4
天。1983 年威优 64 在全国南方各省试种、示范的面积已达 10 万亩以上，一般亩产千斤左
右，湖南省桂东县农科所的中稻丰产田，亩产高达 1 496 斤。

（二）抗性较广，能抗 5 种病虫害。威优 64 抗稻瘟病，白叶枯病，青、黄矮病，褐飞
虱和叶蝉等 5 种主要病虫害。对稻瘟病，根据广东省植保所的鉴定，威优 64 抗中 A_{45}、中
B_1、中 B_5、中 B_{13}、中 B_{15}、中 C_9、中 E_9 等生理小种；在湖南省各地试验中，仅新晃县凉伞
公社有中度发病，其他地方都未曾感染稻瘟病。据湖南省安仁县农科所接种鉴定，该组合对白
叶枯病表现高抗；1982 年在澧县的诱发试验中，仅有零星发病，发病株率为 0.16%。对青、
黄矮病属中抗，湖南农学院常德分院鉴定，威优 64 的发病株率为 10.83%，而威优 6 号为
29.50%。此外，威优 64 对 I 型褐飞虱和叶蝉均表现为较高的抗性。

（三）株叶型良好，分蘖力强，属多穗、中穗、中粒型。威优 64 的分蘖力强，生长势
旺，株叶型良好，株高适中（90~100 厘米），松紧适度，分蘖成穗率较高，一般在 70% 左
右。在中上肥力水平条件下，有效穗 25 万穗左右，每穗粒数 110 粒上下，谷粒长形，千粒
重 27.5~28.5 克，结实率一般为 80%~85%。1982 年作晚稻，虽在整个抽穗扬花期间遇
上连续低温阴雨，但结实率仍在 70% 以上，比对照威优 6 号高 10%。

（四）谷草比值高。威优 64 谷草比值一般高达 1.4 左右。由于谷草比值高，随之带来一
个弱点，即进入黄熟期以后，茎秆的抗倒力不强。

（五）不落粒、易脱粒。不抗纹枯病和小球菌核病。

栽培技术要点

（一）威优 64 的适应性广。在湘南、赣南、粤北、桂北、闽中北可作双季早稻或晚稻。
苏北、皖北、陕南可作麦茬稻。长江流域各省可作早熟中稻或一季早稻，但更宜作插秧期较迟
（7 月 25—31 日）的双季晚稻。

（二）威优 64 作双晚栽培，适宜的播期幅度在湖南省为 6 月末至 7 月初，湘北要适当早
播，湘南可略推迟，每亩秧田播种量 20~25 斤，秧龄 20~25 天，最长不超过 30 天。

（三）栽培密度因季节迟早、土壤肥力而定。一般以每亩插 2.5 万蔸为宜。每蔸 3~6 苗
（包括分蘖）或"两粒谷"的秧。本田用种量 3.0~3.5 斤。

（四）威优 64 适合中等肥力田栽培，一般亩施纯氮 20 斤，就可获得千斤左右的稻谷产
量。但要注意氮、磷、钾配合，特别应增施钾肥壮秆，以防倒伏。作双季晚稻，前期要轰得

起，长好苗架。在田间管理上，一方面要注意适时排水晒田防治纹枯病；另一方面又不要在齐穗以后使田面长期无水层而发生小球菌核病。此外，威优 64 不适宜在排水不良的烂泥田、深泥田种植。其他栽培技术与一般杂交水稻相同。

制种技术要点

测 64 的花粉量大，因此制种产量高，这是威优 64 的又一个主要优点。湖南省 1983 年制种 5 万亩，平均亩产达 265 斤，最高的丘块亩产达 723.5 斤，创造了籼型杂交水稻制种产量的新水平。

威优 64 在长江流域夏制或秋制均可。夏制父本的播种期在 4 月或 5 月，要因地制宜。秋制时，第一期父本的适宜播种期在湖南为 6 月 20 日以前，秧龄不超过 30 天，这样才能保证在 8 月末或 9 月初完全始穗。播差期以叶龄为准，即第一期父本的叶龄达 5.5~6.0 叶（夏制）、3.5 叶（秋制）、4.1~4.5 叶（海南制）时播母本 V20A。由于父本测 64 的分蘖力强，一般只播两期父本，第一期播后 7~8 天播第二期。

测 64 的植株较矮，穗子较小，因此对父本要适当多喷"九二〇"，父母本的行比以 1∶8 为好，在条件好的地方也可扩大到 1∶10。其他措施与一般制种相同。

作者：袁隆平　孙梅元

注：本文发表于《农业科技通讯》1984 年第 5 期。

杂交水稻超高产育种探讨

1980 年，日本制定了一个水稻超高产育种计划，要求在 15 年内育出比现有品种增产 50% 的超高产品种[1]。面对这一国际上的育种新动向和在 20 世纪末把我国农业产值翻两番的任务，我们认为，我国的水稻育种在注意提高品质的同时，也须制定超高产育种研究计划。特别是杂交水稻，从产量育种看，具有很大的潜力和优越性，主要是杂种优势利用能把形态改良同生理功能的提高密切而有效地结合起来，使生物学产量和经济系数都得到提高，既可增加"源"又可扩大"库"。这就比一般的形态育种能产生更好的效果。因而对实现超高产，相对较易。本文试就杂交水稻的超高产育种，提出一些设想和探讨性意见。希望起到抛砖引玉的作用。

什么叫超高产育种，目前并没有一个统一的标准和严格的定义，其产量指标，当然要随时代和地区的不同而异。日本的目标，要求到 1995 年在原亩产 667 ~ 867 斤糙米的基础上达到 1 000 ~ 1 300 斤[2]（折合稻谷亩产 1 250 ~ 1 625 斤）。日本是一季稻。我们针对湖南省以双季稻为主的特点，初拟如下的超高产育种指标，争取 5 年内即 1990 年达到。

1. 双季早稻：以培育全生育期 110 ~ 115 天的中熟组合为主。要求在生产水平较高的大面积条件下亩产 1 200 斤左右，即每亩日产量 11 斤。

2. 双季晚稻：也以培育全生育期 110 ~ 115 天的中熟组合为主，晚季的温光条件优于早季。要求亩产 1 300 斤左右，即每亩日产量 12 斤。

根据杂交水稻育种的特点，原则上，超高产育种要从两个方面着手：一是充分利用双亲优良性状的互补作用，在形态上作更臻完善的改良；二是适当扩大双亲的遗传差异，以进一步提高杂种的生理功能。二者密切结合，相辅相成。

一、形态上的改进

根据选育威优 64 的经验[3]和新近看到的某些很有苗头的中间试验

材料，我们得到一些重要启发，感到要实现杂交水稻上述指标的超高产育种，在形态上不一定要搞那种难度大、费时长且把握不大的特大穗、特大粒类型。倒是以培育多穗、中穗、中粒型的组合为主，使穗、粒、重三个产量因素协调和均衡的发展，反而比较容易在较短的时期收到良好效果。

模式：所谓多穗、中穗、中粒型。参照威优64及某些优良新组合。我们设想了如下的穗粒结构和主要形态特点：

（一）穗粒结构

穗数/亩　25万~30万穗

粒数/穗　120~140粒

结实率　85%左右

千粒重　28克左右

按照上述穗粒结构，各取其中间值算，理论上每亩约1 700斤。大面积生产上按七至八折算，可产1 200~1 300多斤，即每亩日产11~12斤。若配合以超高产栽培，产量当然会更高些（在较好的条件下，威优64每亩有效穗数为22万~27万穗，每穗粒数100~110粒，结实率80%左右，千粒重27~28克，亩产1 000斤左右）。

（二）形态特点

除要求矮秆、株型紧凑、叶片挺直等基本的优良农艺性状外，我们还强调四个特点：

1. 株型呈倒钟形：即在齐穗后，基部各茎秆略为开散，中部直生，上部又略散开，整个植株像一个倒置的古钟。这种株型在多穗情况下，株行间和株内的通风透光条件都优于其他株型紧凑的类型（如扫帚形、一柱香形等）。基本上可做到"封行不封顶，封顶不封行"。如V20A×IR9852-39-2就属这种倒钟形。在每亩30万穗的密度下，齐穗时行间的泥面仍清晰可见。

培育这种株型的杂交稻，最重要的一点是亲本之一要具有很强的分蘖力，同时叶片比较窄。

2. 叶片为瓦片状：即叶片向内微卷，这种瓦状叶的优点是能增加群体的受光率，增大光饱和点（4.5）。由于透光率高，在单位面内能有较大的有效叶面积指数。如南优2号的上部叶片就是瓦状（但株型不紧凑）。不少籼粳杂交后代即具有这种性状，其遗传规律也较简单[4]。

3. 冠层属半叶下禾：所谓半叶下禾是指在抽穗至齐穗期间，剑叶与稻穗都显露在上（前者比后者略高或齐平，而且挺直）。灌浆以后，稻穗下垂，变成叶下禾。如威优64就属于这种类型。半叶下禾的优点是在抽穗至齐穗期扩大了光合面积，把稻穗前期可制造养分的能力利用起来。当灌浆之后，其光合能力衰退时，便已下垂，把空间腾出来，让剑叶充分发挥作用。

4. 谷草比值要在1.4左右：在杂交稻的选育中，往往容易获得营养体优势强大的组合，超高产育种最重要的研究任务之一，就是要通过株型改进，把这种强大的杂种优势尽量发挥在稻谷上，育成谷草比值要高达1.4左右的组合。但谷草比值很高的，一般茎秆较细，穗重草轻，会带来易倒伏的弱点，这是超高产最大的矛盾之一。如威优64，就是因谷草比值高（1.4）和茎秆较细（第8节间为4毫米左右），而抗倒力不强，难以超高产。解决这个矛盾最简易而有效的办法，是把株高再适当降低一些，使重心下降，增强抗倒力。参照威优64，我们认为谷草比值在1.4左右，穗重为3克的多穗、中穗、中粒型超高产组合，株高80～90厘米为宜。

二、选育强优势超高产组合的途径

杂种优势的本质是生理功能旺盛，具体表现在生长势旺、分蘖力强，根系发达。对光能和肥料的利用率高，养分运转流畅，不早衰以及抗逆性强等方面。提高杂交稻的生理功能，主要有两种方法：一是适当扩大双亲的遗传差异；二是在亲本中加入某些在生理功能上具有高效作用的基因。依据这个原则，并以培育多穗、中穗、中粒类型为模式，我们主要从以下四个方面进行强优势超高产杂交组合的选育。

1. 继续利用IR系统的材料，特别是近年育成的早熟品系，从中测选优良恢复系（地理远距离和不同生态类型的品种间杂种优势）。

实践证明：IR系统是优良恢复系的主要来源，与我国现有早籼不育系的配合力相当好。尤其近年国际水稻研究所育成了一批早熟、优质、多抗、分蘖力强、株叶形很好的新品系，从中测选出超高产早熟组合的可能性很大。例如，我们最近测出的IR25898-69-2-2、IR25846-83-3-2等早熟和早中熟恢复系，与我国早籼不育系所配的杂种。根据初步观察，凡属多穗、中穗、中粒型的，日产量均显著高于目前生产上应用的杂交组合。

值得一提的是，在测64（IR9761-19-1）这个早中熟恢复系中还有潜力可挖。近两三年来，我们从中选出不少早熟单株，其中又可分早、中早、迟早三类。与V20A等相配的初步结果表明，杂种的性状与威优64基本上相同，但成熟期分别比威优64提早了15天、10天和6天。小区产量等于或显著高于威优64，每亩日产量则均显著高于威优64。下表是其中一个迟早株系（测 49）与 V20A 所配的杂种与威优 64 在大田栽培的性状比较。

表1　威优 64 与 V20A× 测 49 性状比较

组合	播种期	全生育期 / 天	株高 /cm	每亩穗数 / 万穗	每穗粒数 / 粒	结实率 /%	千粒重 /g	实际亩产 / 斤	每亩日产 / 斤
威优 64	4 月 16 日	120	97	28.0	107.7	86.2	27	1 113.2	9.27
V20A× 测 49	4 月 15 日	113	87	25.3	130.5	72.2[*]	28	1 206.9	10.70

注：* 实际上测 49 的恢复力比测 64 强，但因乳熟期叶面追肥不当，使部分剑叶受伤，同时还有 2% 的黑粉病粒。因此空秕粒大大增加。

2. 恰当地利用籼粳中间型偏籼或偏粳的材料（籼粳杂交优势）。

籼粳杂种具有极强的杂种优势。但因结实率低，在生产上无法直接利用。现阶段我国的杂交粳稻一般都是采用所谓"架桥"的办法，即从籼粳杂交再回交的后代中选育偏粳的中间类型作恢复系，以解决因籼粳不亲和性而引起的杂种结实不良的问题，并已取得较好的成绩[5]。但是，从杂交粳稻的优势和结实率来看，还不能令人十分满意。鉴于粳型不育系农虎 26A 因有籼稻血缘而配合力较好。我们设想，如果不仅对父本，而且对母本也加入籼稻成分，并使恢复系带有 IR 系统的成分，不育系带有我国早籼的成分，那么，很可能是进一步发掘杂交粳稻增产潜力的重要方法。因为这样既可增加双亲的亲和性来提高结实率，同时又可避免亲缘重叠而影响杂种优势。

在杂交籼稻方面，利用籼粳优势来获得超高产杂交组合大有前途。如南朝鲜[①] 的密阳、水源系统，多数属于籼粳中间型偏籼的品种。其中有不少是优良恢复系。与我国早籼不育系的配合力良好。从中筛选超高产组合是有可能的。再从我们的试验材料来看，粳稻同 IR 系统恢复系杂交和回交所育成的偏籼型品系，特别是粳质籼核的核质杂种（只带有很少粳稻成分）。从中测出几个优良恢复系，与 V20A 配组，杂种优势十分明显，生育期与威优 64 相同。但小区产量高 30% 左右。可惜米质不良和有少量分离，还不能进入高级试验。但通过这些材料我们看到了利用籼粳杂种优势的潜力和前途。看来，造成籼粳杂种强大优势所涉及的互作基因数目并不多。只要父本带有很少的粳稻成分，就能产生良好的效果。同时，我们还设想，若在籼型不育系中，导入一些粳稻成分，有可能会取得更好的效果。

3. 在恢复系中导入野生稻的某些优良性状（种间杂交优势）。

湘潭地区农科所用印度野生稻和意大利 B 杂交，育成分蘖力特强的意印 6 号恢复系，与

---

① 南朝鲜：现称韩国。

早籼不育系所配的组合，属于多穗、中穗型，杂种优势非常显著。1979 年在该所的丰产试验田，每亩有效穗数达 30 万穗以上，每穗粒数 130 粒以上，在结实率不到 60% 的情况下，亩产仍超过千斤[5]。我们通过东乡野生稻与 IR91115 及圭六三〇等进行复式杂交，亦初步获得很有希望的恢复系。与 V20A 配组，在 1983—1984 年海南南繁的小材料试验中，小区产量名列首位。上述两种通过栽野杂交选育的恢复系，前者因恢复力欠佳还不能投入生产应用，后者还未在较大面积上试验。但由此可看到一些很有希望的线索，即把野生稻某些优良性状，如特强的分蘖力，快速的生长势和某些抗逆性等，导入恢复系中加以利用，很可能会对选育超高产早熟组合起重要作用。

4. 选育优良的配子体籼型不育系。

从遗传特点上分，我国现有的籼稻不育系有两大类。一类是孢子体不育系，如野败型、冈型。由于其不育性稳定而在生产上得到广泛应用。另一类是红莲型等配子体不育系，由于育性不够稳定而难以在生产上应用。但从超高产育种战略看，今后则应以选育优良的、育性稳定的配子体不育系作重点。因为配子体不育系的恢复谱较广，选到强优组合的机率较高；同时其可恢复性较好，有利于进一步提高杂交稻的结实率。这项工作，我们正在积极进行中。

— References —

参考文献

[1] 佐藤尚雄. 水稻超高产育种研究.《国外农学水稻》1984. Ⅱ .1～6

[2] 星野孝文. 日本暖地水稻产量分析及超高产的可能性.1984. Ⅰ .1～5

[3] 袁隆平等.《农业科学技术通讯》1984. Ⅰ .1

[4] 沈福成. 水稻卷叶性状遗传初探.《贵州农业科学》1983. Ⅲ .9～12

[5] 沈福成. 关于水稻卷叶性状在育种中利用的几点看法.《贵州农业科学》1983. Ⅴ .6～8

作者：袁隆平

注：本文发表于《杂交水稻》1985 年第 3 期。

杂交水稻研究的现状与展望

【摘　要】中国在世界上首先培育成功杂交水稻并用于商品化生产。中国杂交水稻的产量比常规推广品种增产 20% 左右。中国的成功鼓舞了国际水稻所和其他国家的育种家去探索使用这种技术，进一步提高水稻产量的潜力，特别是在灌溉条件下的增产潜力。初步取得的结果表明：种植杂交稻每公顷能增产稻谷 1 吨左右。现已拥有一批选育杂交稻的遗传工具（即细胞质雄性不育系、保持系和恢复系），但目前主要是使用"野败"不育细胞质，胞质质源的这种单一性使杂交水稻潜伏着感染某种流行性病虫害的危机。现在，中国和国际水稻所的科学家正在选育各种新的细胞质雄性不育系。中国培育的细胞质雄性不育系由于对主要病虫害的抗性不强而不适于在热带地区种植。因此，国际水稻研究所的科学家正致力于在适应热带条件的品系的遗传背景上选育新不育系的工作。籼稻中恢复系的频率相当高（20%），但粳稻中却还没有找到恢复系。

杂交水稻制种技术在中国已趋成熟，制种产量达到每公顷 1～3 吨。这些技术正被应用到中国以外的国家。

杂交稻的米质在中国和其他国家已引起重视。这个问题并非技术上的难关。

杂交水稻的主要问题是种子成本高，每季都需换种，长期依靠农场以外的来源供种以及需要一个高效率的基础来生产和供应高质量的杂交稻种子。在中国，杂交稻的发展速度已表明上述问题并未成为杂交稻发展的制约因素。但是，在中国以外的国家，这种技术的应用将取决于杂交稻与最佳常规稻相比较所存在的产量优势程度以及该国是否有能力在政府部门或私人公司组织有效的种子生产、加工和销售。

前　言

现在中国已广泛地应用于生产的水稻杂种优势，是美国的琼斯（Jones）在 1926 年首先报道的。后来，其他一些研究报告又相继报道了水稻的各种农艺性状也存在明显的杂种优势（Chang et al., 1973; Davis and Rutger, 1976; Virmani, 1981; Virmani and Edwards, 1983）。关于选育水稻 F_1 代杂种将杂种优势应用于生产的建议也由印度（Richharia, 1962; Swaminathan et al., 1972）、中国

（袁隆平，1966）、美国（Stansel and Craigmiles, 1966; Craigmiles et al., 1968, Carnahan et al., 1972）、日本（Shinjyo and O'mura, 1966）以及国际水稻研究所（Athwal and Virmani, 1972）的科学家们先后提了出来。但是，由于杂交稻制种困难而使大多数研究人员丧失信心，只有中国的科学家在继续努力。

中国是世界上第一个将水稻杂种优势应用于生产的国家。杂交水稻的研究工作始于1964年。发展杂交水稻的基础遗传工具（即细胞质雄性不育系、保持系和恢复系）在1973年培育成功，1974年鉴定出具有强优势的杂种，1975年杂交稻制种技术基本配套。

1976年杂交水稻开始向农民推广。从此，中国的杂交水稻种植面积逐年迅速扩大。

继中国成功之后，国际水稻研究所在1979年恢复了对杂交水稻的研究，其目的在于通过这种技术来探索水稻植株的生理产量潜力。由于到20世纪末，稻米消费国对大米需求量的增长，研究重点是培育具有高产潜力的水稻。同时，估计土地、人口之比将会下降。所以，现在不仅仅是国际水稻所，印度、印度尼西亚、南朝鲜、日本、美国、巴西、墨西哥和越南也在进行杂交水稻研究。另外，两个私人种子公司，即美国的圆环种子公司和菲律宾的卡捷尔种子公司也正在对从中国引进的杂交稻进行试验和评价，并研究制种技术。

现状与发展

本文概述了杂交水稻在中国和其他国家的研究现状与发展。由于杂交水稻的研究现状与发展在中国和其他国家之间差距甚大，本文将分别讨论。

（一）杂交稻在中国的研究与发展情况

1. 面积和分布

现在，中国约有3 300万公顷水稻，其中杂交水稻占1/4以上（表1）。在具有不同农业气候条件的各水稻主产地，从辽宁省（低温区，北纬43°）到海南（热带区，北纬18°），从上海（东经125°）到云南（东经95°）都种植有杂交水稻。现有许多不同类型的杂交稻组合能适应各种气候和土壤条件。

（1）生育期长的感光杂交籼稻，如珍汕97A×IR30，一般在中国南部作晚稻种植。

（2）熟期中等的杂交籼稻，如珍汕97A×IR24和V20A×IR26，在中国南部可作早、晚稻种植。在中国中南地区（长江流域）的双季稻区被广泛用作晚稻种植或在单季稻区作中稻种植。

（3）早熟杂交籼稻，如V20A×26窄早和V20A×IR9761-19-64，在长江以南的双季稻区作早稻种植。

（4）早熟杂交粳稻分布在黄河以北地区。

（5）中迟熟杂交粳稻在长江流域作晚稻种植。

表1　杂交稻在中国的收获面积（1976—1985）

年份	公顷／百万
1976	0.15
1977	2.13
1978	4.33
1979	5.07
1980	4.93
1981	5.10
1982	5.60
1983	6.75
1984	8.84
1985	8.43
总产	51.33

2. 杂交水稻的产量潜力

实践证明，只要采用适当的栽培技术措施，杂交水稻的产量比常规稻纯系高20%～30%。表2列出了1985年种植杂交水稻的增产幅度。近年来，四川省和江苏省分别种了190万公顷和80万公顷杂交中稻。两省的平均产量都为7.5吨／公顷；湖南省约有100万公顷杂交晚稻，其平均产量为每公顷6吨，而常规水稻的平均产量每公顷只有4.5吨（常规稻与杂交稻的种植面积几乎相等）；辽宁省的杂交粳稻产量范围在每公顷7.5～8吨。另外还有一些小面积高产纪录。例如：

（1）1983年，威优6号在湖南省醴陵县作晚稻种植有24 000公顷，平均产量为每公顷7.6吨。

（2）1983年和1984年，北京南部的一个农场将汕优2号作中稻种植62公顷，其平均产量达每公顷11.3吨。

（3）在辽宁省，杂交粳稻黎优57的最高产量达每公顷13.7吨。

（4）在江苏省，籼型杂交中稻的最高产量为每公顷14.4吨。在福建省种植的双季杂交水稻产量达每公顷22.6吨。

表2 1985年水稻的面积和产量

项目	面积 / 百万公顷	总产 / 百万吨	单产 /（公斤 / 公顷）
总水稻	31.8	166.9	5 248
常规稻	23.4	112.3	4 799
杂交稻	8.4	54.6	6 474
杂交稻 / 总水稻 / %	26.4	32.7	123.4
杂交稻 / 常规稻 / %	35.9	48.6	134.9

从1976年1985年，由于种植杂交水稻而增产的累计值超过9 400万吨。以上事实表明，在中国发展杂交水稻是提高粮食生产的战略措施。

3. 三系选育

自20世纪70年代初期和中期"野败"型、"冈型"和"BT"型细胞质雄性不育系、保持系、恢复系相继问世以后，又发现了许多其他类型的细胞质雄性不育系及其保持系和恢复系。但其中仅有少数用于生产，原因在于许多来自其他细胞质源的不育系都有一些缺点。如开花习性不好，或找不到较好的恢复系，或在不同环境条件下表现育性不稳定等。现在有7种细胞质雄性不育系被用于生产（表3）。

表3 生产上应用的各种细胞质源的不育系

类型	细胞质源		代表不育系	面积（万公顷）	注
	种	品种或品系			
WA	*O.Sativa f. Spontanea*	Male sterile Wild rice	Zhen Shan 97A V20A	810	*Indica sporophytic*
GA	*O.Sativa L.*	Gambiaca	Chao Yang 1A	7	*Indica sporophytic*
Di	*O.Sativa L.*	Dissi	Di Shan A	7	*Indica sporophytic*
DA	*O.Sativa f. Spontanea*	Dwarf wild rice	Xie Qing Zao A	5	*Indica sporophytic*
HI	*O.Sativa f.Spontanea*	Common wild rice	Qin Si Ai A	3	*Indica.gametophytic*
BT	*O.Sativa L.*	Chinserah Boro II	Li Ming A	11	*Japonica.gametophytic*
TI	*O.Sativa L.*	O-Shan-Ta-Bai	Lu I Qian Xin A	1	*Japonica.gametophytic*
Ohtres				1	

我们在"三系"选育方面的经验可归纳如下：

（1）远缘杂交是选育胞质遗传雄性不育系的有效途径。用原始类型作胞质源，用改良型作核源。

（2）在籼稻中，通过测交筛选现有优良品种，较易获得恢复系。

（3）恢复基因以及雄性不育诱导胞质通过各种育种程序可转移到任何理想的品种中。

4. 杂交水稻制种

杂交水稻种子生产与常规稻不同，它包括两个步骤：

（1）不育系繁殖。

（2）杂种一代的种子生产。

现在，不育系繁殖、制种和大田生产之间的土地面积之比约为 1：50：2 500。

目前在中国，杂交稻制种田的总面积约 15 万公顷，平均产量为每公顷 1.6 吨。在湖南省，2 万公顷制种田的平均产量为每公顷 2.1 吨，1985 年最高产量接近每公顷 6 吨。杂交稻的用种量每公顷 30 公斤，而常规稻则需 110 公斤。

1981 年前，制种平均产量很低（每公顷仅约 0.75 吨）。

近年来制种产量的大幅度提高是由于制种技术的不断改进。其要点是：

①不育系单穴插植数由过去的单苗改双苗。

②扩大父母本行比，由 1：6 至 2：8 改为 1：8 至 2：12，甚至更大。

③大剂量使用"九二〇"（每公顷 75 克）使不育系穗部高于剑叶，从而免去割叶。

④通过良好的和特殊的田间管理促进前、中期植株生长，控制后期剑叶生长。

5. 近年来的进展

最近，中国的水稻科学家正在以下几个方面作进一步的努力：

（1）选育高产早熟或极早熟的杂交稻。中国现有的籼稻组合都是中、迟熟类型，不能在长江流域作双季早稻。而长江流域是水稻主产区，其产量多年徘徊在每公顷 6 吨左右。选育早熟和极早熟组合始于 20 世纪 70 年代中期，但只是在最近两年才取得突破（表 4）。目前已有几个新组合的生育期都在 110 天内，大面积产量可达每公顷 8～9 吨，现已在长江流域作早稻推广种植。例如：1986 年在湖南省双峰县种植的 70 公顷威优 49 示范田平均产量达每公顷 8.7 吨，而邻近的常规稻只有 6.7 吨／公顷。预计今后这些新组合在提高粮食总产方面将起重要作用。

表4　早熟杂交稻 V20A/ 测 49 的产量潜力

年份	品种	生育期/天	产量/（公斤/公顷）	为对照产量的百分数 /%	生产率*	为对照产量的百分数 / %
1984	威优 49	120	9 053	108	80.1	115
	威优 64（CK）		8 382		69.7	
1985	威优 49	112	10 092	116	90.0	119
	湘矮早 9 号（CK）	115	8 700		75.7	

注：*播种至收获的日产量。数据采集于湖南省区域。

（2）选育不同细胞质源的雄性不育系。目前生产中所有的不育系，有 95% 以上属于"野败"型。从长远的观点看，细胞质单一可能使杂交水稻遭受病害的毁灭性危害。所以，有必要用其他不同细胞质源的优良雄性不育系来取代部分"野败"系统的不育系。近年来，已培育出一些新质源的不育系并已用于生产。如"DA"型、"Di"型等（表 3）。

（3）培育异交率高的雄性不育系。杂交稻种子价格较高限制了推广。提高杂交稻制种产量、降低种子生产成本的最有效途径是培育异交率高的不育系。如最近选育出来的一些新的不育系就具有较好的开花习性，柱头较大、外露率高。这些新不育系的自然异交结实率比原来的不育系高出 30%～50%。

（4）中国大面积生产的杂交稻产量较高，但米质欠佳。问题在于所用的不育系米质都不好，因此导致杂交稻的米质下降。为了提高杂交稻的米质，双亲都应具有较好的米质，并且某些性状要一致。基于以上观点，去年已培育出了一个新的杂交稻组合，其优良的米质达到了美国的一级标准（表 5）。这个成功事例表明，通过精心设计的育种计划，是能获得既优质又高产的杂交稻的。

表 5　L301A/R29 的米质

粒型	碱消化值	直链淀粉含量	精米产量		腹白度
			精米率	整米率	
长	2.0	23%	70%	57%	0～1

（5）在某些地区由于生理小种的变化，一些优良的杂交稻组合已丧失其对病虫害的抗性。例如在四川省，由于稻瘟病 B 群生理小种的发生和流行，约 7 000 公顷汕优 2 号（抗 G 群生理小种）遭受稻瘟病侵袭，损失严重，因此必须由新组合来接替。幸运的是，大多数抗性性状是由显性或部分显性基因控制。所以，只要亲本选择得当，特别是利用国际水稻研究所最

近育成的、具有多抗性状的品系，就能育成高产、多抗的杂交水稻。威优64就是这样育成的一个新组合。它不仅具有高产稳产潜力，而且抗5种主要病虫害——稻瘟病、白叶枯病、黄矮病、褐稻飞虱和稻叶蝉（表6）。

表6　威优64的抗性鉴定

年份	季别	稻瘟	白叶枯	纹枯	黄矮	褐稻虱
1982	晚季		S	MR	MR	
1983	晚季	MR	MR	LS	R	MR
1984	晚季	R	S	LS		

注：数据来源于国家区试资料。

（二）其他国家的杂交稻研究

在中国以外的其他地方，杂交稻的研究仍处于探索阶段。主要在以下几方面进行研究：

（1）产量和其他性状的优势和配合力；

（2）细胞质遗传雄性不育和育性恢复；

（3）胞质雄性不育系的自然异交性和杂交稻制种技术；

（4）杂交稻的米质。

另外，还开展了一些与上述主要研究领域有关的辅助性基础研究，以提高杂交稻育种计划的效率。

1. 杂种优势和配合力研究

这方面的研究工作做得最多。现已有充足的依据证明：产量和与产量有关的农艺性状都存在显著的正优势和超亲优势。但在中国以外的地方，这些研究结果仅在近七年才被应用并进行深入研究，以培育出适应于生产的F_1杂交种。国际水稻研究所、印度、印度尼西亚、南朝鲜和美国都在开展这项研究。

1979年，三个从中国引进的优良杂交水稻组合（即汕优2号、汕优6号和威优6号）在国际水稻研究所试种时，由于对热带病虫害的抗性不够，并不比最好的常规稻品种（IR36和IR42）增产。因此，从中国直接引进F_1杂交组合的试种工作就停止了。而是用一批品种间杂交组合以及从中国引进的胞质雄性不育系，与来自国际水稻研究所、印度尼西亚、印度及南朝鲜的恢复系所配组育成的杂交组合同最佳地方对照品种进行品比试验。

在国际水稻研究所的试验中，参试的400多个杂交组合，有几个产量显著高于相应的对照品种。表 7 列出了最佳杂交组合的平均产量与最佳常规对照品种的平均产量。总的来说，最佳杂交稻组合的产量比最佳地方对照品种增产 20%~30%。

表7　1980—1986 年国际水稻研究各种试验中的最佳杂交稻的产量与最佳常规对照相比较的增产幅度

季别	杂交稻	年份	产量 /（吨/ 公顷）	为对照产量百分数 /%
	IRI1248-242-3/IR15323-4-2-1-3	1980	5.9	122
	珍汕 97A/IR3420-6-3-3-1	1981	6.2	123
	珍汕 97A/IR54	1982	4.4	113
雨季	IR46828A/IR54	1983	5.3	112
	IR46828A/IR54	1984	4.5	140
	IR54752A/IR13419-113-1	1985	5.1	107
	平均		5.2	120
	IET 3257/IR2797-105-2-2-3	1981	10.4	132
	IET 3257/IR2797-105-2-2-3	1982	8.9	135
	IET 3257/IR42	1983	9.6	124
旱季	IR29799-17-3-1-1A/IR2797-125-3-2-2	1984	7.2	108
	IR46828A/IR13524-21-3-3-3-2-2	1985	5.4	123
	IR54754A/IR46R	1986	7.4	119
	平均		8.2	123
	总平均		6.7	122

表8　1980—1985 年，印度尼西亚、南朝鲜和印度进行的各种试验中，最佳杂交稻的产量和与最佳常规对照相比的增产幅度

国家	试验数目	最佳杂交稻的产量 /（吨／公顷）		为对照产量百分数 %	
		变幅	平均	变幅	平均
印度尼西亚	11	4.2~8.9	6.2	102~170	123
南朝鲜	9	8.1~11.5	9.0	97~142	113
印度					
亚热带	5	6.2~9.8	8.1	103~130	120
热带	14	3.3~7.3	5.6	91~132	112
总数	39	3.3~11.5	7.1	91~170	116

　　表 8 表明，1980—1985 年在国际水稻研究所以外的各个国家进行的试验中，一些试种杂交组合比当地最佳对照增产 13%～23%，其中 V20A/ 密阳 46、V20/IR54、珍汕 97A/IR2307-247-2-3 和 IR45831A/IR54 在一个以上试验地表现很好，但它们对病虫害的易感性和主要受母性遗传影响而形成的较差米质，阻碍了推广。近来，用国际水稻研究所新近育成的胞质雄性不育系所配的组合已有较好表现（表 9）。

表 9　一些用国际水稻研究所新培育的不育系配出的有希望的杂交稻及对照品种的产量和生育期（国际水稻研究所，1986 年旱季）

组别	杂交稻 / 对照	生育期 / 天	产量（吨 / 公顷）	为对照产量百分数 %
第一组	IR54754A/IR46R	126	7.4	119*
	IR64（对照）	126	6.2	100
第二组	IR54754A/ARC 11353R	133	7.9	141*
	IR54753A/IR46R	126	7.0	125*
	IR54752A/IR46R	126	7.0	125*
	IR54752A/ARC 11353R	131	6.8	121*
	IR64（对照）	126	5.4	96
	IR54（对照）	130	5.6	100

注：* 表示与对照相比，差异达 5% 显著水平。

　　在美国，1983 年和 1984 年从中国引进一些杂交稻品种，作为圆环种子公司的专卖品种在得克萨斯密西西比、路易斯安那、阿肯色斯和佛罗里达进行多点试种。1984 年，杂交稻的产量比籼稻对照增产 14%，比南方品种增产 19%。这些组合的米质却不能适应美国市场需要。

　　南朝鲜水稻科学家 Kim 在 1985 年报道了不同肥力水平的杂交稻的产量优势（表 10）。

表 10　在不同肥力水平下，4 个杂交稻的平均优势、超亲优势和竞争优势

组合名称	N 水平 /（公斤 /10 公亩①）	产量 /（公斤 /10 公亩）		平均优势 /%	超亲优势 /%		竞争优势 / %（与水源 294 相似）	
		1983	1984	1984	1983	1984	1983	1984
V20A/ 水源 287	12	—	924	56**	—	22**	—	17**
	18	999	963	64**	16**	23**	—	20**
	24	1.005	883	70**	17**	32**	-2	20**
	平均	1.002	923	63	17	26	-3	19

① 公亩：1 公亩＝ 100 平方米，后同。

续表

组合名称	N 水平/(公斤/10公亩)	产量/(公斤/10公亩)		平均优势/%	超亲优势/%		竞争优势/%(与水源294相似)	
		1983	1984	1984	1983	1984	1983	1984
V20A/密阳46	12	—	1.070	86**	—	48**	—	36**
	18	1.025	1.148	92**	21**	43**	16**	43**
	24	1.203	989	84**	12*	40**	17**	34**
	平均	1.204	1.069	87	17	44	17	37
珍汕97A/水源287	12	—	900	55**	—	19**	—	14**
	18	1.085	956	63**	26**	22**	4	19**
	24	1.079	877	67**	25**	31**	5	19**
	平均	1.082	911	62	26	24	5	18
珍汕97A/密阳46	12	—	1.038	84**	—	43**	—	32**
	18	1.103	1.086	81**	11*	36**	6	35**
	24	1.082	988	82**	1	41**	5	34**
	平均	1.037	1.093	82	6	40	6	34

注：* 表示达 5% 显著水平；** 表示达 1% 显著水平。

关于影响产量的其他农艺性状的杂种优势也有报道。这包括干物质生产、收获指数、叶面积、作物生长率、叶绿素含量、每穗粒数和每平方米粒数以及粒重。另有报道杂种优势亦表现在早熟、根长、根重和根的抗拉强度等方面。根系的杂种优质对于杂交稻在某些"天水田"环境种植是很有益的。单位面积日产量图显示单位面积的日产量较高（图1），这表明杂交稻具有很高的效能。配合力研究表明：在产量方面存在着显著的一般配合力和特殊配合力效应（表11）。在某些试验中，特殊配合力产生的方差大于一般配合力，这说明是由显性基因优势所致，尽管也有研究证明加性基因作用同等重要。国际水稻研究所的一个试验结果是：大多数特殊配合力高的组合其亲本至少有一个一般配合力是高的（表12）。只是在少数情况下，双亲一般配合力较低，而杂种的特殊配合力高。所以，在选配亲本时，至少应有一个亲本的一般配合力是高的。

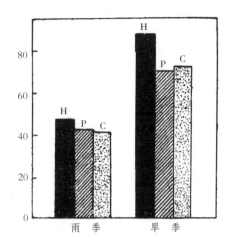

图 1　杂交稻（H）及其亲本（P）和对照品种（C）的
单位面积日产量［公斤/（日·公顷）］
（1980—1981 国际水稻所雨季、旱季）

表 11　产量配合力分析的研究总结

配组设计	配合力效应		变异		方差期望值 特殊 / 一般	参考文献
	一般配合力	特殊配合力	一般	特殊		
4×4Dᵃ	—	—	**	NS	—	Sivasubramanian and Menon（1973）
5×5D	***	***	321.9	133.3	2.4	Rahman et al.（1981）
4×4D	**	**	40.5	143.7	3.5	Mohanty and Mohapatra（1973）
6×6Dᵃ	**	NS	—	—	—	Ranganathan et al.（1973）
6×6D	**	**	5.2**	12.5**	2.4	Singh and Nanda（1976）
8×8D	NS	**	8.9	22.4	2.5	Singh et al.（1977）
6×6Dᵃ	**	**	118.0***	118.0***	1.0	Singh（1977）
7×7D	**	**	1.5**	53.5**	35.7	Maurya and Singh（1977）
7×7D	**	**	—	—	1.0	Rao et al.（1980）
5×5D	**	NS	—	—	1.0	Hapue et al.（1981）
6×6Dᵃ	**	**	23.34	26.09	1.1	Zhao and Rui（1982）
6×6D	***	***	295.47***	736.70***	2.5	Singh et al.（1980）
15×15D	**	**	144.11***	39.49***	0.27	Shrivastava and Seshe（1983）
6×6D	***	***	361.24***	315.49***	0.87	Kumar et al.（1975）
8×2L	**	**	49.23***	31.47**	0.64	Anandakumar and Rangasamy（1983）
10×10D	***	***	91.88***	11.58***	0.13	Subramanian and Rathinam（1984）

注：①a 表明双列杂交包括正反交，其余没有正反交。
　　②*、**、*** 分别表示达 10%、5% 和 1% 显著水平。

表12　具有不同类型一般配合力及相应特殊配合力的亲本所配组合的数目与百分频率
（国际水稻研究所，1985旱季）

亲本一般配合力	表现特殊配合力效应的组合数			总组合数
	+	0	−	
+/+	6 (55)	2 (18)	3 (27)	11
+/− or −/+	8 (36)	− (0)	14 (64)	22
+/0 or 0/+	5 (26)	3 (16)	11 (58)	19
−/0 or 0/−	5 (42)	4 (33)	3 (25)	12
0/0	− (0)	2 (50)	2 (50)	4
−/−	1 (14)	1 (14)	5 (72)	7
总数	25	12	38	75

2.细胞质遗传雄性不育及育性恢复

细胞质导致水稻雄性不育，最先是由 Weeraratne 以及 Sampath 和 Mohanty 在 1954 年报道的。第一个栽培稻细胞质雄性不育系是由日本的 Shiniyo 和 O'mura 1966 年用籼稻包罗Ⅱ和粳稻台中 65 育成。随后，Erickson 和 Carnahan 分别在 1969 年和 1972 年用一个粳稻品种和另一个籼稻 Birco（PI279120）作细胞质源育成了细胞质雄性不育系。1972 年在国际水稻研究所，Athwal 和 Virmani 用一个细胞质源为台中（本地）1 号的籼稻与 Pakhari203 作核源育成细胞质雄性不育系。Cheng 和 Huang 在 1979 年也用 Oryzg rufipoqon 作细胞质源与台中 65 育成细胞质雄性不育系。所有这些不育系中，以 Shiniyo 和 O'mura 1966 年育成的不育性最稳定。此雄性不育系已在中国被广泛用于培育杂交粳稻。中国以外地方育成的其他细胞质雄性不育系或由于不育性不稳定，或由于找不到理想的恢复系而未能应用。

中国通过用 *Oryza sativa f.spontanea* 培育出细胞质雄性不育系后，将此细胞质源命名为"野败"（WA）。

国际水稻研究所和其他几个国家应用回交育种法，将"野败"质源转移到他们选育的优良品系中，从而培育出新的不育。所用的野败型不育系有：V20A、珍汕 97A、二九南 1 号 A、

V41A 等。之所以有必要开展这项工作，主要是因为中国的不育系对主要病虫害的易感性和较差的米质而不能培育出适合中国以外地区的杂交水稻。表 13 列出了国际水稻研究所和其他国家用自己的优良品系转育成的"野败"型不育系。其中一些不育系非常适合热带地区，米质优良，配合力好，异交结实率高。因此有希望用这些不育系选配出适合中国以外地区的杂交水稻。

在恢复系方面，已鉴定出"BT"型、"野败"型和"Gam"型恢复系。而 Pankhari 203A 的有效恢复系则尚未鉴定出来。有效恢复系品种主要分布在热带籼稻生长区。在国际水稻研究所培育的品种和品系中约有 20% 对"野败"型不育细胞质具有恢复能力，粳稻品种中的恢复系频率则微乎其微。南朝鲜的许多 Tongi l 水稻品种（籼 × 粳的后代）恢复能力较好，其恢复基因应当是由它们的籼稻亲本遗传而来的。表 14 列出了在国际水稻研究所等地鉴定出的有效恢复系。来自中国的报道认为：恢复系出现的频率，迟熟品种高于早熟品种，这也许是因为迟熟品种的亲缘关系较接近野生稻。但是考虑到国际级和国家级育种计划中，不同生育期的品种间杂交的规模不同，这种相关性在改良品种或品系中已不复存在。

表 13　国际水稻所和其他国家优良品系 / 品种从野败转育的新不育系

不育系	选育地	来自优良品系 / 品种	来源
IR 46826A	国际水稻研究所	IR 10154–23–3–3	国际水稻研究所
IR 46827A	国际水稻研究所	IR 10176–24–6–2	国际水稻研究所
IR 46828A	国际水稻研究所	IR 10179–2–3–1	国际水稻研究所
IR 46829A	国际水稻研究所	IR 19792–15–2–3–3	国际水稻研究所
IR 46830A	国际水稻研究所	IR 19807–21–2–2	国际水稻研究所
IR 46831A	国际水稻研究所	Jikkoku Seranni 52–37	印度
IR 48483A	国际水稻研究所	MR 365	印度
IR 54752A	国际水稻研究所	IR 21845–90–3	国际水稻研究所
IR 54753A	国际水稻研究所	IR 19657–87–3–3	国际水稻研究所
IR 54756A	国际水稻研究所	Iri 356	南朝鲜
IR 54757A	国际水稻研究所	Suweon 310	南朝鲜
IR 54758A	国际水稻研究所	PAU 269–1–8–4–1–1–1	印度
Madbu A	印度	Madhu	印度
CRMS 1	印度	–	印度
CRMS 2	印度	–	印度
TNMS 31A	印度	–	印度
TNMS 37A	印度	–	印度
TNMS 47A	印度	–	印度
HR 7017A	南朝鲜	Samkangbyeo	南朝鲜
HR 7019A	南朝鲜	Hankagchalbyeo	南朝鲜

表14　在国际水稻研究所、南朝鲜、印度和印度尼西亚鉴定出的一些优良恢复系

恢复系	来源	恢复系	来源
IR 24	国际水稻研究所	水源　294R	南朝鲜
IR 26	国际水稻研究所	水源　332R	南朝鲜
IR 36R	国际水稻研究所	水源　333R	南朝鲜
IR 42R	国际水稻研究所	Iri　347	南朝鲜
IR 48R	国际水稻研究所	Iri　371	南朝鲜
IR 50R	国际水稻研究所	Sadang	印度尼西亚
IR 52R	国际水稻研究所	Sumeru	印度尼西亚
IR 54R	国际水稻研究所	Krueng Aceh	印度尼西亚
IR 56R	国际水稻研究所	CO　39	印度
IR 62R	国际水稻研究所	CR　1009	印度
IR 64R	国际水稻研究所	Intant mutant	印度
IR 9761-19-1R	国际水稻研究所	Pankaj	印度
IR 13146-243-2-3R	国际水稻研究所	Radha	印度
IR 19661-63-1-2-3-1-3R	国际水稻研究所	Vani	印度
IR 28228-119-2-3-1-1R	国际水稻研究所	Pusa　33	印度
IR 29723-143-3-2-1R	国际水稻研究所	ADT　36	印度
IR 33043-46-1-3R	国际水稻研究所	Krishna	印度
密阳　46R	南朝鲜	PR　106	印度
密阳　54R	南朝鲜	PAU　102-8-3-10	印度
水源　287R	南朝鲜	PAU　50-9	印度

3. 细胞质雄性不育系的自然异交结实与制种技术

花器结构、开花和花药开裂的特有方式，使水稻成为严格的自花授粉作物。栽培品种的自然异交结实率变幅0%~6.8%。至于野生稻（O.perennis Moench.），曾观察到其异交结实率为16.5%~100%。雄性不育系植株的自然异交结实率已有报道，其变幅为0%~44%。Azzini 和 Retger 观察"Birco"不育系的自然异交结实率为5%~32%。在国际水稻研究所，我们观察到不育系的自然异交结实率变幅为3%~43%。在印度，观察到一些不育系的自然异交结实率为10%~25%。

水稻开花的高峰期，晴天从上午9时到11时。国际水稻研究所的研究表明：开花高峰时间有种的差别：O.glaberrima 约有60%的颖花在上午9时开花，而此时 O.sativa 栽培种 IR36 只有5%的颖花开花。开花习性和特性的品种间差别对异交结实率的影响已有大量报道。Parmar 等发现品种的生育期长短与开花时间长短有直接关系。另外发现授粉延迟或失败使开花延长，结果导致不育系的开花时间比可育的保持系开花时间长。高温能使开花延迟并略微延长开花时间。

Virmani 和 Edwards 在所发表的文章中报道了对水稻异交结实率有影响的开花特性的变异范围和理想的开花特性资源。遗传研究表明，花药长度、柱头长度和柱头外露率等性状是由多基因控制的。在其遗传过程中，加性和非加性基因都有重要作用。国际水稻研究所现在正试图将 *Oryza longistaminata* 的大柱头和大花药特性转育到具有改良株型的栽培品种中，此工作已取得了进展。

中国发展的杂交水稻制种技术已在国际水稻研究所和印度尼西亚以及南朝鲜进行了小区试验。制种产量范围在国际水稻所为 10~1 578 公斤 / 公顷，在印度尼西亚为 600~1 800 公斤 / 公顷。1985 年，朝鲜的科学家得到 750~1 530 公斤 / 公顷的产量。制种产量取决于父母本花期相遇、季节、农艺性状和开花特性。一般地讲，制种产量将随经验积累而增加。

4. 杂交稻的米质

由于稻米在大多数情况下是以米粒形式供食用的。所以人们很关心杂交稻的米质是否可能由其化学特性方面的遗传分离而致蒸煮品质下降。

目前，中国以外地方的杂交稻研究人员主要致力于探索杂交稻的增产潜力，而对其米质注意不够。不过，也并没有完全忽略这个问题。现在看来杂交稻米质的化学特性的遗传分离并不影响米质。只要亲本选择适当，是可以配出具有理想米质的组合。

结　论

杂交水稻在中国大面积栽培以及在国际水稻研究所和其他国家进行的试验表明，杂交水稻技术使水稻产量更接近其生理产量潜力水平。杂交稻已显示出比改良的半矮秆品种的极限产量增产 15%～20%（约每公顷 1 吨）。中国已经非常广泛地应用了这一技术。预期在中国以外地区的前景也很光明，特别是在农民已经得到很高产量的灌溉地区更是如此。由于杂交稻发达的根系以及前期秧苗和营养优势，有可能适应低洼天水田及各种不良土壤和气候条件。巴西科学家正在探索其在旱土种植的可能性。

现在已拥有一批发展杂交稻的遗传工具（即不育系、保持系和恢复系），适合不同产稻国家采用的三系亲本材料不久即将问世。因而有可能选育适合所有国家的杂交水稻。

目前种植的杂交稻 90% 以上属"野败"质源，这是因为"野败"在各种条件下不育性最稳定而且恢复性也较好。因而，"野败"正被转育到各种遗传类型中，以育出适合其他国家的雄性不育系。这种状况有可能使杂交稻容易招致与"野败"细胞质有关的病虫害。所以，中国和国际水稻研究所现在都在鉴定新的、多种多样的不育细胞质，并用之选配新的组合。一种有效的化学杀雄剂在这方面非常有用。中国已鉴定了一批杀雄剂，但因其安全问题及效果不稳定而用途非常有限。

中国已非常成功地发展了制种技术，并在继续努力以不断提高制种产量。从小面积试验看，此技术也适合其他国家采用。尽管存在劳动强度大的问题，然而在大多数劳力充足和低廉的水稻生产国是适用的。事实上，杂交稻制种技术将有助于增加农村就业机会。选育具有理想开花习性的亲本也将有助于提高异交结实率，进而增加制种的单位面积产量。

杂交稻的主要问题是种子价格高，每季都须换种以及依靠外源供应种子。中国已建立了有效的基地进行种子生产、加工、鉴定和销售。但很难说其他国家也能建立这样的基地。一个国家如果没有一个有效的种子生产体系，杂交稻技术就难以获得成功。其他国家能否应用这种技术取决于：①在单位面积上需要生产更多粮食的压力大小。②杂交稻比最佳常规稻增产的幅度。③国家能否以公有或私有形式组织一个有效的体系来生产、加工、鉴定和销售种子。

作者：袁隆平　费马尼

注：本文汇编于《首届杂交水稻国际学术讨论会（中国长沙，1986 年 10 月 6—10 日）论文集》。

中国的杂交水稻

一、引言

杂交水稻的研究成功，是水稻育种上的一项重大突破，也是水稻生产上的一项技术革新，它为大幅度提高水稻产量提供了有效的新途径。1981年，这项科研成果获得了中国第一个特等发明奖。

中国是世界上第一个在生产上利用水稻杂种优势的国家。1964年开始杂交水稻研究，1973年实现了"三系"配套，1974年选育出强优组合，1975年研制出一整套制种技术，1976年开始在生产上大面积推广，从那以后，杂交水稻的种植面积逐年迅速扩大。目前，全国水稻种植面积约为4.95亿亩，而杂交水稻将近占四分之一（表1）。

表1　全国杂交水稻收获面积

年份	面积 /（百万亩）
1976	2.25
1977	31.95
1978	64.95
1979	76.05
1980	73.95
1981	76.50
1982	84.00
1983	101.25
1984	123.00
合计	633.90

多年来的生产实践证明：只要栽培措施得当，杂交水稻一般比常规良种增产20%左右。从1976年到1984年，因推广杂交水稻而累计增产稻谷近700亿斤。由此可见，发展杂交水稻对于提高粮食产量具有十分重要的战略意义。

二、中国杂交水稻的分布

目前，杂交水稻已在中国各地广为种植。从北纬 43° 的辽宁（温带）到北纬 18° 的海南（热带），从东经 125° 的上海到东经 95° 的云南，处处都种植有杂交水稻。与此同时，还选育出适合在不同气候和土壤条件下种植的各种杂交水稻类型和组合。比如：

1. 生育期长、对光照敏感的籼型杂交组合，如珍汕 97A×IR30，可在珠江流域作晚稻栽培。

2. 中熟籼型杂交组合，如珍汕 97A×IR26、V20A×IR26、珍汕 97A×明恢 63。可在珠江流域作早稻和晚稻栽培，在长江中游的双季稻区普遍作晚稻栽培，或在一季稻区作中稻栽培。

3. 早熟籼型杂交组合，如 V20A×IR9761、V20A×二六窄早，在长江以南的双季稻区作早稻栽培。

4. 早熟粳型杂交组合适合在黄河以北地区作一季稻或麦茬稻。

5. 生育期长的粳型杂交组合可在长江流域作晚稻栽培。

中国杂交水稻种植面积最大的三个省是四川、湖南、广东，每个省的种植面积都在 1 800 万亩以上；江苏、江西、福建次之，其种植面积在 900 万到 1 200 万亩之间。

杂交水稻组合的产量优势是很明显的，表 2 中列举 1983 年全国因推广种植杂交水稻而增产的情况。

表2　1983 年的水稻种植面积和产量

项目	面积 /（百万亩）	总产量 /（百万吨）	单产量 /（斤 / 亩）
全部水稻	495	168.9	682.4
杂交水稻	101.25	43.3	849.4
杂交水稻 / 全部水稻	20.5%	25.6%	124.5%

四川和江苏分别种有 1 950 万亩和 1 200 万亩杂交中稻。这两个省的平均产量为 1 000 斤 / 亩。湖南省大约种植 1 500 万亩杂交晚稻，平均产量为 800 斤 / 亩，而常规品种（其种植面积与杂交水稻的种植面积接近）的平均产量为 600 斤 / 亩。最近几年，辽宁省粳型杂交组合的产量达 1 000~1 100 斤 / 亩。另外，各地还创造了许多小面积的高产记录，如：

1. 湖南省醴陵县 1983 年种植 36 万亩杂交晚稻（威优 6 号）的平均产量为 1 013.4 斤 / 亩。

2. 位于北京以南的一个军垦农场，1983 年和 1984 年种植 930 亩杂交中稻（汕优 2 号），每亩平均产量达 1 500 斤。

3. 辽宁省的粳型杂交组合（黎优 57）的最高产量为 1 826.6 斤 / 亩。

4. 江苏省作中稻栽培的籼型杂交组合的最高产量为 1 920 斤 / 亩，福建省的双季杂交水稻的最高产量为 3 013.4 斤 / 亩。

三、杂交水稻研究的历史背景

在过去 30 年中，中国的水稻高产育种工作可分为三个主要阶段。20 世纪 50 年代，育种目标主要是通过系统选择和品种间杂交改良本地高秆品种。60 年代，由于化肥使用量增加和使用密植，倒伏成为水稻高产的重要障碍，于是，育种目标转向半矮秆品种的选育，并在产量上取得突破。一般说来，半矮秆品种要比高秆品种平均增产 20%～30%。但是，半矮秆品种很快就达到了它的产量极限。在十几年当中，尽管不断地选育和推广了许多新的半矮秆品种，但是其产量潜力始终没有明显增加。就日产量而言，新品种的产量水平几乎与第一批推广的半矮秆品种相差无几。后来，许多科学家认识到：半矮秆品种的选育仅仅是通过改良形态来提高产量的一种方法，要想进一步提高水稻产量潜力，必须把形态改良和生理功能的提高结合起来。70 年代，育种家们探索了多种非常规方法，如花粉培养、远缘杂交、杂种优势利用，高光效植株的筛选等。实践证明，迄今为止，其中水稻杂种优势利用，是最有效和最成功的方法之一。

四、杂交水稻的研究工作

要使杂交水稻在生产上大面积推广使用，必须实现三项必不可少的技术要求：

1. 选育适用的细胞质雄性不育系、保持系和恢复系（简称"三系"）。

2. 培育出在产量和其他性状方面远比最佳常规品种优越的 F_1 代杂种。

3. 通过异交授粉能够获得大量的杂交种子。

五、三系选育

为了在生产上大面积利用水稻杂种优势，首先必须生产大量的 F_1 代杂交种子。但水稻是自花授粉作物，花器又小，特别是每朵颖花只能结一粒种子。因此，通过人工去雄生产大量杂交种子是不可能的。到目前为止，在其他自花授粉作物中，克服这一困难最有效的方法业已证明是利用细胞质雄性不育系。

1964 年，我们开始了水稻雄性不育系的研究，试图按照杂交玉米育成的方法，利用雄性不育突变体作为培育杂交水稻的原始材料。我们在抽穗期对稻穗逐个进行仔细观察，发现了几株具有不同形态特征的雄性不育株。这些不育株就被用来作为杂交水稻育种的第一批材料。

然而，所有这些雄性不育类型的不育性，均受隐性核基因的控制，通过各种不同的育种方法，都无法获得满意的保持系，于是，我们在1970年改变了育种策略，开始采用远缘杂交，由核不育育种改为核质互作雄性不育育种。

1970年，在为远缘杂交收集野生稻资源过程中，我的助手李必湖在海南发现了一株雄性不育野生稻（*O. sativa*.L.f. *spontanea*），后被命名为"野败"。这个发现在杂交水稻育种上是一个突破。

1971年，杂交水稻研究课题被列为全国协作项目，由中国农科院和湖南农科院负责主持，这个"野败"材料被分发到其他水稻科研单位，以便协同攻关。通过广泛测交和连续回交，1972年，我和其他育种工作者选育出了第一批野败型雄性不育系和保持系。到现在为止，已有200多个来自野败质源的雄性不育系，其中在生产上普遍采用的珍汕97A和V20A已回交了30~35代，在不同的环境条件下始终保持雄性不育稳定性。

1973年，从已有品种中筛选出第一批恢复系，如普遍采用的IR24和IR26恢复系引自国际水稻研究所。近年来，采用不同的育种方法还选育了许多优良的恢复系，从而配制出很多早熟、高产、多抗和优质的优良杂交组合。

继"野败"型之后，还发现了其他几种雄性不育细胞质可用作选育新型雄性不育系的材料。目前在生产上广为采用的，除了野败型外，还有另外两个细胞质源的不育系。一个是冈型（G型）（其不育细胞质来源于迟熟的籼稻品种）。在四川省，用这个类型的不育系配制的杂交组合的种植面积约150万亩。另一个是1972年引自日本的"包台"型（BT型）。由于BT型三系属高秆，同时F₁代杂种不具有优势，因而不能用于生产，这个类型的不育系遂被转育而成黎明A和农虎26A等。另外，还成功地选育了一些较好的恢复系，如C-57、培C115、77302，它们与粳稻不育系配组表现出明显的杂种优势。近两年，广东省还在生产上大面积采用另一种籼稻不育系——青四矮A（属于配子体雄性不育）。

我们选育三系的经验归纳为以下三点：

1. 选育细胞质遗传雄性不育系最有效的方法是以进化程度较低的类型为母本进行远缘杂交。

2. 通过测交从已有优良品种中筛选恢复系是一条较为便捷的途径。

3. 采用不同杂交方法，可以把恢复基因和不育细胞质转育到任何有希望的品种中。

六、水稻的杂种优势

1972年到1974年，湖南农科院以雄性不育系为母本，配制了53个杂交组合，并对水稻的杂种优势进行了研究，在分析和测定这些材料之后，我们得出五点结论：

1. 95% 以上的杂交组合的产量比其亲本增加 30%～70%。

2. 其中半数组合比对照常规优良品种增产 20%～30%。

3. 就稻谷产量而言，生育期长的杂交组合高于生育期短的，籼型高于粳型。

4. 用亲缘关系相近的亲本配制的杂交组合的产量与其亲本类似。

5. F_2 代的产量比 F_1 代低 50%。

1975 年，湖南农科院曾从已有的组合中选出一批优良组合（如南优 2 号和南优 3 号）进行产量试验（试验面积为 105 亩），取得了很大的成功。杂交组合的平均产量为 1 044 斤 / 亩，而最好的常规品种（对照）为 860 斤 / 亩。另一个例子是：1976 年在江苏省练湖农场曾对高产品种 IR24 和杂交水稻（二九南 A×IR24）进行一次大面积产量比较试验，两者种植面积均约为 10 000 亩。尽管 IR24 的生育期比该杂交组合长 5 天，IR24 的平均产量为 1 009.4 斤 / 亩，而该杂交组合为 1 140 斤 / 亩。1977 年又进行了重复试验，该组合的产量比 IR24 每亩高出 100 多斤。这些结果使大多数人，包括那些不相信水稻存在杂种优势的人完全信服了。

水稻杂种优势可以概括为四个方面：

1. 形态特征好，如根系发达，穗大粒多，粒重，秆硬，株型好。

2. 生理效能高，如光合强度，根系活力，分蘖力等均优于常规稻。

3. 抗多种主要病虫害，因为大多数抗性基因属显性和部分显性。

4. 遗传背景广，适应性好。

七、杂交水稻制种

雄性不育系作为一种工具仅仅提供了通过自然异交生产杂交种子的可能性。想要获得高产，实际上还有许多技术问题需要解决。例如，1973 年首次杂交水稻制种的产量是极低的，每亩仅 12 斤。后来，通过许多试验，我们才取得了很大进展，从而在 1975 年研制出杂交水稻制种的基本技术。1976 年，在海南进行大规模的制种，制种总面积达 3 万多亩，平均制种产量为 80 斤 / 亩。近些年来，制种技术逐渐完善和标准化，所以产量逐年增加。1983 年，全国制种面积为 210 万亩，总产 3.2 亿斤以上，湖南省制种面积为 37.5 万亩，平均产量为 227 斤 / 亩，最高的达到 720 斤 / 亩。由于制种的单位面积产量增加，杂交稻种子的价格减少了 30%～40%（从每斤 1.5～2.0 元降到 1.0～1.5 元），这有利于杂交水稻的进一步推广。

杂交水稻制种的技术要点是：

1. 调节父母本的播种期，使其花期相遇。

2. 利用时间或空间进行隔离，避免串粉。

3. 适宜的行比和行向。

4. 割掉剑叶，排除传粉的主要障碍。

5. 喷洒赤霉素，使不育系的稻穗从叶鞘中完全抽出。

6. 拉绳或用棍竿震动正在开花的父本稻穗，进行辅助授粉。

八、基本理论研究

在选育杂交水稻的同时，一些高等学校和科研单位进行了有关杂交水稻的基础理论研究，主要成果有以下几个方面：

1. 雄性不育系的分类：我国的水稻雄性不育系，根据遗传特点，可分为配子体不育系和孢子体不育系；按恢保关系可分为"野败"型、红莲型和包台型（BT 型）。

2. 遗传分析揭示：普遍采用"野败"型不育系的育性恢复受两对显性基因控制，三系的基因型是：不育系为 $S(r_1r_1r_2r_2)$，保持系为 $F(r_1r_1r_2r_2)$，恢复系为 $F(R_1R_1R_2R_2)$。

3. 同工酶分析表明：强优势组合在过氧化物同工酶和脂酶同工酶的酶谱方面具有以下三个特征之一：

（1）双亲的酶带；

（2）新酶带或多个酶带；

（3）高活性的酶带区。

4. 细胞学观察显示：不育系败育花粉从形态上可分为三类，即典败型、圆败型和染败型。第一类属于孢子体不育，其恢复系较少，但不育性稳定。第二类和第三类主要属于配子体不育，其恢复系较多，但不育性不如孢子体不育系稳定。

九、化学杀雄

广东和江西省正在进行化学杀雄研究。现有两种杀雄剂（杀雄剂 1 号和 2 号）已用于制种。在生产上应用较广的组合有"赣化 2 号"和"钢化青兰"等。化学杀雄的制种产量和纯度低于用不育系配制的。1983 年，用化学杀雄配制的杂交组合推广面积为 10 万亩左右。

十、杂交水稻制种的组织和体系

杂交水稻种子的生产过程不同于纯系品种，它包括两个步骤：

1. 不育系繁殖（不育系 × 保持系）。

2. 杂交种子生产（不育系 × 恢复系）。

目前，不育系繁殖田、制种田和大田栽培面积的比例为 1∶50∶2 500，即：繁殖 1 亩不育系，可供 50 亩制种田用，50 亩制种田生产的种子可供 2 500 亩大田栽培。

在繁殖和制种过程中，由于不育系具有异交结实的特性，杂交稻种子和不育系种子容易因串粉而导致不纯。为了确保杂交稻种子的纯度和质量，除了采用正确的技术外，还必须组织好种子生产体系。

当前，我国杂交水稻制种体系由三个层次组成：

1. 省种子公司负责三系提纯和原种生产。例如，湖南省就有四个由省种子公司管理的原种生产基地。三系原种的标准见表 3。

表 3　三系原种标准

项目	不育系 /%	保持系 /%	恢复系 /%
纯度	99.8	99.8	99.8
发芽率	93	98	13.5
全量水分	13.5	13.5	85
不育度	100		
恢复率			

2. 地区种子公司负责不育系的繁殖。一般由国营农场繁殖不育系，为县种子公司制种提供不育系种子。用于制种的不育系的纯度应达到 99%。

3. 县种子公司负责杂交稻种子生产，一般选择 1 到 2 个适合制种的地区，组织制种区周围的农民按照与县种子公司签订的合同进行制种，种子纯度应达到 97% 以上。

十一、杂交水稻的经济效益

对于制种者和栽培者来说，从以下计算数字可以看出，种植杂交水稻的经济效益是很明显的。例如在湖南省，制种田面积与大田栽培面积之比一般为 1∶50（即一亩制种田可供 50 亩大田用种），这与常规稻的种子生产情况大致相当（1∶50）。所以，杂交水稻的制种成本较高仅仅表现在要比常规稻多花 1/3 的劳动日以及施用赤霉素和较多的化学肥料（需要多投资约 10 元 / 亩）。

种植常规稻，每亩可生产 800 斤稻谷，需费 16 个劳动日，即一个劳动日生产 50 斤稻谷。杂交水稻制种，一亩制种田，需要 24 个劳动日，可生产 266.7 斤杂交稻种子（按目前杂交稻种子的价格相当于 1 867 斤常规稻）和 200 斤父本种子，其总产量为 2 067 斤/亩，即一个劳动日可生产 96 斤稻谷。

栽培杂交水稻，除每亩需 4 斤杂交种子的成本外，不需要另外的投资。杂交种子成本为常规种子的 2 倍，但其产量要较后者每亩增产 160 斤左右（表 4）。

表 4　栽培常规水稻、杂交水稻以及杂交水稻制种经济效益比较

项目	栽培		制种
	常规水稻	杂交水稻	
每亩需劳动日 / 个	16	16	24
每亩产量 / 斤	800	960	2 067
每个劳动日产量 / 斤	50	60	96
总产值 / 元	128	160	310
种子成本 / 元	2.4	4.7	6.60
额外成本 / 元			10.0
每亩收益 / 元		29.7	77.8

由于杂交水稻制种收益大，近年来，许多制种者，特别是那些有丰富制种经验的农民已越来越富裕了。

十二、杂交水稻在我国取得成功的基本经验

1. 各级政府领导人非常重视这项研究。国家科委和农牧渔业部在制定全国农业科研规划时总是把杂交水稻研究放在优先地位，这样，研究中所需的人力、仪器设备和资金能够及时得到提供。

2. 我们组织了一支强有力的专家队伍，专门负责此项研究工作。研究人员不仅要求其知识面宽，具有较丰富的育种实践经验，而且应对杂交水稻育种充满信心和热情。

3. 不同领域的科研单位与专家进行密切合作。

4. 海南冬天气候温暖，为全年种植育种材料（一年 3 代），加速世代进程，提供了良好条件。

5. 培训大批推广杂交水稻和从事杂交水稻制种的技术人员。

6. 在新组合鉴定合格以后和推广以前，应安排高产示范田，召开现场会，以增强农民和各级领导对种植杂交稻的信心。

7. 推广和制种体系要配套。

十三、问题和展望

尽管杂交水稻在中国取得了巨大的成功，但是还有一些问题需要继续研究和解决。目前，中国水稻科学家正在从下述几个方面对杂交水稻育种进行改进和提高，预计杂交水稻无论在栽培面积和单产上，还是在经济效益上，都将会有更大的增长。

1. 选育和推广不同质源的雄性不育系。目前，在生产上应用的籼型不育系，95% 以上属"野败"型。从长远来看，细胞质太单一，潜伏着意外的危机，很可能招致某种毁灭性病害的大发生、大流行。因此，有必要因地制宜选育一些其他质源的不育系，特别是配子体不育的优良籼型不育系，来取代一部分"野败"型不育系。最近几年，已经选育出了一批具有新质源的不育系，并正在向生产上逐步推广。

2. 选育异交率高的不育系。杂交水稻种子的价格高，不利于杂交水稻的推广，选育异交率高的不育系是提高制种产量和降低制种成本的治本之道。例如，最近已选育出一批开花习性好、柱头大、柱头外露率高的新不育系。它们的异交率比现在应用的当家不育系高30% ~ 50%。

3. 超高产育种。杂交水稻的育种技术具有把改良株型和提高生理机能结合起来的优越性。所以，超高产育种已成为一些水稻研究单位的主攻目标之一。计划在五年内育成在大面积生产条件下每亩日产量达 12 斤的强优势杂交组合。这意味着在不久的将来，长江流域大部分地区的双季杂交水稻的平均每亩总产量将达 2 500 斤左右。事实上，新选育成的个别优良组合的产量潜力已非常接近这个指标。所以超高产育种的前景十分令人鼓舞。

4. 优质米育种。用国际稻米市场的标准来衡量，现有杂交稻品种在米质上还存在一定缺陷（如直链淀粉含量较高，透明度欠佳，整精米率较低）。近几年来，中国的杂交水稻育种在重视进一步提高产量的同时，还注意到米质改良，以提高其经济价值。杂交水稻育种具有易于获得高产的优点，但要改良品质则存在一定难度，这是因为杂种 F_1 代植株上所结谷粒的某些品质性状常常发生分离现象。因此，在杂交水稻优质育种中，要求杂交的父母本双方均是优质品种，而且在品质性状方面基本相同。

目前，中国已有较多的优质籼型恢复系，而绝大多数不育系品质较差，故应把选育优质米籼型不育系作为发展杂交水稻的重点之一。

5. 多抗性育种。目前，在中国一些地区，一些当家杂交组合的抗性正在逐渐丧失，这是因为病虫害生理小种和生物型发生了变化，必须用新组合来取代。幸运的是，大多数抗性受显性或部分显性基因控制。因此，选择适当的亲本，特别是利用国际水稻所最新育成的多抗品系，可以选育出高产多抗的杂交组合。威优 64 就是这类新组合，它不仅高产、稳产，而且抗五种主要病虫害——稻瘟病、白叶枯病、黄矮病、褐飞虱和叶蝉。

总之，上述杂交水稻育种计划，正在中国各地执行，并已取得很大进展，预计一批熟期短、超高产、优质、多抗、低成本的新组合即将问世。

作者：袁隆平

注：本文原系英文，为作者于 1985 年 6 月在国际水稻研究所国际水稻科研会上作的报告。后翻译发表于《杂交水稻》1986 年第 1 期（创刊号）。

杂交水稻育种的程序

【摘　要】本文概述了利用细胞质遗传性不育和育性恢复系统进行杂交水稻育种的程序，也对开展杂交育种研究提出这样一个原则，即杂交育种并不是与自交育种进行竞争，而是相互补充配合，以促进水稻的遗传改良。

一、引言

大面积栽培的自花授粉作物的杂交种选育程序与自交品种的选育过程迥然不同。自交育种的目的是累加产量基因，这种基因在同质的遗传背景下表现良好。杂交育种方法是将这些在异质背景（F_1）中表现好的基因综合在一起。

杂交水稻育种是一个相对较新的领域。除中国和国际水稻研究所外，只有少数几个国家在探索这方面的潜力，其他一些国家现在尚抱等待和观望的态度。如果国际水稻研究所和中国以外的其他国家在这方面取得了可观的进展，他们也可能会将力量转移到这上面来。到目前为止，中国和国际水稻研究所在杂交水稻上所获得的成功经验已形成了一种选育程序模式。本文论及的这种模式对有志于着手杂交水稻研究的各国水稻改良计划将有所帮助。

二、模式

这个模式的基点是利用细胞质遗传雄性不育和育性恢复系统。对水稻这种严格的自花授粉作物来说，这种系统已被认为是进行杂交水稻育种的最有效的遗传工具。因为三系亲本都是自交种，所以常规育种和杂交种必然有着密切联系。通过常规育种选育的优良品系可以被用来选育具有优势的杂种，从而使其得到进一步的遗传改良。因此，杂交育种应该与常规育种互相配合，而不是互相排斥。

三、程序

这个模式涉及的杂交水稻育种程序分为两个阶段：①三系亲本的选育；②杂种优势的鉴定。

（一）三系亲本的选育

这个阶段包括雄性不育系（A）、保持系（B）和恢复系（R）的选育。在中国即称之为"三系选育"。三系亲本的本质特点是：①具有对当地条件的适应性；②一般配合力好；③理想的花器结构，从而有利于异花授粉，提高制种产量。中国和国际水稻研究所已有一系列可以被利用的不育系、保持系和恢复系。这些亲本材料可以在当地进行适应性、配合力等方面的测定，以便直接用来选育杂种。另一条途径是：可通过专门的方法选育优良的三系亲本。

从现行常规育种中选出具有良好农艺性状（例如，高产、抗病虫害、米质好等）的优良品系或品种，然后将这些品种（或品系）与已有的最好的不育系进行单株成对测交，其 F_1 代和相应的父本，以及当地最好的品种（对照）分行单本种植（每行 10～20 株）。

如果测交后代的花粉表现 95%～100% 的不育，那么它相应的父本可以认为是有前途的保持品种（品系）。这种情况下，将测交 F_1 代与原父本回交。回交 F_1 代中的完全不育株再次分单株同原父本回交。通常经过 5～7 代的连续回交，就育成了一个新的稳定的不育系。这样的不育系的核遗传组成是原父本的，这个父本也就成为了新不育系的保持系。

如果测交 F_1 代表现花粉正常或颖花达 80% 以上的可育，那么用于测交的父本就可以看作是有前途的恢复品种（品系）。这样的品种（品系）还须通过单株复测以进一步证实其恢复力且使之纯化。

在这同一个测交试验中，比对照品种具有优势的杂交组合可以进行鉴定，并对它们有希望的不育系和恢复系进行配合力测定，以后还需进一步鉴定组合优势的强弱。

测交 F_1 代中表现部分花粉和颖花不育的组合就被淘汰了，因为它们的父本不能转育成不育系，也很难育成恢复系。

（二）不育系和恢复系的配合力测定

当有好几个不育系和恢复系被认为可以利用时，就有必要测定它们的配合力了。通过将供试品种与相应的品系杂交，配制一批杂交组合。F_1 杂种与父本一起在设重复的小区比较试验中进行鉴定，然后按照Kempthorne（1957）程序分别测定一般配合力和特殊配合力。那些表现出一般配合力高的不育系和恢复系，以及由它们配组表现出特殊配合力强的杂种后代即选择出来作进一步的测定。

（三）杂种优势鉴定

选择性状好，一般配合力高的不育系和恢复系配组。新配制的组合与最好的当家品种要进行连续几次的产量比较试验，即初、高级的产量鉴定和多点试验。试验中比对照品种具有明显优势的组合被选出来进行更高级的区域产量比较试验。多年表现优势强的组合可以进行大面积生产试验，从中择优推广。

四、各种鉴定圃的设置

如果有组织地经过以下几个鉴定圃来筛选育种材料（即①原始材料圃；②测交圃；③复测圃；④回交圃），那么三系选育将更有效。

原始材料圃包括可利用的不育系和经本地选育或从国内外引进的优三系亲本。测交圃是用已有的不育系和原始材料圃中的亲本作测交，它包括测交F_1相应的父本，以及最好的对照品种，这样就能通过比较选出育性正常的强优势组合。

复测圃用来进一步观察鉴定和纯化恢复系。复测F_1代和用于做测交的恢复系的单株选择后代种植在该圃中，也包括有规则地相间种植的对照品种，以便鉴定组合的优势。

回交圃是供选育稳定不育系和保持系用的。回交F_1代及父本后代种植在这个圃中。

五、制种和产量试验

如前所述，杂种优势鉴定需要通过有组织的制种来进行配组比较。用于配合力测定和产量观测圃中的杂种可通过手工杂交获得，也可以通过在隔离笼罩内将花期相遇的父母本配组获得。小区设重复的鉴定圃和区试的杂种是在小规模或中等规模的隔离小区内制得的。各鉴定圃之间的相互关系如图1所示。这个选育程序完全不同于系谱育种的程序。

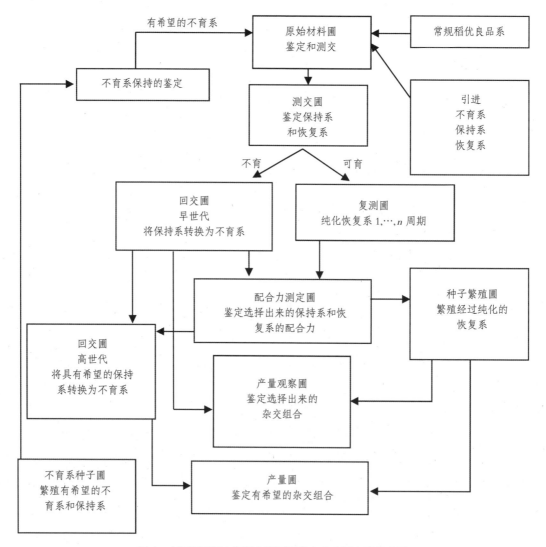

图1　中国和国际水稻研究所采用的杂交水稻育种程序图

　　以上描述的选育程序是杂交稻育种中的一般程序，但在实践中必须灵活掌握。一些好的组合可以越级参加试验。例如，如果一个在复测圃中表现异常好的组合，育种者就可以将它推荐参加小区产量鉴定，甚至区试。这样，育种年限就能缩短，优良组合也就可以比较快地投入大面积生产。

作者：袁隆平　乌马利

注：本文汇编于《首届杂交水稻国际学术讨论会（中国长沙，1986年10月6—10日）论文集》。

杂交水稻育种的战略设想

一

农业生产和作物育种的历史表明，凡在育种上有所突破，就会给农业生产带来一次飞跃。如杂交玉米和高粱、矮秆水稻和小麦等的育成和应用，都大幅度地提高了这些作物的产量。

杂交水稻培育成功，属于水稻育种上的一项突破。生产实践证明，在相同条件下，杂交水稻一般比普通良种增产20%左右。近年来，杂交水稻的种植面积约占全国水稻面积的四分之一，其产量则占稻谷总产的三分之一左右（表1）。由此可见，进一步发展杂交水稻，对提高我国粮食产量具有十分重要的战略意义。

表1　1987年全国水稻种植面积和产量

	面积 /（百万公顷）	总产 /（百万吨）	单产 /（公斤/公顷）
全部水稻	31.8	166.9	5248
杂交水稻	8.4	54.6	6474
杂交水稻/全部水稻/%	26.4	32.7	123.4

事物的发展是无止境的，特别是在人为控制下，事物的发展规律是螺旋形上升的。我国杂交水稻的研究和利用虽然成绩巨大，但从育种上分析，现在的杂交水稻只是处于发展初级阶段，它还蕴藏着巨大的增产潜力，具有广阔的发展前景。根据国内外对水稻杂交优势利用研究的新进展、新动向和发展趋势，我个人认为，杂交水稻的育种，无论在育种方法上还是杂种优势水平上，都具有三个战略发展阶段，而每进入一个新阶段都是一次新突破，从而会把水稻的产量推向一个更高的水平。

二

从育种方法上说，杂交水稻的育种可分为三系法、两系法和一系法三个战略发展阶段，朝着程序上由繁到简而效率越来越高的方向发展。

1. 三系法　即培育和生产杂交水稻必须做到不育系、保持系和恢复系"三系"配套。这是行之有效的经典方法，当前正处于方兴未艾时期，预计在今后 5～10 年内仍将继续作为选育杂交水稻新组合的主要方法。但是，三系法的育种程序和生产环节较复杂，以致选育新组合的周期长、效率低，推广环节多、速度慢，同时种子成本高、价格贵，这些都是三系法杂交水稻在种植面积上和提高产量上受到限制的内在不利因素。所以，从杂种优势育种的长远战略上考虑，三系法应设法用其他较先进的方法来取代。三系法迟早要被淘汰，这是杂交水稻育种科学发展的必然趋势。

2. 两系法　分光敏核不育系法和化学杀雄法两种。本文所讨论的两系法仅指前者，因为此法最有希望，预计在不久的将来便可逐步取代三系法，在生产上应用。

由石明松发现继而由湖北省协作组基本上培育成功的水稻光敏核雄性不育系，是水稻育种上的又一次突破，将使杂交水稻的发展跨入一个新阶段。这种核不育系的雄性不育主要受一对隐性核基因控制而与细胞质无关。在长日照条件下，表现完全雄性不育；在短日照条件下，则表现可育，能自交结实。利用这种遗传工具于杂交水稻育种和制种具有三大优越性：

（1）配组自由，选育出优良组合的概率比三系法大大增加。由于不育性是受隐性基因控制的，所有正常品种都能使其育性恢复正常，都是恢复系。育种工作者在选配组合时，不必去考虑父本有无恢复基因的问题。而在三系法中，往往出现这样的情况：有些组合表现出很强的杂种优势，而父本却无恢复力；或者结实率良好的组合却不表现杂种优势。

（2）一系两用。在长日照下（夏季）可用于制种；在短日照下（春、秋）可用于自身的繁殖，并不需要借助保持系。因而能简化繁殖、制种程序，使种子成本下降。

（3）避免不育细胞质的负效应。

光敏核不育的研究成绩很大，然而尚未达到实用阶段。今后还须对其遗传规律和转换机制作深入研究；要在不同的地理位置和生态条件下鉴定其适用范围；要加强和扩大协作研究，加快选育步伐，使之尽早应用于生产。

3. 一系法　即培育不分离的 F_1 杂种，将杂种优势固定下来，免除制种。从理论上讲，固定 F_1 杂种优势的可能途径有多种，我个人认为，其中以培育无融合生殖系最有前途。

无融合生殖是指以种子形式进行繁殖的无性生殖方式（无性种子繁殖），它可使世代更迭但不改变核型，后代的遗传结构与母体相同，因此，通过这种生殖方式可将 F_1 杂种优势固定下来。育种工作者只要获得一个优良的 F_1 杂种单株，就能凭借种子繁殖，迅速在大面积生产上推广。

无融合生殖育种是一项新的育种方法，成败的关键在于能否获得可资利用的无融合生殖基因。水稻在这方面的研究刚刚开始，目前尚处于探索阶段。禾本科是拥有无融合生殖的属和种最多的科之一，因此，从理论上推测，在稻属中很可能存在无融合生殖基因。同时，还可通过远缘杂交和遗传工程的方法，把异属的无融合生殖基因导入水稻。总之，利用无融合生殖对水稻杂种优势育种是一项价值极大、颇有希望但难度很大的研究课题，应将其列为长远的战略目标。

三

从杂种优势的水平上分，杂交水稻育种可分为品种间、亚种间和远缘杂种优势的利用三个发展阶段。

1. 品种间杂种优势　　目前生产上应用的杂交水稻属此范畴。由于品种间的亲缘关系较近，杂种优势是有限的。新杂交组合的选育可以继续在生育期、米质、抗性和不育系的异交率等方面有所改良和提高，但若仅依靠品种间杂交，想在现有基础上再取得产量上的突破，看来很难办到。

作物育种的实践和成就表明，只有采用崭新的材料和方法，才能在产量或其他方面取得突破。因此，杂交水稻育种必须冲出品种间杂交的框框，寻找新的优势途径，才能使产量有新的飞跃。就当前的育种科学和技术水平来看，最有希望且能在较短的时期内取得成效的途径，是利用亚种间杂种优势。

2. 亚种间杂种优势　　籼、粳和爪哇稻为普通栽培稻的三个亚种。众所周知，亚种间杂种尤其是籼粳杂种具有巨大的产量潜力。这里仅举湖南杂交水稻研究中心栽培研究室1986年的一个观察试验例子，便以说明（表2）。

表2　籼粳 F_1 杂种的产量潜力

类型	株高 /厘米	每穗颖花数 / 个	每株颖花数 / 个	结实率 /%	实产 /（斤 / 厘[①]）
城特232（粳）×26窄早（籼）	120	269.4	1 779.4	54.0	11.0
威优35（对照）	89	102.6	800.3	92.9	11.5
优势%	34.8	162.8	122.4	−41.9	−4.3

注：小区面积0.01亩，7月29日插秧，株行距4寸[②]×6寸。

从表2可知，城特232×26窄早的每亩颖花数比对照多122%，可达4 400万粒以上，

① 1厘≈6.67平方米。
② 1寸≈3.33厘米。

尽管结实率仅 54%，但产量几乎与品种间强优组合威优 35 相当。如果能将结实率提高到 80%，作双季晚稻栽培，其产量潜力每亩可超过 1 600 斤。

直接利用强大的籼粳杂种优势是育种工作者多年来梦寐以求的愿望，但一直未能实现，主要难关在于无法解决杂种结实率低的问题。直到最近，经日本池桥宏等人的研究，终于揭示了籼粳不亲和性和由此引起的杂种结实率低的本质，表面上看起来很复杂的不亲和性现象基本上只受一对主基因所控制，这个基因存在于许多广亲和性品种中，称为广亲和基因。这项卓越研究是理论上的一大贡献，为解决籼粳杂种 F_1 结实率低的问题指明了方向。现在，我们已掌握了攻克这个难关的基本方法和材料，预计在不久的将来，就能实现利用籼粳杂种优势的愿望。

3. 远缘杂种优势　远缘杂交水稻特别是有异属基因的杂交水稻，可能具有人们今天还难以想象的强大优势，现在提出这个目标，似乎带有科学幻想的性质，但从生物技术的发展来看，它的实现并非没有可能。利用无融合生殖和借助遗传工程，可能是培育远缘杂交水稻最有希望的途径和方法。

四

上述三种育种方法和三种优势水平之间存在着一定的内在关系。三系法主要适用于选育品种间杂交组合，选育亚种间组合固然可以，但难度较大。两系法对选育品种间和亚种间组合均适用，但用于亚种间杂交则能更好地发挥其优越性。至于远缘杂种优势，三系法和两系法均能利用个别有利的远缘基因，但通过一系法来利用，则可能产生更好的效果。这样，就可把杂交水稻的育种综括地划分三个战略发展阶段：

（1）三系法为主的品种间杂种优势利用。

（2）两系法为主的亚种间（籼粳、籼爪、粳爪）杂种优势利用。

（3）一系法为主的远缘杂种优势利用。

（参考文献 6 篇略）

作者：袁隆平

注：本文发表于《杂交水稻》1987 年第 1 期。

利用无融合生殖改良作物的潜力

一、无融合生殖的概念

无融合生殖是一个新的研究领域。无融合生殖是不通过受精作用而产生种子的生殖方式，它包括单倍体和二倍体两种无融合生殖。对作物改良意义最大的是二倍体无融合生殖。被子植物的二倍体无融合生殖可以有3种不同的途径：

1. 二倍体孢子生殖　胚囊由孢原细胞或大孢子母细胞不经减数分裂发育而成。

2. 无孢子生殖　由胚珠中的体细胞直接发育成胚囊。

3. 不定胚　由胚珠的体细胞（多为珠心）直接发育成胚。

现有性生殖和二倍体无融合生殖的途径图解如下（图1）：

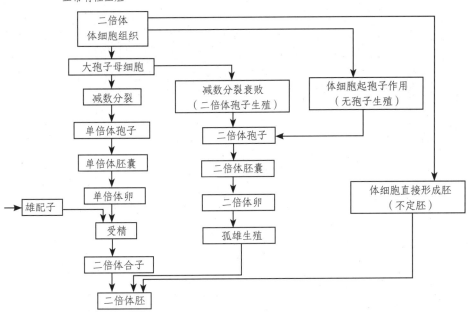

图1　有性生殖和二倍体无融合生殖的途径

二、作物育种利用无融合生殖的意义

二倍体无融合生殖可使世代更迭但不改变基因型，后代的遗传结构与母本相同，将这种特性用到育种上，具有巨大的优越性。

1. 无融合生殖可以固定杂种优势，育成不分离的杂交种（True breeding hybrid）。因此，只要获得一个优良的杂种单株，就可凭借种子繁殖，迅速地在大面积生产上应用其杂种优势。

2. 简化杂交种的制种程序和方法。不需专门的制种田和隔离区，不需大量繁殖父、母本，也不会因发生串粉而引起生物学混杂，因而能降低杂交种子成本和提高其纯度。

3. 扩大杂种优势利用的领域。对难以育成三系的作物和靠异花传粉不能生产大量 F_1 代杂交种的作物（如花生等闭花受精植物），通过无融合生殖育种，亦能达到利用杂种优势的目的。

4. 增加选到优良基因组合的机会。通过无融合生殖这一特殊机制，能将任何基因型固定下来，而不论其杂合性的复杂程度如何。所以，无融合生殖育种的意义不仅仅在于固定杂种优势，而还在于它能快速固定任何高度杂结合的基因型，甚至包括发生疯狂分离的远缘杂种。

5. 减轻或避免某些杂交种由于遗传上的不协调而导致的结实不良问题。

6. 对以营养体进行繁殖的作物如马铃薯等，利用无性种子繁殖也有很多好处。如能免除病毒病传递，复壮更新，降低种子成本，减轻贮藏和运输负担，以及便于推广等。

三、无融合生殖育种的现状

对植物无融合生殖的研究，过去在理论方面作了大量工作并取得很大的成绩。如通过细胞学和胚胎学的观察研究，将各种无融合生殖方式进行了分类并弄清了它们的发生机制和发育过程；通过遗传学的分析研究，肯定了无融合生殖是受基因控制的遗传性状，但因物种和无融合生殖类型不同，所涉及的基因数目有多有少，并有显隐性之分，但是，直到 20 世纪 70 年代，这项研究才逐步转向应用方面，其主要动向是：

1. 在无融合生殖作物（主要是禾本科牧草）中寻找有性生殖的基因型。

2. 在有性生殖作物或其近缘野生种中寻找无融合生殖的基因型。

3. 对无融合生殖进行遗传操作。

现在已知，在大多数无融合生殖种中具有少数的有性或部分有性生殖的植物；同时在高粱和珍珠粟中发现了兼性的无融合生殖植物，这些发现预示其他有性生殖植物也可能存在无融合生殖的个体或类型。迄今，粟草、早熟禾（Bluegrass）、草莓和柑橘等四种作物已育成无融合生殖品种并在生产上应用。目前正在取得进展的作物有：

（1）玉米。无融合生殖基因来自摩擦禾属（*Tripsacum*）。

（2）珍珠粟。基因来自狼尾草属（*Pennisetum*）。已初步育成兼性无融合生殖品系，这种远缘杂种具有极强的杂种优势，生物学产量比母本珍珠粟高一倍以上，籽粒产量虽不高，但可看出其巨大的潜力。

（3）高粱。已获得几个兼性的无融合生殖类型，借鉴早熟禾的成功经验，正在按如下方案进行选育（图2）。

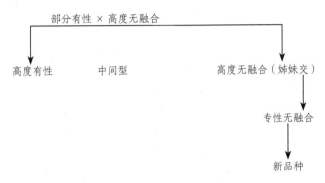

图2　兼性无融合生殖材料的选育方案

开始在探索的作物有：小麦和水稻等。

四、无融合生殖育种的可能途径和方法

作物遗传育种的经验告诉我们，凡是在产量或其他方面具有突破性的新品种的诞生，都是由于采用了新方法和新材料（基因）的结果。矮秆基因的发现和利用，使水稻和小麦育种起了革命性的变革，大幅度提高了产量。无融合生殖育种是一项新的育种方法研究，成败的关键在于能否获得可资利用的无融合生殖基因。从理论上推测，很多栽培作物都可能存在无融合生殖类型。然而，迄今仅在为数很少的作物中有所发现。其原因，一方面作物的无融合生殖育种一直很少有人涉猎；另一方面，大多数现行的育种方法易使人们忽略可能是无融合生殖的材料。比如，杂交后代表现整齐一致或母性遗传时，就把它们当作自交后代淘汰。因此，育种家只要把注意力集中到这个领域，就可能在更多的作物中发现无融合生殖材料。

理想的无融合生殖基因要具备四个条件：①显性遗传；②简单遗传；③无融合生殖性稳定，表现充分，受环境影响少；④专性无融合生殖。现以水稻为例，介绍目前正在进行或将要开展的关于获得无融合生殖基因的研究方案：

1. 在现有品种资源和近缘野生种中寻找

（1）通过显隐性遗传关系筛选

①用具有3个隐性性状（雄性不育、矮秆、光稃）的品种作母本同相对显性性状的父本杂交，在F_2（按穗行播插）中逐行检查表现型大大偏离理论数的穗行。

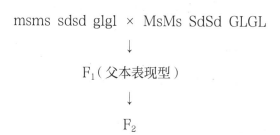

msms sdsd glgl × MsMs SdSd GLGL

↓

F₁（父本表现型）

↓

F₂

②用雄性不育、糯性品种作母本同雄性正常、非糯品种杂交，在母本穗上找糯性种子。

msms wxwx × MsMs WxWx

↓

F₁

（2）多胚苗的筛选。一粒种子（不包括复粒稻）长出两个以上的苗叫多胚苗。多胚现象是不定胚的主要表现形式。因此，从大量品种资源中筛选多胚苗，也是寻找无融合生殖材料的途径之一。此法简单易行，每个材料只需数百粒种子，发芽后一周左右，用肉眼观察就能识别出来。用此法已获得几个能稳定遗传的多胚苗材料。其中最好的一个，多胚苗的遗传率为30%左右。获得遗传的多胚材料后，需进行胚囊发育的细胞学观察，判断其是否属于无融合生殖。同时，还要配合遗传学方面的试验加以确证。最近，湖南杂交水稻研究中心与美国加州大学合作研究，通过胚囊发育的细胞学观察，发现不同双胚苗品系的胚囊含有2.6%~5.1%的不定胚，初步判断这些不定胚起源于体细胞组织（珠心或珠被）。另一方面，利用带有显性遗传标记的紫色稻与双胚苗杂交，后代出现相似比例的绿色苗，从而肯定了细胞学观察的结果。这项结果，首次揭示了在水稻中存在二倍体无融合生殖类型，虽然这些材料是频率很低的兼性无融合生殖，离实用阶段还很远，但在科学研究上具有重要意义。

为了提高不定胚的遗传频率，可用不同来源的多胚苗材料互交；按图2方案进行选育。染色体加倍，辐射和其他诱变剂处理，对提高多胚苗的遗传率，也可能是有效的手段。

（3）近缘野生种胚囊的细胞学观察。正在对国际水稻所的500份各种野生稻样本进行胚囊发育的细胞学观察，从中寻找无融合生殖材料，现已发现其中有胚位异常的材料。

2. 异属远缘杂交　禾本科是拥有无融合生殖的属和种最多的科之一。因此，通过远缘杂交，把异属的无融合生殖基因导入水稻，也是一条颇有希望的途径。但常规的远缘杂交很难成功，必须借助原生质体融合、组织培养和基因工程等生物技术才能最终解决这个难题。

五、结语

无融合生殖基因是具有奇特活力的遗传工具。然而利用这种奇特的工具培育品种，提高育种效率的工作，仅仅才是开始。绝大多数的作物尚待开拓，而无融合生殖育种的成败关键在于能否获得这些基因。因此，锲而不舍，广开思路，奋发开拓，通力协作，才有希望取得成功。

作者：袁隆平

注：1986 年 4 月 21—25 日，本文作者应洛氏基金会邀请，参加了意大利贝拉吉奥召开的 "The Potential Use of Apomixis in Crop Improvement" 学会讨论会，本文根据会上有关育种部分的发言，整理成摘要，其中包括作者在水稻方面的工作和看法。该文曾于《作物杂志》1986 年第 3 期发表；重刊于《杂交水稻》1988 年第 4 期，且作者作了部分修改和补充。

水稻广亲和系的选育

【关键词】广亲和系

水稻广亲和基因的发现及其遗传研究，揭开了籼粳稻亚种间杂种一代半不育现象的遗传本质，从而为利用水稻亚种间杂种优势指明了方向。但是，已知的广亲和材料，绝大多数为古老的农家品种，农艺性状不良，如植株太高，株叶形披散和生育期很长等，不能直接作为选配优良组合的亲本加以利用。因此，培育具有优良农艺性状和经济性状、可供实用的水稻广亲和系就显得非常重要，这是开展亚种间杂种优势育种的首要任务之一。

普通栽培稻有籼稻、粳稻和爪哇稻三个亚种，所以亚种间的杂交可分为籼粳交、籼爪交和粳爪交三种形式。初步研究表明，水稻杂种优势强弱的程度，具有籼粳交＞籼爪交＞粳爪交＞籼籼交＞粳粳交的一般趋势，也就是说，亚种间的杂种优势一般要强于品种间的。为了能充分利用各种形式的亚种间杂种优势，提高育成强优组合的概率，就有必要建立籼、粳、爪三个亚种类型的优良广亲和系。这三类广亲和系的应用方式，有如下几种：

1. 在两系法中，可直接作为恢复系与不同亚种的光敏核不育系配组选优；

2. 导入光敏核不育系基因，育成广亲和的光敏核不育系，从理论上讲它能与任何亚种的正常品种配组选优；

3. 将它们转育成细胞质雄性不育系或其恢复系，以便在三系法中利用水稻亚种间杂种优势。

基于上述认识，我们在以前工作的基础上，近年来开展了水稻广亲和系选育的研究，育成了籼型广亲和系培 C311，爪哇型广亲和系轮回 422，籼型广亲和系培矮 64S 以及广亲和光敏核不育系培矮 64S。现将研究结果报道如下。

一、材料和方法

1. 广亲和亲本：用印度尼西亚地方品种培迪（Paddy）作为广亲和基因的供体。培迪属典型的爪哇稻，株高 110~120 厘米，生育期在湖南 140 天左右，不感光；谷粒细长，颖壳上具粳稻类型的绒毛，颖壳在温度较低时呈淡紫色；开花时间早，似籼稻；叶色深绿，似粳稻；米粒透明无腹白，直链淀粉含量 16% 左右，似粳稻。

早在 20 世纪 70 年代，用培迪作父本，与粳稻不育系 BT-C 和黎明 A 测交，F_1 的结实率在 70% 以上，表明培迪含有 BT 型配子体雄性不育的恢复基因。用它与籼稻矮黄米和糯籼谷杂交，F_1 的结实率亦在 70% 以上，表明培迪具有广亲和性。

2. 籼、粳稻亲本：供试的三个中籼品种为矮黄米、芦苇稻和测 64（即 IR9761-19-1-64，野败型不育系的恢复系）；粳稻品种有 C57（BT 型不育系的粳稻恢复系）。以上品种均无广亲和性。

3. 光敏核不育亲本：原始农垦 58S。

二、选育过程

粳型广亲和系培 C311 的选育是以培迪作母本与粳稻恢复系 C57 杂交，按一般系谱法在 F_2 至 F_7 中选性状倾粳单株，并将中选单株与 BT 型不育系测交，淘汰不具有恢复力的父本。到 F_7 育成几个性状稳定的恢复系，其中的培 C311 具有广亲和力（图 1）。

爪哇稻轮回 422 的选育：在培迪与 C57 的杂交后代中选性状倾爪哇型的恢复单株，到 F_7 育成了培 C422 恢复系，为了提高它的配合力，选用一个大穗籼稻品种芦苇稻与它杂交，然后又用培 C422 作轮回亲本回交两次而育成（图 1）。

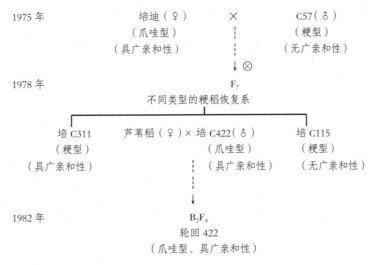

图 1 "培 C311"（粳型）和"轮回 422"（爪哇型）的选育

籼型广亲和系培矮 64 及培矮 64S 的选育过程如图 2 所示。从培迪与矮黄米杂交的 F_7 代中育成了 BT 型的恢复系培矮，再用测 64 作父本与培矮杂交，经多代选择育成性状为籼型的培矮 64。1985 年用农垦 58S 与培矮 64 杂交，在 F_2 中选择不育度高、性状倾父本的单株与培矮 64 回交，在 B_1F_1 中继续选择性状似父本的单株，在 B_1F_2 中选完全雄性不育株，进行短光照处理，收取自交种子，到海南加代后，1988 年在长沙为 B_1F_4，共有 30 个株系，其中 2 个株系（共 536 株）不育株率和不育度均达 100%，且性状似父本，无分离，暂定名为培矮 64S。

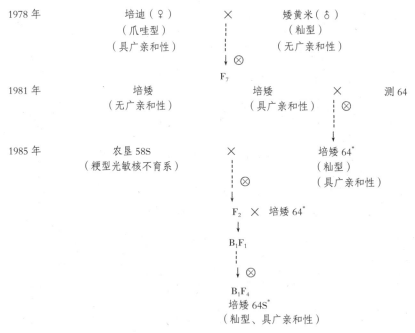

图 2　籼型培矮 64 及培矮 64S 的选育

注: S 表示光敏核不育系; * 具粳稻绒毛性状。

三、结果和分析

（一）特征特性和植物学分类

培 C311 叶色深绿，剑叶直立，籽粒短圆，颖壳上有密集的绒毛，不易脱粒，抗寒性强，花时晚，这些都是粳稻的特点。因此，把培 C311 归为粳型广亲和系。

培矮 64 的亲缘以中籼成分较多，外部形态与测 64 很相似，叶色较淡，分蘖力强，谷粒长形，易脱粒，抗寒力弱，花时较早，这些都属于籼稻的典型性状，因此，培矮 64 为籼型广亲和系。

轮回 422 的性状与母本培迪相似，是兼有籼、粳特点的中间类型（表 1），属于爪哇型广亲和系。

表 1　轮回 422 的籼粳性状

性状	籼	粳
株叶形态		叶片深绿、直立、偏粳
粒形	细长，偏籼	
米粒外观	似籼	
颖壳		绒毛密集，似粳
落粒性		不易，似粳
抗寒性	中等偏弱，似籼	
直链淀粉含量		15% 左右，偏粳
亲和性	72.7（59.0～80.0）	76.9（67.9～82.7）
开花时间	比籼稻晚	比粳稻早

除广亲和性外，这 3 个品系还有 4 个主要的共同特点，即矮秆、早中熟、株叶形态良好和对细胞质雄性不育系（包括野败型和 BT 型）都有恢复力。因此，具有较大的实用价值。

（二）广亲和性表现

上述 3 个广亲和系与目前国内外三种不同类型的 4 个优良广亲和系——CPSLO17（爪哇型）、02428（粳型）、真系 8544（粳型）和真系 8539（籼型）进行了亲和性比较，用作测交的有 7 个籼稻品种、3 个粳稻品种（包括不育系），将 F_1 的结实率作为亲和性的指标，结果见表 2。

1. 表 2 结果初步表明，爪哇型品种 CPSLO17 的广亲和性优于其他被测品种，测交的 F_1 平均结实率为 78.94%，以下依次为培 C311（76.18%），培矮 64（74.21%），轮回 422（74.12%）和 02428（72.88%）。真系 8539 和真系 8544 较差，总平均结实率未到 70%。籼型真系 8539 对早籼稻的测交一代结实率高，而对粳稻的亲和性最差，粳型真系 8544 却相反，与粳稻的测交一代结实率高而对籼稻的亲和性差。看来，直接利用这两个从日本引进的广亲和系的可能性较小。

2. 除真系 8544 和真系 8539 外，其余 5 个广亲和系的亲和性经显著性测定差异不显著，测交 F_1 的结实率都在 75% 左右。

3. 粳型广亲和系 02428 与培 C311 比较，二者对粳稻的亲和力都好，对中籼的亲和力接近，但 02428 对早籼的亲和力较差（测交 F_1 平均结实率为 60.5%，居第 6 位），而培 C311 对早籼的亲和力较好（测交 F_1 平均结实率为 73.5%，居第 3 位）。

轮回 422 与 CPSLO17 比较，两者对籼和粳型的亲和力都好，表现为明显的中间型（爪哇型）。

培矮 64S 用指定的测验种南京 11 号（籼）、IR36（籼）、秋光（粳）和巴利拉（粳）测交，F_1 结实率分别为 86.44%、66.35%、77.96% 和 79.92%。表明培矮 64S 已组合了广亲和基因与光敏核不育基因。

4. 测交 F_1 结实率的变幅以 02428 和真系 8544 最大，分别为 31.5%～86.6% 和 40.0%～89.3%，表明二者的亲和谱不广或者有专一性。培 C311、培矮 64、轮回 422 的变幅大体与 CPSLO17 接近。

表 2　几个广亲和品种的亲和性比较（以测交 F_1 结实率来衡量）　　单位：%

测交品种		CPSLO17	培 C311	培矮 64	轮回 422	02428	真系 8544	真系 8539
早籼	V20B	80.0	74.5（V20A）	79.8	70.0	63.4	69.9	75.0
	W6154S	75.0	70.0	71.0	70.1	31.5	40.0	80.0
	26 窄早	71.6	75.9（TF115）	64.7	72.6	86.6	58.7	75.0
	平均	75.5	73.5	71.8	70.9	60.5	56.2	76.7
	变幅	71.6～80.0	70.0～75.9	64.7～79.8	70.0～72.6	31.5～86.6	40.0～69.9	75.0～80.0
	位次	2	3	4	5	6	7	1
中籼	明恢 63	84.8	74.7（2-2）	86.1	77.9	80.0	74.8	
	9024	74.8	78.5		80.1			
	南京 11 号			86.4		84.7	76.3	
	IR 系统	81.2	71.7	66.4	59.0	54.0	30.3	
	密阳系统	83.9	66.5	75.0	80.8	75.0	75.0	
	平均	81.2	70.9	78.5	74.5	73.4	66.4	
	变幅	74.8～84.8	66.5～74.7	66.4～86.4	59.0～80.8	54.0～84.7	30.3～76.3	
	位次	1	5	2	3	4	6	
粳稻	秋光	85.3	82.9	78.0	80.1	84.3	85.0	55.8
	农虎 26A	84.8	85.5	66.0	82.7	31.5*	15.0*	74.5
	农虎 26B			79.9*（巴利拉）		85.0	75.0	

续表

测交品种		CPSLO17	培C311	培矮64	轮回422	02428	真系8544	真系8539
粳稻	城特232A	68.0	86.0	68.7	67.9			57.0
	城特232B					84.3	89.3	
	平均	79.4	84.8	70.9	76.9	84.5	83.1	62.4
	变幅	68.0～85.3	82.9～86.0	66.0～78.0	67.9～82.7	84.3～85.0	75.0～89.3	55.8～74.5
	位次	4	1	6	5	2	3	7
总平均数		78.94	76.18	74.21	74.12	72.88	67.43	69.81
总变幅		68.0～85.3	66.5～86.0	64.7～86.4	59.0～82.7	31.5～86.6	30.3～89.3	55.8～80.0
总位次		1	2	3	4	5	7	6

注：此表为1986—1988年长沙夏播试验资料。有"*"的未参加平均数的计算。

四、结语

1. 培迪具有广亲和基因，广亲和性与CPSLO17接近。通过转育，可以把其广亲和基因转育到粳型、爪哇型和籼型不同品种中。

2. 培迪的广亲和基因可以与BT型不育系的恢复基因或野败型不育系的恢复基因重组在一起，为三系法利用亚种间杂种优势提供了可能性。

3. 培迪的广亲和基因可以与矮秆基因相结合，因而有望克服亚种间杂种株高超亲的问题。

4. 培迪的广亲和基因可以与光敏核不育基因重组，从而可能选育出籼粳亚种优势利用中的最佳遗传工具——粳质、籼核、广亲、光敏核不育系。

5. 以培迪转育的广亲和系具有较好的农艺性状，有较好的配合力，为配制生产用种提供了良好的亲本。

注：本文引用了李新奇、肖金华、邱趾忠的试验资料，特致谢。

080

References

参考文献

[1] Eke hashi，H and H.A rauki，1984. Varietal Screening for Compatihitity type rerealed in F_1 fertility of distant crosses in rice Japan L. Breed 34（3）:304-213.

[2] 李新奇，袁隆平. 广亲和爪哇品种与籼、粳稻的亲和力遗传及杂种优势初步研究. 湖南农业科学，1987（3）：4-9.

[3] 罗孝和. 籼粳杂种优势及"培迪"广亲和性研究. 四川冈 D 型杂交水稻协作组成立十周年纪念刊.1987: 47-54.

[4] 廖翠猛，袁隆平. 水稻籼爪交，粳爪交杂种优势及其与亲本关系的研究. 研究生毕业论文（未发表）.1988.

[5] 袁隆平. 杂交水稻的育种战略设想. 杂交水稻，1987（1）: 1-3.

[6] 袁隆平等. 杂交水稻育种栽培学. 湖南科技出版社，1988: 126-131.

[7] 彭科晋. 籼粳亚种间杂交组合表现出很高的产量优势. 杂交水稻，1988（2）: 25.

[8] 曾世雄，杨秀青等. 栽培稻、籼粳亚种间杂种一代优势研究. 作物学报，1980，6（4）: 193-202.

[9] 顾铭洪.1987. 水稻广亲和基因的遗传及其利用. 江苏农学院.

[10] 肖金华.6 个广亲和品种的广亲和性研究简报. 杂交水稻，1988（2）: 30.

作者：袁隆平

注：本文发表于《杂交水稻》1989 年第 2 期。

"无融合生殖"水稻84-15还有待科学验证

【摘　要】从已有资料看，还缺乏科学证据说明水稻84-15属于无融合生殖水稻。84-15不是固定杂种第一代的优势；迄今没有胚胎学证据说明84-15属于何种无性繁殖方式；84-15与具标志性状品种杂交后，杂种第二代出现一定比例的分离；84-15属于兼性无融合生殖的解释将导致84-15的"无融合生殖"个体以每代近14.3%的速度递减以致消失；以胚乳游离核染色体数目异常为依据亦缺乏说服力。

《人民日报》1988年10月5日第三版有关无融合生殖水稻84-15选育成功的报道，在国内外农业生物科学界引起了强烈的反响。最近，在北京召开的国家"863"计划生物技术领域年会上，该水稻的育成者，中国农科院作物所陈建三同志又对84-15的选育和遗传研究作了进一步阐述。但是，从已有的资料看，还缺乏科学证据可以说明84-15属于无融合生殖水稻。

第一，从遗传学理论和植物杂种优势利用的实践看，生物杂种优势主要只表现于杂种第一代。因此，固定杂种优势概念应该是指固定杂种第一代的优势。据陈建三同志介绍，84-15是以野生稻为母本，与粳稻品系806195杂交，在严重分离的杂种第二代群体中通过选择，在杂种第三代开始稳定的品系，这怎么能说是固定了杂种优势呢？

其次，无融合生殖是植物的一种特殊的无性繁殖方式，它有多种类型，最常见的为无孢子生殖，二倍性孢子生殖和不定胚等几种。84-15属于哪一种类型，迄今没有拿出任何胚胎学证据来说明。在这种情况下，怎么能武断地说它是无融合生殖水稻呢？

第二，据介绍，水稻84-15在与显性标志性状品种杂交后，杂种一代全表现为隐性，以此推断它属于无融合生殖并受隐性基因控制。但是，在杂种第二代却出现了一定比例的显性性状个体。这一结果只能说明 84-15 为正常的有性生殖品种，否则，在杂种第二代就不会出现分离。或者，他所说的显性标志性状很可能是隐性性状。

陈建三同志根据另一个杂交组合的第一代也出现显性性状个体的现象，又提出 84-15 属于兼性（即部分）无融合生殖类型的解释，这同样不能解决前后资料间的矛盾。如果如其所说，84-15 有 14.3% 的有性生殖，那么，这么个体由于隐性无融合生殖基因的表达受显性基因的掩盖，以后将很少有恢复无融合生殖的可能。这样，84-15 在逐代繁殖过程中，无融合生殖的个体将以每代近 14.3% 的速度递减以至消失。因此，兼性无融合生殖论点也不能成立。

第四，在细胞学上，陈建三同志提供的证据是，84-15 胚乳游离核中的染色体数目为 $2N=24$（即二倍体），而正常品种的为 $3N=36$（即三倍体）。但是以胚乳游离核染色体数目为依据是不可靠的。因为水稻胚乳形成初期，是以无丝分裂为主的方式进行的，而无丝分裂不能保证染色体均等地分配到两个子核中，所以在游离核中可存在从单倍（$1N$）至多倍（$8N$）的染色体数。因此，这个证据也就缺乏说服力。

我们认为，水稻杂种优势的研究在我国已经取得了巨大的成就，在生产上发挥了明显的经济效益。但是，目前农作物的杂交种子，需要年年制种，耗费的人力物力较多。为此，开展无融合生殖固定杂种优势的研究具有深远的意义。但是，无融合生殖在农作物中的发生频率极低，选育无融合生殖品种的难度很大，因此，在论证某种作物品种是否具有无融合生殖特性时，更应持严肃的科学态度。这样做会更有利于这项研究的健康发展。

水稻 84-15 来自于野生稻与栽培稻的杂交后代，它在遗传上可能存在着某些特点，作为一种育种材料，值得进一步研究，但在未获得足够的证据之前，不宜对其作过早、过头的结论。

作者：袁隆平　顾铭洪

注：本文发表于《杂交水稻》1989 年第 4 期。

Hybrid Rice: Achievements and Outlook

【Abstract】Extensive research on and development of hybrid rice in China have clearly demonstrated its usefulness in significantly increasing rice yields beyong the levels of improved semidwarf varieties. Hybrid rice is currently planted on 9 million ha in China. Several IRRI-bred cultivars are involved as male parents of commercial varieties. Research at IRRI and in several collabor ating countries shows that hybrids can increase yields 15%~20% more than those of the best semidwarf varieties. Hybrid varieties suitable for other countries are still in the pipeline. In addition to higher yield, heterotic rice hybrids show higher productivity per day, adaptability to certain stress environments, and better utilization of applied nitrogen fertilizers. Hybrid rice seed production technology has been well developed in China, and prospects for its adoption outside China appear promising. A number of the cytoplasmic male sterile (CMS) lines developed and used in China are not adaptable outside China for lack of adequate disease and insect resistance and acceptable grain quality. Several cytoplasmic male sterile lines developed at IRRI are not being evaluated in collaborating countries. Both China and IRRI are involved in research to diversify CMS sources to prevent genetic vulnerability problems in hybrid varieties. Sufficient numbers of restorer lines are available among indica rice cultivars, but restorer numbers are negligible in japonica rices. By selecting appropriate CMS and restorer lines, hybrids possessing multiple disease and lisect resistance and desired grain quality disease and insect resistance and desired grain quality can be developed. Constratints and the outlook for hybrid rice technology also are discussed.

Development during the 1960s and 1970s of semidwarf rice varieties possessing multiple resistance and/or tolerance for biological and physicochemical stresses helped nationjal programs achieve high and stable rice yields, particularly under irrigated and favorable rainfed lowland environments. On the basis of available technology and projectious of infrastructural developments for rice production, Barker et al (1985) predict that rice production in the world will increase by more than 3% a year, against a projected rice demand of 3.5% a year, to the end of the century. That shortfall prediction makes it logical to look at the prospects of those technologies which can help increase rice yields per unit area per unit time. Experiences in China over the last 10 years indicate that hybrid rice is a prospective technology to meet this goal. China's success in developing hybrid rice is not only a major breakthrough, but also a technological innovation in rice breeding. China is the first country in the world to put hybrid rice into commercial use.

Research on hybrid rice in China began in 1964. The cytoplasmic male sterile (CMS), maintainer, and resterer lines, essential for producing F_1 hybrids were developed by 1973; hybrid seed production techniques were essentially developed by 1975. By 1976, hybrid rice had been released commercially and was planted on 140,000 ha. In 1986, hybrid rice was grown on 9 million ha (Fig. 1), yielding about 20% more than improved rice varieties. To cover such a vast area with hibrid rices, thousands of technicians and farmers were trained.

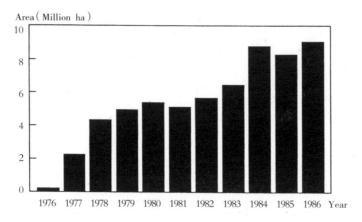

Fig.1   Area grown to hybrid rice in China, 1976—1986.

Encouraged by the developments in China, in 1979 IRRI took initiatives to explore the potentials and problems of hybrid rice in countries outside China. In collaboration with the Chinese Academy of Agricultural Sciences, IRRI organized training courses on hybrid rice technology at Changsha, Hunan, China, during 1980 and 1981, to train rice scientists from several Asian countries. Those scientists initiates hybrid rice research in their home countries. In 1980, China transferred hybrid rice technology through private seed companies to the USA and some other countries.

This paper highlights the achievements and outlook of hybrid rice technology in increasing rice yield beyond the levels possible with improved semidwarf rice varieties.

GRAIN PRODUCTION, YIELD POTENTIAL, AND YIELD STABLITY OF F_1 RICE HYBRIDS

The maximum yield from a single crop of an indica rice hybrid has been recorded at 14.4 t/ha, in Jiangsu Province. That compares to 10.4 t/ha from an improved variety. In Jiangsu, the average yield of hybrid rices in 1986 was 8.01 t/ha (from 0.77 million ha). In Hunan Province a 15 t/ha average yield was obtained from 2 crops of hybrid rices (0.14 million ha). Among the japonica hybrids, Li-You 57 yielded the highest (13.7 t/ha) in Liaoning Province. Japonica hybrid variety: iu You 57 yielded an average 28% more than improved japonica variety Jin Yin 39 in Ningxia Province (Table 1). Average yield of hybrid rice in China 1983—1986 was 6.5 t/ha (Table 2). In 1985, the hybrid rice growing area was 26.4% of the total rice area; it contributed 32.7% of total rice production. The cumulative production increase due to cultivation of hybrid rice in China, 1976—1985, has been estimated as 94 million t.

Outside China, hybrid rice still in the experimental stage. Hybrids developed in China generally are not adaptable in the thropics, primarily because of their susceptibility to major diseases and insects. A number of experimental rice hybrids, mostly developed at IRRI from China-or IRRI-bred CMS lines and IRRI-bred restorers, were compared to the best available improved rice varieties. The best hybrids yielded an average 16% more than the best improved varieties (Table 3). The growth durations of heterotic hybrids ranged from 104 to 133 d, indicating that higher-yielding rice hybrids can be developed in different varietal malturity groups.

Table 1　Yields of hybrid and improved rice varieties, Ningxia, China,1982—1986.

Year	Yield/ (t/ha)		Hybrid yield Advantage/%	Hybrid rice growing area/ha
	Hybrid rice (ha)	Improved rice Jing-yin 39		
1982	5.2	4.1	26.6	3.5
1983	5.2	3.9	34.8	420.0
1984	5.1	3.8	33.0	3 597.1
1985	4.8	3.8	4.3	4 665.8
1986	4.7	3.8	21.8	10,097.0

Table 2　Total growing area, yield, and average yield ha of hybrid rice in China 1983—1986.

Year	Total area /million ha	Total yield /million t	Average yield / (t/ha)
1983	6.74	42.97	6.4
1984	8.84	56.62	6.4
1985	8.27	53.53	6.5
1986	9.13	60.24	6.6

Table 3　Yield of the best experimental F_1 rice hybrids and the best check varieties. IRRI, 1981—1986.

Season	Trials (No.)	Hybrids Evaluated (No.)	Yield/ (t/ha)		Percent of check	
			Range	Mean	Range	Mean
Dry	14	207	5.4—9.6	7.8	86—141	116
Wet	16	202	2.6—5.6	4.2	100—140	116

During 1986—1987, IRRI also evaluated some promising experimental rice hybrids, along with more than 370 elite breeding lines and varieties developed through conventional rice breeding program at IRRI. The highest yielding experimental hybrid was only slightly superior (0.3 t/ha) to the best breeding lines (Table 4). The difference from the best commercial variety IR64 was higher (0.7 t/ha). The hybrid was strikingly superior to all other lines in productivity per day.

Table 4　Performance of an F_1 rice hybrid, best elite lines, and best commercial variety. Replicated yield trials, IRRI 1986—1987.

Hybrid/ variety	Yield/ (t/ha)				Percent of IR36 (CK)				Productivity / (kg/day per ha)				Mean Growth duration/d
	1986		1987		1986		1987		1986		1987		
	DS	WS	DS	$\bar{X}$	DS	WS	DS		DS	WS	DS	$\bar{X}$	
IR46830A/ IR9761–19–1	7.7	4.2	6.9	6.3	108	125	111	115	86	51	75	71	109
IR35366–90–3–2–1–3	6.5	4.5	7.0	6.0	90	157	122	123	65	48	71	61	118
IR64 (CK)	6.3	4.2	6.3	5.6	93	135	107	112	66	41	64	57	120
IR44707–31–1–3–2	7.5	3.5	6.9	6.0	132	113	144	130	68	32	62	54	131
IR44668–85–1–2–2–3	–	3.7	7.2	5.5	–	98	118	108	–	42	78	60	111

The best experimental rice hybrids identified in the hybrid rice breeding program at IRRI are being eualuated with the best improved lines developed through the conventional breeding program, to identify hybrids qualified for on-farm multilocation testing outside IRRI.

A number of experimental rice hybrids also were evaluated in replicated yield trials in collaboration with the national programs of the Philippines, Indonesia, Korea, India, Malaysia, and Vietnam. On average, the highest yielding hybrids yielded 16% higher than the best improved varieties (Table 5).

Table 5　Yield performance of best experimental F_1 rice hybrids and best check varieties in international trials, 1980—1986.

Country	Trials (No.)	Yield of best Hybrids/ (t/ha)		Percent of check	
		Range	Mean	Range	Mean
Indonesia	15	4.1–8.9	6.2	102–143	117
Korea	11	8.1–11.5	9.1	97–142	113
India	21	3.3–9.8	6.2	91–143	116
Malaysia	2	4.2–5.0	4.7	89–127	108
Philippines	8	4.8–7.4	5.4	92–133	114
Vietnam	4	5.3–6.6	6.0	91–122	108
Overall	61	5.3–11.5	6.6	89–143	116

Commercial rice hybrid Wei You 64 developed jointly by Chinese and IRRI scientists (Yuan et al.,1985) was evaluated in the 1986 International Rice Yield Nursery-very Eearly, coordinated by the International Rice Testing Program (IRRI,1987). This hybrid yielded the highest (5.4 t/ha) among 24 entries evaluated at 25 locations, with a CV less than 25%. The second ranking line yielded 5.3 t/ha. The hybrid ranked number 1 among the 24 entries at 5 locations (4 in China-Guanzhou, Fuzhou, Luzhou, and Lian Tang and 1 in Bangladesh-Comilla). The hybrid was among the top five entries in Chainat, Thailand: Sakha, Egypt: Amol, Iran: and Menemen, Turkey. Wei You 64 is not suitable for tropical countries because it does not possess adequate disease and insect resistance.

At certain locations (Namyang and Milyang, Korea), and Bangalore, Adhuturai, and Cuttack, (India), none of the IRRI-bred hybrids tested so far have performed better than the checks. For such situations we need to develop hybrids derived from locally adaptable CMS and restorer lines which possess acceptable grain quality. Such lines are being developed at IRRI and in national programs.

These results indicate that F_1 hybrids that have helped increase rice varietal productivity in China beyong the levels of improved semidwarf varieties/also can help increase varietal yields by 15%–20% in countries outside China. Suitable hybrids adapted to these countries are in the pipeline and should be available within a year or two.

Hybrid show a general tendency to possess higher productivity per day than parents and commercial check varieties (Virmani,1986). Higher yield potential combined with higher per day productivity should make hybrid rice technology the way to meet the challenge of producing more rice on less land and provide opportunities for crop diversification without risking reduction in rice production.

ADAPTABILITY OF RICE HYBRIDS IN STRESS ENVIRONMENTS

Hybrid rices have been adapted to various climatic (tropical, subtropical, and temperate), topographical (plain, coastal area, and hilly regions), and cultural (irrigated, drought-prone, and upland) conditions in China. Hybrids are cultivated in single crop, double crop, and ratooned fields. They have been adapted to semidry irrigated areas in North and Northeast China, including Liaoning, Henan, Beijing, and Tainjing provinces, where average yields of 5–6 t/ha are obtained from 67,000 ha. Rice hybrids have been grown with 50% less water (4,500 m) compared to 9,000 m^3 in irrigated fields. In Northern China, growing hybrid rice has been more profitable than growing maize (Table 6).

Table 6　Benefit of hybrid rice over and maize under semidry irrigated conditions. Baode county, Tianjing, China, 1984.

Site	Hybrid rice			Maize			Benefit for rice / ($/ha)
	Yield / (t/ha)	Input / ($/ha)	Net income / ($/ha)	Yield / (t/ha)	Input / ($/ha)	Net income / ($/ha)	
1	4.8	266	518	4.20	59	277	241

Continued

Site	Hybrid rice			Maize			Benefit for rice / ($/ha)
	Yield / (t/ha)	Input / ($/ha)	Net income / ($/ha)	Yield / (t/ha)	Input / ($/ha)	Net income / ($/ha)	
2	5.25	266	574	4.50	54	305	269
3	4.86	248	534	3.00	46	194	340
4	5.52	268	615	5.25	85	335	280
AV.	5.13	262	562	4.24	61	278	284

IRRI results also indicate F_1 superiority for root number and root diameter (R. Peiris, G. Loresto, and T. T. Chang, 1982, unpubl.) and root-pulling resistance (the force required to pull a plant from the soil) (Ekanayake et al., 1986). Root-pulling resistance and tolerance for moisture stress during vegetative growth have been found to be correlated (O'Toole and Soemartono, 1981). We asume the F_1 hybrids would have better adaptation than their parents in drought-prone rainfed area.

In Liaoning Province, China, hybrid rices showed tolerance for soil alkalinity (Table 7) and area, being grown in coastal areas. At IRRI, we also observed better survival of hybrids than parents under saline conditions (EC = 7 ds/m) (Senadhira and Virmani, 1987).

Table 7　The Alkali-tolerant abilities of different rices Liaoning, China, 1986.

Rice type	Productive panicles/plant	Relative alkali tolerance/%
Conventional rice	6.83±1.47	82
Hybrid rice	7.66±0.90	92
Alkali – tolerant rice	8.31±1.01	100

Kaw and Khush (1985) report heterosis for various traits related to low temperature tolerance. We also observed better performance of certain F_1 hybrids under high-altitude irrigated conditions at Banaue, Philippines (Table 8).

Table 8　Yield and growth duration of some promising experimenttal rice Hybrids, elite breeding lines, and commercial varieties at high altitude location. Banaue, Philippines, 1986 wet season.

Hybrid/line	Yield/(t/ha)	Growth duration/d
Observational Yield Trial		
IR54752A/IR14753–120–3 F_1	7.4	168
IR54752A/IR19392–211–1 F_1	7.4	160
V20A/Milyang 46R	7.3	145
IR40094–4–5–5	7.2	160

Continued

Hybrid/line	Yield/（t/ha）	Growth duration/d
Barkat（early check）	3.1	140
Pinidua（local check）	4.4	180
Replicated Yield Trial		
Wei You 64 F_1	6.6	140
IR40094–1–5–2	7.3	159
IR9202–5–2–2–2	4.5	148
Barkat（early check）	4.0	140
Pinidua	4.1	180
LSD（0.5%）	0.91	–

These results indicate the usefulness of F_1 rice hybrids, not only under favorable conditions but also under certain stress environments.

PERFORMANCE HYBRID UNDER DIFFERENT FERTILITY LEVELS

Chinese scientists report that hybrid rices utilize applied fertilizer more efficiently than improved varieties. Korea-IRRI collaborative studies showed yield advantages of F_1 hybrid V20A/Milyang 46 at different N fertilizer levels. IRRI evaluated a number of experimental rice hybrids, their parents, and check varieties under three N levels. Selected hybrids showed significant superiority over parents, but only numberical superiority over chack varieties（Table 9）. Hybrid IR54752A/IR19392–211–1 yielded the highest (7.3 t/ha) in the trial at 75 kg N; best check variety/IR64 yielded the highest (6.6 t/ha) at 150 kg N. These results indicate that not every hybrid will utilize applied N better than its parents. Perhaps only heterotic combinations will possess such capability.

Table 9　Yield of rice hybrids, their parents, and check varieties at 3 N levels. IRRI, 1986 dry season.

Parents/Hybrid[a]	Grain yield/（t/ha）			
	0 kg N	75 kg N	150 kg N	Mean
IR54752B	4.9	4.8	4.7	4.8
F_1	4.1	5.7	4.0	4.6
IR54	5.1	5.9	6.3	5.7
IR54752B	4.9	4.8	4.7	4.8
F_1	5.1	6.1	6.3	5.8

Continued

Parents/Hybrid[a]	Grain yield/ (t/ha)			
	0 kg N	75 kg N	150 kg N	Mean
IR29512–81–2–1	4.4	5.4	4.7	4.8
IR54752B	4.9	4.8	4./	4.8
F₁	5.2	7.3	7.0	6.5
IR19392–211–1	4.3	5.9	4.9	5.0
IR54752B	4.9	4.8	4.7	4.8
F₁	4.1	5.7	5.0	4.9
IR20933–68–21–1–2	3.4	3.9	3.3	3.5
IR54752B	4.9	4.8	4.7	4.8
F₁	4.9	5.4	3.9	4.7
IR4422–480–2–3–3	4.1	4.9	4.4	4.4
IR46830B	2.4	4.3	5.0	3.9
F₁	3.3	5.3	5.5	4.7
IR50	4.4	5.2	5.4	5.0
V20B	0.9	1.2	1.3	1.1
F₁	2.0	3.5	3.2	2.9
IR9761–19–1	3.5	5.5	5.9	5.0
IR64 (Check)	3.8	6.0	6.6	5.5
LSD (0.05)	0.9	0.9	0.9	

Note: a.In a group, line 1=female parent; line 2=hybrid; line 3=male parent.

ADVANCES IN HYBRID SEED PRODUCTION TECHNOLOGY

In the 1970s, hybrid seed yield in China were very low. Yields have improved tremendously. It is not uncommon to get 2–2.5 t hybrid seeds/ha. The highest hybrid seed yield in China was 6.1 t/ha for hybrid Wei You 64 in Xu-Pu county, Hunan Province, during 1986 summer.

With increasing seed yield, the field area ratio between 'A' line multiplication, hybrid seed production, and commercial production of hybrid rice has changed, from 1 : 30 : 1,000/ha in the 1970s to 1 : 50 : (3,000–5,000) /ha today. Increased seed yields have been primarily responsible for the expansion of the rice area planted to hybrids. That reduced seed prices and farmer investments in hybrid rice cultivation.

The increase in hybrid rice seed yield and quality in China has been attributed to the following factors:

1. Shitting seed production from autumn to summer .

2. Planting two seedlings CMS lines per hill instead of the one seedling per hill planted previously. This increased the number of total effective panicles of a CMS line per nit area.

3. Increasing the row ratio of A : R lines. Currently, row ratios of 1 : (8–10) or 2 : (14–16) are used for indica rice and 1 : 6 or 2 : 8 for japonica rice.

4. Increasing the Gibberrellic acid (GA_3) dosage to enhance spikelet and panicle exsertion.

5. Discontinuing the reduction of flag leaf clipping to increase seed weight and quality.

6. Developing CMS lines showing higher outcrossing rates (up to 50%) — II 32A, Chang Hui 22A, Xiu Qing Zao A, and L 301A.

7. Increasing the efficiency of the organization set up for hybrid seed production, certification, and distribution. The provincial, perfectural, and county level management system for hybrid rice seed production has been found very effective. Under this system, the provincial seed company is in charge of purification and foundation seed production; the perfectural seed company responsible for large-scale CMS seed multiplication; and the county seed company for commercial seed production. Earlier, CMS multiplication was carried out mostly by county seed companies, which resulted in some problems with seed purity. Also, in recent years, provincial governments have enacted some laws regulating seed production and seed quality.

At IRRI, we have attempted to adopt the basic seed production techniques developed in China. In experimental seed production plots (size 10–1,250 m^2), seed set on cms lines ranged from 0.1% to 43% and seed yield ranged from 5 to 1,510 kg/ha. Outcrossing on a line was generally lower in the wet season than in the dry, because of unfavorable weather conditions (heavy rains and typhoons) and higher incidence of diseases and insects during the wet season. Ootcrossing rate and seed yield depended on synchronization of flowering in the male female parents and the agronomic and floral traits of the parental lines.

Indonesian and Korean scientists also have tested the adaptability of the hybrid seed production techniques developed in China, in collaboration with IRRI. They have obtained seed yields of 60–1,800 kg/ha in Indonesia and 750–1,530 kg/ha in Korea.

We recognize that hybrid rice technology will not succeed outside China unless appropriate and economical packages of practices for hybrid seedproduction, along with suitable F_1 rice hybrids, can be develop. IRRI, in cooperation with China and other collaborating countries, is actively involved in developing such pakages of practices and suitable CMS and restorer lines.

DIVERSIFICATION OF USABLE CMS SYSTEMS

Chinese scientists have identified seven different CMS systems among the CMS lines used to develop commercial hybrids. Five are sporophytic (WA, Gam, Indonesian Paddy (IP), Dissi (Di), and dwarf wild rice (DW). The two others are gametophytic BT (CMS-boro) and Hong Lien (HL).

Two or three years ago, more than 95% of the CMS lines used in developing commercial hybrids in indica rice were of the WA CMS system. This situation made hybrid rice in China potentially vulnerable to a disease or insect outbreak, which might be genetically associated with this CMS system. Recently, new CMS lines with different CMS sources [Ⅱ −32A (IP), Xiu Qing Zao A (DW), Zhen Shan 97 (DI)] have been released, which should help diversify the CMS sources in China.

Outside China, mostly the WA cytosterility system has been transferred to elite maintainer lines to develop new CMS lines. During the last 8 years 29 new CMS lines have been developed at IRRI (Table 10). Some of them have been found suitable for the development of rice hybrids for the tropics.

Table 10   CMS lines developed at IRRI 1987.

CMS Line	Elite breeding line Used as maintainer	Origin of maintainer	Varietal group	Growth duration/d
IR46826A	IR10154−23−3−3	IRRI	I	105
IR46827A	IR10176−24−6−2	IRRI	I	105
IR46828A	IR10179−2−3−1	IRRI	I	110
IR46829A	IR19792−15−2−3−3	IRRI	I	110
IR46830A	IR19807−21−2−2	IRRI	I	110
IR46831A	Jikkoku	India	I/J	115
IR48483A	MR 365	India	J	110
IR54752A	IR21845−90−3	IRRI	I	135
IR54753A	IR19657−34−2−2−3−3	IRRI	I	135
IR54754A	IR19657−87−3−3	IRRI	I	135
IR54756A	Iri 356	Korea	I/J	125
IR54757A	Wuweon 310	Korea	I/J	130
IR54758A	PAU 269−1−8−4−1−1−1	India	I	125
IR58019A	IR19809−12−3−2−1	IRRI	I	105
IR58020A	IR17525−278−1−1−2	IRRI	I	130
IR58021A	IR19805−12−1−3−1−2	IRRI	I	110
IR58022A	IR19774−23−2−2−1−3	IRRI	I	100
IR58023A	PY 2	India	I	110
IR58024A	Suweon 161	Korea	J	110
IR58025A	Pusa 167−120−3−2	India	I	120
IR58026A	IR31787−24−3−2−2	IRRI	I	110

Continued

CMS Line	Elite breeding line Used as maintainer	Origin of maintainer	Varietal group	Growth duration/d
IR58027A	IR15795–151–2–3–2–2	IRRI	I	120
IR58052A	IR4763–73–3–11	IRRI	I	115
IR58053A	IR19746–27–3–3–1–3	IRRI	I	115
IR58054A	IR22103–26–6–2	IRRI	I	117
IR58055A	IR19728–9–3–2–3–3	IRRI	I	120
IR58056A	IR25474–41–2–3–2	IRRI	I	130
IR58057A	IR19661–283–1–3–2	IRRI	I	140
IR58058A	IR12979–24–1	IRRI	I	130

Concurrently, attempts have been made to identify new sources of CMS. Interspecific, intraspecific and intervarietal crosses have been made. We have developed a stable CMS line, IR54755A, possessing the CMS source of Oryza sativa cultivar ARC13829–16 introduced from Assam, India, Preliminary results indicate that the CMS system of IR54755A is different from WA (Virmani and Dalmacio, 1987).

Although no evidence so far associates any disease or insect susceptibility in rice with the WA or any other cyto-sterility system, hybrid rice breeders in and outside China cannot afford to be complacent. Cytoplasmic and genetic diversity must be assured when parents are choosen to develop commercial hybrids. Also, the relationship between the available CMS sources and resistance or susceptibility to major diseases and insects found in the tropics must be monitored.

Lu and Wang (1986) reported a genetic male sterility system in rice which was photoperiod sensitive. It showed complete male sterility under long daylengths but reverted to fertility under short daylengths. Jin et al. (1986) proposed using this system in hybrid rice breeding. If efective, it would simplify hybrid seed production. The CMS line can be multiplied by selfing and growing under short daylength conditions. Moreover, the choice of male parents is not restricted to restorers; any fertile cultivar can be used as the male parent. Hybrid development would involve only two lines the male sterilie and a fertile male parent, instead of the three lines required in the CMS system.

DEVELOPMENT OF RESTORER LINES AND GENETICS OF FERTILITY RESTORATION

Chinese and IRRI scientists have used a large number of restorer lines from among cultivated rice varieties and elite breeding lines for WA, Gam, BT and newly identified CMS systems. The restorer lines widely used in developing commercial rice hybrids in China were mostly introduced from IRRI.

Restorer lines are more often found among rice cultivars originating at lower latitudes than among cultivars from higher latitudes. The frequency of restorer lines found among japonica rices is negligible. Consequently, japonica F1 hybrids in China have been developed from restorer lines bred by transferring restorer gene (s) from indica rices.

Work at IRRI also has indicated that several commercial rice varieties (IR24, IR26, IR28, IR36, IR42, IR46, IR50, IR54, IR56, IR58, IR60, and IR64) are effective restorers. In fact, of about 3,000 elite lines and varieties screened at IRRI to date, 20% are effective restorers for the WA CSM system. Therefore, IRRI has linked hybrid rice breeing programs with the line breeding program so that the best available conventional lines can be used as parental lines to develop heterotic hybrids. IRRI is constantly supplying elite restorer lines possessing multiple disease and insect resistance and early maturity to China to develop heterotic rice hybrids.

A number of Tongil (indica/japonica derivatives) rice varieties from South Korea (Suweon 287, Suweon 294, Milyang 46, and Milyang 54) also have been found to be effective resotrers. Their fertility restoration ability may be due to the restorer genes inherited from indica parents.

We have found that the identified restorer lines are usually mixture of restorer, partial restorer, and nonrestorer genotypes. The test F_1 progenies derived from single plant selection of restorer lines showed differential behavior in fertility restoration (Table 11) . Purification of restorer lines is essential for breeding F_1 hybrids.

Tale 11　Fertility restoration behavior of single plant selections of resotrers in purification nursery.

Restorer	Tester	Single plants (No.)	Behavior of testcross F1 progenies[a]						
			F	F/PF	F/PS	F/P	S/S	FS	S
IR54	IR465831A	32	15	14	3				
IR9761–19–1	IR46830A	56	43	7	1		1	3	1
IR64	IR54752A	56	23	23			10		
IR13419–113–1	IR54752A	78	39	21	13	4			1

Note: a.F= fertilie; PF = partially fertile, PS = partially sterile, S = sterile.

Results obtained in China (Wang, 1980; Yang and Hao Ran,1984; Zhou,1983) and at IRRI (Govinda Raj and Virmani, unpublished) (Virmani et al.,1986; Young and virmani,1984) indicate that the restoration ability of the WA CMS system is governed by two dominant genes, one of which appears to be stronger than the other. The mode of action of the two genes varied with different CMS/restorer combinations (Govinda Raj and Virmani, unpubl.) . At IRRI, four groups of restorers possessing different pairs of restorer genes were identified.

It appears that there are a number of restorer genes present in rice, which results in the wide-spread occurrence of restorer lines among elite breeding lines. Hybrid rice breeders have the opportunity to choose genetically diverse restorer parents to develop heterotic rice hybrids. Govinda Raj and Virmani (unpubl.) also found that certain CMS lines possess inhibitory genes that affect fertility restoration.

DISEASE AND INSECT RESISTANCE IN RICE HYBRIDS

The genes in rice that control resistance to major diseases and insects in rice are mostly dominant or partially dominant (Khush, 1977; Khush and Virmani, 1985) . If one of the parents has dominant resistance gene (s) , the F_1 hybrid is also resistant. The hybrid breeding approach facilitates expeditious pyramiding of dominant major genes or minor polygenes. The hybrid Shan You 63 released in China is resistant to blast, bacterial blight, and brown planthopper. The resistance genes are contributed by both parents. In studies at IRRI, certain F_1 rice hybrids showed a wider spectrum of blast resistance than their parents (IRRI, 1983) . The vigor of hybrids also may contribute to their field tolerance for diseases and insects. This aspect of rice hybrids needs to be studied critically.

Grain QUALITY OF RICE HYBRIDS

The economic product of rice hybrids is the seed borne on F_1 plants. Although F_1 plants are uniform the grains they produce/represent the F_2 generation and are expected to segregate on some important grain characteristics. We must consider the effect of this segregation on grain quality, because consumer acceptance of the grain of F_1 rice hybrids will have an important bearing on the adoption of this technology.

There are three major determinants of grain quality in hybrid rice: 1) milling and head rice recovery; 2) size, shape, and appearance; and 3) cooking and eating quality compared to parents. Khush et al. (1986) studied the effect of genetic differences for grain characteristics of the parents on the quality of grain borne on the F_1 rice hybrids derived from them. They drew the following conclusions:

1. Physical properties (differences in length, breadth, shape, and weight of grains of parents) do not pose any problem because the seeds borne on hybrid plants do not vary from each other. F_1 hybrids with desired milling and head rice recovery can be obtained.

2. Crosses among parents of different endosperm appearance result in grains with different types of endosperm in bulk samples from hybrid plants. Such variation in grain appearance would affect market acceptability. Using a waxy or dull endosperm parent with parents possessing translucent grains should be avoided in hybrid rice breeding program.

3. When parents differ widely in amylose content, F_2 single grains are clearly classifiable into 2–4 categories. However, segregation for amylose content does not have any adverse effect on cooking and eating quality.

4. Parents possessing grain with different gelatinization temperatures (low, intermediate, or high) do not cause any detectable difference in cooking quality of grain borne on their F_1 hybrids.

We concluded that rice hybrids with desired physical, chemical, and cooking characteristics can be developed with appropriate selection of parents.

The first set of hybrids developed in China (Wei You 2, Wei You 6, Shan You 2, Shan You 6, etc.) possessed bold, chalky grains that are not acceptable in several countries outside China. The grain quality of these hybrids was inherited from the female parent developed in China. Chinese scientists are now critically evaluating the parental lines used to develop new rice hybrids. Some CMS lines (L 301A, U-1A, Qiu Guang A) and restorer lines (R 29, IR9761-19-1-64) possessing good grain quality have been developed. The hybrid L 301A/R 29 has been found to possess excellent grain quality acceptable in the U.S. market, where Wei You 6 and some other first-generation hybirds were rejected.

CONSTRAINTS

Despite the tremendous success of hybrid rice in China, some constraints need to be overcome. These are discussed below:

1. The indica hybrids now used in China mainly are of medium and long growth duration. They cannot be cultivated as the first crop in the double rice cropping system in the Yangtze valley, which is the major rice production region in China (6 million ha). Very few short-duration hybrids suitable for this region are available so far. Development of early-to very early-maturing hybrids possenssing high yield potential is extremely important if the hybrid rice area is to be expanded to 13.5 million ha by the end of 1990, as per the China national plan.

2. The commercial F_1 hybrids in currently grown China have very little cytoplasmic diversity, which makes rice production potentially vulnerable to a disease or insect outbreak. New hybrids with diverse CMS systems are needed. Development of CMS lines with diverse CMS sources is a high priority research area. For japonica hybrids, the available CMS lines are not stable for complete sterility under high temperature. That causes impurity problems. We need to develop stable CMS lines for japonica hybrids.

3. Effective restorer lines are lacking to develop heterotic japonica rice hybrids in japonica rice.

4. The high cost of hybrid rice seed is a major constraint to further expansion of hybrid rice. The best way to improve hybrid seed yield and to reduce seed production costs is to develop CMS lines possessing high outcrossing rates.

5. Most of the available commercial rice hybrids do not possess good grain quality. To improve grain quality, both male and female parents must have good grain quality and be uniform in certain grain characteristics. Breeding rice hybrids for good grain quality is an important objective of the hybrid rice breeding program in China.

6. The leading rice hybrids in China are becoming susceptible to diseases and insects in certain regions,due to changes in physiological races and biotypes.For example, about 7,000 ha in Sichuan Province were seriously damaged by blast disease in 1985. Breeding rice hybrids possessing multiple disease and insect resistance is a continuous objective of hybrid rice breeders.

7. Rice hybrids derived using chemical emasculation techniques have shown higher yield potential than those derived from a CMS system.But the chemical emasculants have not been used extensively to deveiop commercial rice hybrids because of the lack of effective and safe gametocides. Research to

identify suitable chemical male gametocides needs to be intensified.

Major constraints to hybrid rice research and development in countries outside China are:

1. Inadequate numbers of trained scientist and lack of infrastructure and government support for hybrid rice research and development.

2. An attitude among research managers and policymakers that hybrid rice technology cannot be adopted in developing countries with market economies and that efforts should be directed improving rice production in unfavorable rice environments rather than to increasing production in favorable environments.

3. Lack of outstanding and stable male sterile lines, especially in the tropics.

4. Lack of strikingly superior and stable hybrids for commercial use.

5. Low yields in hybrid seed production plots, resulting in high costs for hybrid seed.

6. Inadequate information on the economics of hybrid rice cultivation and hybrid seed production in China and its applicability in countries outside China.

The technical constraints could be overcome with additional research at IRRI in collaboration with national rice improvement program interested in exploring the potentials and problems of this technology. The nutional programs need to establish multidisciplinary teams to work full time on hybrid rice research if they are to participate effectively in this collaboration. Those teams should work closely with ongoing programs on improved variety development.

OUTLOOK FOR THE FUTURE

In China, research and utilization of heterosis in rice have made tremendous advances during the last 20 years. Preliminary results at IRRI and in some national rice improvement programs indicate that hybrid rice technology can help to increase rice varietal yields 15%–20% beyond the levels obtained with semidwarf improved rice varieties. However, packages of technology suitable for rice farming in countries outside China have yet be developed.

The classical, three-line breeding method use to develop heterotic rice hybrids, ijvolving CMS, maintainer and restorer lines, is expected to remain effective for the next decade. But it is more complicated than is necessary and may in the long run, be replaced by two line or one line systems. Two-line systems may involve photoperiod-sensitive genetic male steriles or chemical emasculation. Theone-line method involving apomixis is considered the most worthwhile long-erm goal. It will make posible true breeding hybrids with permanently foxed heterosis. Development of apomictic rice will required biotechnology processes.

To increase heterosis levels beyond the level currently obtained, emphasis may have to be shifted from intervarietal hybrids to intra-specific hybrids (involving indica and japonica cultivars) which exhibit very strong heterosis for spikelet number (Table 12) . However, spikelet sterility in such hybrids at in the way of exploiting this heterosis commercially. The recent discovery of a wide compatibility gene

(Araki et al., 1986; Ikehashi and Arakia, 1986) should make it possible to get indica-japonica hybrids to set normal seed. To make this approach practical, wide compatibility genes will have to be transferred to various CMS and restorer lines.

Table 12　Yield potential of indica-japonica and indica-indica hybrid rices. Hunan Hybrid Rice Research Center, 1986.

Type	Plant height /cm	Spikelets/ panicle	Spikelets/ plant	Seed set /%	Yield / (t/ha)
Indica/japonica (I/J) Chengte 232/26 Zhai zao	120	269.6	1779.4	54	8.2
Indica/indica (I/J) V20A/26 Zhai zao (CK)	89	102.6	800.3	93	8.6
% increase of decrease		102	12	12	−4

Using of hybrid vigor from wide crosses is hard to imagine today. But with the help of genetic engineering tools, it may be possible to develop elite lines from interspecific crosses which have unique gene blocks, resulting in greater heterosis in F_1 hybrids.

To develop approprizte hybrid rice technology for countries outside China, the following research needs to be undertaken:

1. Develop suitable CMS and restorer lines that will result in F_1 rice hybrids that will yield 15%–20% (0.75–1 t/ha) more than the best improved lines developed through conventional breeding methods. The hybrids must possess multiple disease and insect resistance and acceptable grain quality.

2. Develop and adapt of suitable seed production techniques to produce good quality hybrid seed economically.

3. Develop optimum management practices for maximum economic yields and adaptability of hybrids.

4. Assess the economic feasibility of hybrid rice.

5. Use growth regulators to modify floral characters and select CMS lines that possess large exserted stigma and multiple pistil to influence outcrossing.

6. Study the effect of hybrid vigor in imparting field tolerance for diseases, insects, and other stresses.

7. Explore prospects of somatic embryogenesis for hybrid seed or seedling production, as an alternative to large-scale seed production in the field.

Critics of hybrid breeding approach contend that heterozygosity is not a prerequisite to high performance, uniformity, and stability of performance of hybrid varieties (Jinks,1983) . They say that heterosis is primarily due to the bringing together of unidirectionally dominant alleles dispersed between the parental lines showing linkage disequilibrium. Therefore, it is contended that conventionally bred lines with performance equal or better than F_1 hybrid can be developed, especially in self-pollinated crops.

However, there are no critical data in rice to show that conventionally bred lines that perform as well as a heterotic F_1 hybrid have been developed from such a cross. We believe that repulsion phase genetic linkages in parental lines would not allow all the useful genes from the parents to combine in the F_2 and subsequent generations through pedigree breeding methods. The hybrid breeding approach would enable to overcome the effects of repulsion phase linkages in the F_1 generation.

IRRI is currently exploring the possibility of working collaboratively with the University of Birmingham to critically study whether or not and how inbred lines with as high yield potential as F_1 hybrids can be developed.

Hybrid rice has already revolutionized rice production in China, and it will continue to do so during the remaining years of this century. Outside China, the rate of this technology is still uncertain. Its future will depend on the resources international and national rice improvement programs provide to develop and use the technology, and the availability of alternative conventiional inbred breeding methodologies to increase rice yields per unit area per unit time.

References

Araki H, Toya K, Ikehashi H (1986) Role of wide compatibility gene (s) in hybrid rice breeding. Paper presented at the International Symposium on Hybrid Rice, 6–10 Oct 1986, Changsha, Hunan, China.

Barker R, Herdt R W, Rose B (1985) The rice economy of Asia. The John Hopkins Univ. Press, Washington, D. C.

Ekanayake I J, Garrity D P, Virmani S S (1986) Heterosis for root pulling resistance in F1 rice hybrids. Int. Rice Res. Res. Newsl. 11 (3) : 6.

Ikehashi H, Araki H (1986) Genetics of F1 sterility in remote crosses of rice. Page 119–130 in Rice genetics International Rice Research Institute, P. O. Box 933, Manila, Philippines.

International Rice Research Institute (1983) Annual report for 1982. P. O. Box 933, Manila, Philippines. p. 125.

International Rice Research Institute (1987) Preliminary report of 1986 IRTP nurseries results, P. O. Box 933, Manila, Philippines, p. 2–5.

Jin D M, Li Z B, Wan J M (1986) Utilization of photoperiod sensivive genetic male sterility in rice breeding . Paper presented at the International Symposium on Hybrid Rice, 6–10 Oct 1986, Changsha, Hunan, China.

Jinks J L (1983) Pages 1–46. Biometrical Genetics of Heterosis in Heterosis, Reapprainsal of Thory and practice. R. Frankel, ed., Spring-Verlag-Berlin, Heidelberg.

Kaw R N, Khush G S (1985) Heterosis in traits related to low temperature tolerance in rice. Phillipp. J, Crop Sci. 10:93–105.

Khush G S (1977) Disease and insect resistance in rice. Adv. in Agron. 29:265–341.

Khush G S, Kumar I, Virmani S S (1986) Grain quality considerations in hybrid rice. Paper presented at the

International Symposium on Hyrid Rice, 6–10 Oct, 1986. Changsha, Hunan, China.

Khush G S, Virmani S S (1985) Breeding rice for disease resistance. Page 239 in Progress in plant breeding. G. E. Russel, ed., Butterworth.

Lu X G, Wang J L (1986) Stability and genetic behavior of a photoperiod sensitive genic male sterile line in rice. Paper presented at the International Rice Symposium on Hybrid Rice, 6–10 Oct 1986. Changsha, Hunan, China.

O'Toole J C, Soemartono (1981) Evaluation of a simple technique for characterizing rice root systems in relation to drought resistance. Euphytica 30:283–290.

Senadhira D, Virmani S S (1987) Survival of some F_1 rice hybrids and their parents in saline soil. Int. Rice Res. Newsl. 12 (1) :14–15.

Virmani S S (1986) Prospects of hybrid rice in developing countries. In Rice: progress assessment and orientation in the 1980s. Int. Rice Comm. Newsl. 34 (2) :143–152.

Virmani S S, Dalmacio R D (1987) Cytogenic relationship between two cytoplasmic male sterile lines of rice. Int. Rice Res. Newsl. 12 (1) :14.

Virmani S S, Govinda Raj K, Casal C, Dalmaeio R D, Aurin P A (1986) Current knowledge of and outlook on cytoplasmic genetic male sterility and fertility restoration in rice. Page 633–647 in Rice genetics. International Rice Research Institute, P. O. Box 933, Manila, Philippines.

Wang S L (1980) Inheritance of R genes in rice and methods of selection of new 'R' lines. Agri. Ssi. And Technol. (Hunan) 4:1–4.

Yang R C, Hao Ran L (1984) Preliminary analysis of R genes in IR24. Acta Agron. Sin. 10 (2) :81–86.

Young J B, Virmani S S (1984) Inheritance of fertility restoration in a rice cross. Rice Genet. Newsl. 1:102–103.

Yuan L P, Virmani S S, Khush G S (1985) Wei You 64–An early duration hybrid for China. Int. Rice Res. Newsl. 19 (5) :11–12.

Zhou T L (1983) Analysis of R genes in hybrid indica rice of 'WA' type. Acta Agron. Sin. 9 (4) :241–247.

作者：Yuan Longping　S.S.Vrmani　Mao Chang Xiang
注：本文汇编于《杂交水稻国际学术讨论会（1989）论文集》

水稻无融合生殖研究的新进展 ※

【摘　要】从水稻双胚苗中提高不定胚的发生频率是研究利用无融合生殖固定 F₁ 代杂种优势的主要途径之一。通过特殊的选育方法，已将原有双胚苗品系的双苗率由 30% 提高到 60% 左右。由于不定胚与双苗的发生率有着一定的内在联系，随着双苗率的提高，内含不定胚的频率也有相应提高的趋势。又据细胞学观察，认为不定胚发生在远离珠孔端的部位，故在种子中部的萌发苗可能是不定胚产生的种子形态学特征。1989 年，我们采用特定苗位为标记的筛选方法，选取种子正中部萌发苗并结合对其他非正中苗的人工切除，使水稻的不定胚频率由原来的 2.6%～5.1% 提高到 21.2%，水稻不定胚频率的大幅度提高，为无融合生殖在杂交水稻育种中的研究应用开拓了诱人的前景。

【关键词】无融合生殖；双胚苗；不定胚；杂种优势固定

农业发展史业已证明，植物育种技术一旦突破，必将带来农业生产的新飞跃。杂交玉米、杂交高粱、半矮秆水稻和小麦以及杂交水稻的巨大成就都是其光辉的范例。

以战略的观点看，杂交水稻育种无论从杂种优势程度以及育种方法等方面，均可分别划分为三个战略阶段。即：品种间→亚种间→远缘杂种优势利用和三系法→两系法→一系法育种。一系法育种其目的即在于固定 F₁ 杂种优势[2]。无融合生殖（Apomixis，Apogamy）系指以种子形式进行繁殖的无性生殖方式。它可使世代更迭而不改变核相，即后代的遗传结构与母本完全相同，可以固定 F₁ 代杂种优势。迄今，在植物界中已有 36 个科 300 多个种中发现无融合生殖突变体[4]。高粱、马铃薯、珍珠粟和棉花等作物也均已发现这种生殖现象[5-7]。在无融合生殖中又以二倍体孢子生殖，无孢子生殖和不定胚生殖三种类型能固定 F₁ 杂种优势。在水稻及其近亲植物中，二倍体无融合生殖材料（特别是兼性无融合生殖

※ ①双 3 和双 13 两个籼稻双胚苗原始品系实验材料系由郭名奇同志提供。
　②参加本项试验工作的还有王桂元、张克明和赵炳然等同志。

102

材料）的发现与利用，可能是杂交水稻一系法育种固定杂种优势的重要途径之一[1]。

1980年，我们在上千份材料中筛选获得了4份具有遗传特性的双胚苗水稻品系（双苗频率在5.0%～32.4%）。自1985年以后，我们便利用这些材料系统地、逐步深入地开展了无融合生殖水稻的研究。1987年，我们首次发现并证实在水稻双胚苗品系中存在着低频率（2.6%～5.1%）的不定胚；并用显性紫色稻与它们杂交从遗传特点上验证了这一现象[3]，从而揭示了走利用不定胚固定水稻杂种优势途径的可行性。但是，由于不定胚频率太低而不能在育种实践上加以利用，因此，研究提高水稻不定胚频率是一个很重要的探索方向。

由于双苗中存在着具有一个中胚轴的双苗和具有独自中胚轴的双苗之别，后者又有双卵（或助细胞）受精苗和不定胚苗等类别[3]。我们在注意到双胚苗（含多胚苗）频率与不定胚频率之间存在着一定的内在相关性的同时，在细胞学检查中也发现了不定胚往往发生在远离珠孔端的部位；同时也注意到合子胚与不定胚在营养生理发育上的特点与差异，1989年采用了选择具独立中胚轴苗，选择种子正中萌发苗和人工切除非正中苗等综合筛选措施，最终使双苗率和不定胚频率获得大幅度提高。现将研究进展报告如下。

材料与方法

（一）双胚苗材料。AP Ⅰ（陆52）和AP Ⅱ（阿里斯思尼）属粳稻类型，双苗率分别为16.1%和5.6%；AP Ⅲ（双13）和AP Ⅳ（双3）属籼稻类型，双苗率分别为32.4%和5.0%。

（二）选育方法。双胚苗品系的种子，在去壳后用2‰升汞消毒，再用冷开水浸种催芽或在黑暗条件下培育7～10天使中胚轴延长，以区分出以下两种双胚苗（图1）：1.两根苗共有一个中胚轴；2.两根苗具有两个中胚轴。每代只选具有自己中胚轴的双苗或三苗留种，淘汰共有一个中胚轴的双苗。并注意筛选从种子中部萌发的秧苗留种（切除非正中苗）。

（三）胚囊分析。每份材料取开花后2～3天的颖花300朵，在卡诺氏固定液中固定24小时，然后移至70%酒精中保存，采用D.M.Stelly（1979）所制定的梅氏苏木精-水杨酸甲酯整体染色透明法进行胚囊分析[7]。

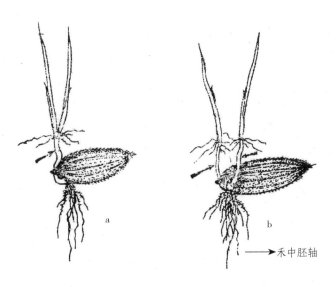

a. 共一个中胚轴的双苗；b. 具两个中胚轴的双苗。

图 1　两种类型的双苗

结果与分析

　　水稻双胚苗品系的种子内含多个胚，位于中部的不定胚，由于受谷壳仅有一个萌发孔的限制，影响直接从中部萌发成苗或仅从萌发孔出苗，因此，除壳后浸种催芽有利于各种胚体的正常萌发并能区分两种类型的双胚苗。通过两代选择，其中两个籼稻双胚苗品系的双苗率已有明显提高（表 1）。

表 1　四个双胚苗品系选择的效果

双胚苗品系	原有双苗率 / %	系选后的双苗率	
		鉴定株数 / 株	平均双苗率 / %
AP Ⅰ	16.1	52	13.8
AP Ⅱ	5.6	20	5.2
AP Ⅲ	32.4	94	59.4
AP Ⅳ	5.0	69	10.9

　　从表 1 可以看出，AP Ⅰ和 AP Ⅱ平均双苗率分别为 13.8% 和 5.2%，与原有双苗率相比，无明显变化。但是，AP Ⅲ检查 94 个双苗单株的后代，平均双苗率为 59.4%，比原有频率（32.4%）提高了将近 1 倍，其中有 6 个单株的双苗率在 80% 以上；AP Ⅳ检查 69 个双苗单株的后代，平均双苗率为 10.9%，比原有频率（5.0%）提高了 1 倍多，其中 1 个单株

的双苗率在20%以上。在选择中只留具两个中胚轴单株的双苗，这种双苗才是真正由两个胚发育产生，其中可能含有来自不定胚的苗，由于两类双苗在遗传上有差异，因而收到明显的选择效果。

随着双苗率的提高，内含不定胚的频率是否也提高呢？为此，我们对这两个籼稻双胚苗品系进行了胚囊分析。对AP Ⅲ系选后双苗率达77%的株系中1个单株检查220个胚囊，共有28个胚囊内含不定胚，不定胚频率为12.7%，比原来的频率（3.6%）提高了2倍多；对AP Ⅳ系选后双苗率达到19.8%的株系中1个单株检查184个胚囊，其中28个胚囊内含不定胚，不定胚频率为15.2%，比原有频率（5.1%）提高了2倍（表2）。

表2　两个籼稻双胚苗品系胚囊分析的结果

双胚苗品系	原有不定胚频率 / %	系选后不定胚的频率		
		检查胚囊总数 / 个	含不定胚的胚囊数 / 个	占胚囊总数的百分比 /%
AP Ⅱ	3.6	220	28	12.7
AP Ⅳ	5.1	184	28	15.2

由于不定胚多位于偏离珠孔端的位置（图2），从种子中部萌发是不定胚的特点。据此，从不定胚频率较高的双胚苗品系（双3）中，采用特定苗位为标记（图3），筛选从种子正中部或偏中部萌发的苗，并以萌发孔和谷粒顶端萌发的苗为对照，对它们的胚囊作了观察。结果表明：珠孔苗和偏中苗内含不定胚的频率很接近，分别为3.7%和3.3%，可能都是由合子胚产生，而正中萌发的苗（CM），内含不定胚高达21.2%，比原来的频率（5.1%）提高了3倍，也比系选后（混收）的频率（15.2%）高，顶苗（T）不定胚的频率为8.2%（表3）。

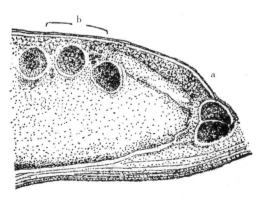

a. 双合子胚；b. 不定胚。
图2　胚囊中部的不定胚

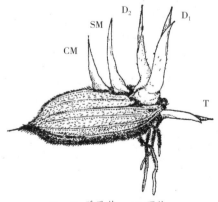

D_1、D_2. 珠孔苗；　T. 顶苗；
SM. 偏中苗；CM. 正中苗。
图3　不同萌发部位的苗

表3　双3各种异位苗胚囊分析的结果

苗位	检查胚囊总数	含不定胚的胚囊数	不定胚的频率 ／ %
珠孔苗（D_1+D_2）	164	6	3.7
顶苗（T）	365	30	8.2
偏中苗（SM）	478	16	3.3
正中苗（CM）	1378	292	21.2

注：珠孔苗（D_1+D_2），由萌发孔长出的苗；顶苗（T），由谷粒纵向顶端长出的苗；偏中苗（SM），偏离萌发孔长出的苗；正中苗（CM），由种子正中部萌发的苗。

上述结果表明，AP Ⅲ和 AP Ⅳ系选后双苗率有显著提高，但由于不定胚与合子胚在组织来源上存在差异，不注意苗位的选择，虽然内含不定胚也随双苗率的提高而有所提高，但提高的幅度比不上正中苗，因此，选用正中苗可以作为提高不定胚频率在苗位上的选择依据。

讨论

1. 种子中部萌发的苗（正中苗）极可能来自不定胚，筛选正中苗繁殖后代能显著提高内含不定胚的发生频率，但中苗后代出现中苗的频率却不高，因此，不定胚并非都能发育成具有发芽能力的胚胎，多数不定胚在与合子胚共存的发育过程中，可能因营养竞争问题以致发育受阻，只有在合子胚败育的胚囊中才有利于不定胚的正常发育（图4），不定胚可能是一种生殖补偿效应。如能在受精前后采用药物杀卵细胞或合子胚，将可能促进不定胚的发育并提高其成苗率。

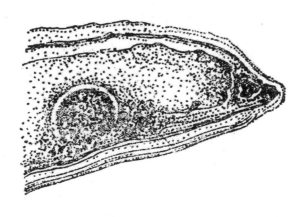

图4　合子胚败育胚囊中的不定胚

2. 现在已选育出内含不定胚达 21% 的双胚苗单株，如果能稳定遗传，那么，这种材料就有应用价值。我们设想，如能筛选出一种专性杀雌配子或合子的化学药物，使每个不定胚都能成苗，则可利用这种不定胚来固定杂种优势。此法在 F_1 的群体中每年可获得 21% 的无性种子供大田生产应用。如果不定胚频率进一步提高到 30%～50%，从而不仅能大大简化繁殖与制种的程序，种子产量也会超过目前三系法和两系法的制种产量。

3. 引进休眠期长的基因，使合子胚在浸种催芽时不易萌发，但不影响不定胚正常出苗，这样也能排除合子胚苗，达到利用不定胚固定杂种优势的目的。

4. 本文提到的不定胚，是用整体染色透明法对水稻双胚苗品系开花后 3 天的子房进行观察，在珠孔附近或胚囊中部所发现的胚状体结构，至于它的起源还有待进一步研究。

References

参考文献

［1］袁隆平.1988.利用无融合生殖改良作物的潜力.杂交水稻,（4）: 1.

［2］L.P.Yuan.1989.Outlook on the development of Hybrid Rice Breeding. Int. Rice Science Conference Hongzhou China.

［3］Yuan-Ching Li, Long-Ping Yuan and J. Neil Rutger, 1989. Searching for Apomixin in Rice via twin Seedlings. Third Annual meeting of the Rockefeller Foundation's international program on Rice Biotechnology.

［4］Bashaw, E.C, 1980. Apomixis and its application in crop impiovement. P.45-62. In Hybridization of Crop plants, W.R.Fehr and H.H.Hadley（eds.）Amer. Soc. Of Agron Madison, WI.

［5］Rao, W.P.G. and U.R.Murty. 1972. Further studies on obligate apomixis in grain sorghum. Ind.J. Genetics and plant breed. 32:379-383.

［6］Rao, N.G.P. and L.L.Narayana, 1968.Apomixis in grain sorghums. Ind.J.Genetics and plant breed.28:121-127.

［7］Murty, U.R., P.B.Kiti, M.Bharathi and N.G.P.Rao, 1983. The nature of apomixis and its utilization in the production of hybrids（"vybrids"）in sorghum bicolor（L.）Moench. Z.Pflanzenzucht.95:113-117.

［8］Stelly,D.M., S.J.Peloquin, R.G.Palmer and C.F.Crane. 1984. Mayer's hemalum-methyl salicylate; A stain-clearing technique for observations within whole ovules, Stain Tech.59:155-161.

作者：黎垣庆　邓鸿德　袁隆平

注：本文发表于《杂交水稻》1990 年第 1 期。

两系法杂交水稻研究的进展

【摘　要】两用核雄性不育系是两系法杂交水稻的基础，我国现已育成 17 个达到规定标准的两用核不育系。根据育性对光温条件的不同反应，它们分为光敏型和温敏型两个基本类型。在生产上均有实用价值，但要因地制宜选用。两系法品种杂交水稻已经研究成功，一些优良组合开始在生产上试种和示范。两系法亚种间杂交水稻的选育也取得了很大的进展。

【关键词】两系法杂交水稻；水稻品种间和亚种间杂种优势利用；光敏核雄性不育水稻；温敏核雄性不育水稻

利用植物的雄性不育性培育雄性不育系，再借助这种遗传工具来大量生产杂交种子，从而能使许多作物特别是自花授粉作物的杂种优势得以在生产上应用，这是现代农业科学最主要的成就之一。按照三型学说，植物的雄性不育分为细胞质、细胞核和质核互作三种遗传类型，其中唯有后者能找到保持系和恢复系，实现三系配套。因此在作物杂种优势育种工作中，选育质核互作类型的不育系一直被视为是正确可行的技术路线，而三系育种法就成了经典的方法。

我国现时在生产上应用的杂交水稻也是循着三系法的技术路线而培育成功的，并且此法还将继续在杂种优势育种中起着重要作用。但是，应当指出，三系法的育种程序和生产环节比较复杂，以致选育新组合的周期长、效率低，推广的环节多、速度慢，同时种子的成本高、价格贵，这些都是三系法杂交种在种植面积和单产提高上受到限制的内在不利因素。所以，从作物杂种优势育种的长远战略上考虑，许多育种家都在设法探索各种新的技术路线，以期采用较简易而效率更高的方法来取代三系法。在这一领域的诸项研究中，现已取得良好结果的首推我国选育成功的光、温敏型核雄性不育水稻。这项成就为杂交水稻由三系法改革为两系法创造了条件，标志着我国杂交水稻的发展跨入了一个新阶段。现将近两年来，我国在两系法杂交水稻育种研究方面所取得的主要进展作一概述。

两用核雄性不育系的选育和研究

两用核雄性不育系（以下简称核不育系）是指既能自交结实繁殖自身又能表现完全雄性不育用作制种工具的水稻品系，不育性由隐性核基因控制，与细胞质无关。这种核不育系是两系法杂交水稻育种的基础，它的优越性除一系能两用、不需要保持系外，还具有以下几个特点。

1. 恢复谱极广。几乎所有同亚种的正常品种都能使其育性恢复正常。根据不完全统计，W6154S 和安农 S-1 分别有 97.6% 和 99.3% 的籼稻品种是恢复系。

2. 遗传行为简单，不育性由 1~2 对隐性基因控制；因而容易转育和稳定，有利于培育多种类型的不育系。

3. 可避免不育细胞质的负效应和细胞质单一化的潜在威胁。

为了选育可供实用的核不育系，特制定了如下标准：

群体 1 000 株以上，性状整齐一致；不育株率 100%，不育度 99.5% 以上；育性转换明显，不育时期连续 30 天以上，可育期的结实率 30% 以上；异交结实率不低于 V20A 或珍汕 97A。核不育系的符号用 S 表示。

核不育系的研究成绩很大，截至 1989 年，已育成并通过省级以上技术鉴定，达到规定标准的核不育系有 17 个（表 1）。其中籼稻 9 个，粳稻 8 个；不育基因来自原始农垦 58 光敏核不育株（HPSGMR）的有 14 个，非 HPSGMR 来源的有 3 个。此外，全国还有不少单位近年来又发现一些新的可两用的雄性不育突变体或育成了核不育系。

表 1　通过省级技术鉴定的水稻核不育系（1989）

不育系名称	亚种	不育基因来源	对光温反应类型	选育单位
5047S			光敏	湖北农科院
3111S			温敏	华中农大
WD-1S			光敏	武汉大学
7001S	粳	农垦 58 不育株（HPSGMR）		安徽农科院
C407S				中国农科院作物所
1541S			温敏	湖北宜昌农科所
AB019S				武汉市东西湖农科所
6334S				华中师范大学

续表

不育系名称	亚种	不育基因来源	对光温反应类型	选育单位
W6154S				湖北农科院
W7415S				湖北农科院
8801S		农垦 58 不育株（HPSGMR）		湖北仙桃市
K7S				广西农科院
K9S	籼			广西农科院
K14S				广西农科院
安农 S-1			温敏	湖南安江农校
衡农 S-1		非 HPSGMR	温敏	湖南衡阳农科所
5460S			温敏	福建农学院

根据育性对光温条件反应的不同，有人将核不育系分为两个基本类型。

1. 光敏型：育性变化主要受光照长度影响，在长日照（13 小时 45 分以上）条件下，表现完全雄性不育，在短日照条件下，表现雄性可育。反应敏感时期为二次枝梗分化到花粉母细胞形成期。温度高低（在水稻正常生长发育范围内）对育性变化基本上不起作用或作用很小。如农垦 58S、5047S 和 7001S 等中晚粳属此类。

2. 温敏型：育性变化主要受温度影响，在较高温度下表现完全雄性不育，在较低温度下表现可育，光照长度对育性变化基本上不起作用或作用很小。诱导不育的临界温度为 25 ℃或 27 ℃（因材料不同而异，下同），诱导可育的临界温度为 23 ℃左右。反应敏感期为花粉母细胞形成期或花粉母细胞减数分裂期。如 W6154S、安农 S-1、衡农 S-1、5460S 等早、中籼核不育系属此类。

若再分出一个光温互作型，它们的育性变化是光温互作效应的结果，也许更有助于解释某些疑惑现象和问题，其中可分两个亚类。①以光为主，温度起协调作用的光温型。当光长在临界值左右时，高温可诱导不育，低温导致可育。②以温为主，光长起协调作用的温光型。当温度在不育临界值和可育临界值之间时，长光照可导致不育，短光照导致可育。

观察和研究还表明，同一核不育基因置入不同遗传背景下，对光温的反应有所不同甚至差异很大。5047S 和 W6154S 的不育基因均为 HPSGM，但前者为光敏，后者为温敏；W6111S 与 W6154S 系来自同一杂交组合的姊妹系，均属早籼稻，但 W6111S 对低温的反应较迟钝。1989 年长江中下游 7 月上旬出现连续 3~5 天日平均温度在 22 ℃左右的异常低温，W6154S 部分恢复正常，而 W6111S 仍基本上保持完全不育。安农 S-1 的不育基因也存在类似情况，应当指出，这种现象对于作物育种无疑是十分有利的。另外，同一个不育系在不同的生态环境条件下，育性对光温的反应也不同。如在沈阳，夏季日照长达 16 小时，1989 年亦遇到异常低温，但各类粳型不育系包括温敏型的 3111S 等均表现完全不育。反之，在贵阳，海拔 1070 米，夏季冷凉，所观察的各类不育系包括光敏的农垦 58S 在内均无明显的能持续 30 天以上的完全雄性不育现象。由此可见，影响核不育水稻育性变化的内外因素颇为复杂，其规律性尚待进一步研究才能完全弄清。

从生产应用角度看，上述几种类型的核不育系，除过分温敏的外，都有实用价值。我国幅员辽阔，不同水稻生态地区的光温条件差别很大，各个地区对核不育类型的要求不一样。华南诸省，温度高、日照相对短，夏季不会出现低温（1989 年的异常低温也未影响到华南），看来，温光型的核不育系比较适合，其次为温敏型。长江流域的夏季，每 10 年左右会出现一次弱低温，每 30 年左右有一次强低温出现，因此采光敏型或光温型不育系比较稳妥。华北地区，夏季日照长，秋季降温快，光敏型固然有利于制种，但繁殖不育系种子有困难。看来，光温型也可能较适用。至于云贵高原以何种类型为宜，尚待研究解决。总之，核不育系的选用，要因地制宜，不能一概而论。

品种间杂交组合的选育

两系法杂交水稻的育种分两个层次，一是品种间优良组合的选育，二是亚种间强优组合的选育。二者在技术上的难度不同，在育种计划中，对其产量指标的要求也有高低之别。

两系法品种间杂交种一般不存在亲和性差而引起的结实率低的问题，同时核不育系不受恢保关系的制约，恢复谱极广，配组较自由，因此选配到优良组合的概率比三系法要高。1989 年不少单位已选出一批两系法品种间杂交稻，且同步进行品比试验、区域试验和生产示范，获得了比较好的结果。湖北省农科院的粳稻组合 5047S/R9-1 在湖北省示范面积达 1 万亩，经多点测产和实收，亩产 350～517 公斤，比常规品种鄂宜 105、鄂晚 5 号增产 20% 以上；广西农科院在广西 20 多个县试种 1 千亩 K14S/03 等 3 个组合作双晚栽培，普遍表现早熟高产，验收产量最高的亩产达 660.4 公斤，创下每亩日产量 6.28 公斤的纪录。同时，在全国以及一些省和单位的区试和品比试验中，均出现比同熟期对照三系杂交稻增产 5% 以上，名列第一的新组合。如培矮 64S/湘早籼 1 号、W6154S/特青、N98S/特青、IR8S/明恢 63 对汕优 63（中、晚籼），W6154S/测 49 对威优 49（早籼），7001S/轮回 422 对当优 C 堡（粳）等。此外，用安农 S-1 还配出几个在长江流域可作早熟早稻的组合，产量比同熟期的常规早稻高 10%～20%，而这是三系法迄今尚未解决的问题。根据初步计划，1990 年全国试种、示范的两系法杂交水稻面积将超过 20 万亩。由此可见，两系法品种间杂交水稻的选育工作进展比较顺利，发展速度很快，可以说这是两系法优越性的具体体现。但要指出，以上所提到的杂交组合尚属初试产品，预计在今后两三年内，新的更好的组合将陆续出现，从而两系法杂交稻会以更快的发展速度投向大面积生产。

亚种间杂交组合的选育

普通栽培稻可分为籼稻、粳稻和爪哇稻 3 个亚种，因而亚种间杂交有籼／粳、籼／爪和粳／爪 3 种形式。研究表明，水稻的杂种优势强度具有籼粳交＞籼爪交＞粳爪交＞籼籼交＞粳粳交的一般趋势。有不少优势很强的亚种间尤其是籼粳交组合，它们的库大源足，理论上的产量可超过现有高产品种间杂交稻 30%～50%。因此，利用强大的籼粳杂种优势一直是两系法杂交水稻育种的战略重点。然而，选育能实用的强优籼粳组合在技术上的难度很大，主要是必须同时解决杂种一代结实率低、植株过高、生育期太长和籽粒充实度不好等四大难题。针对这些问题，育种家们正在从以下四个方面进行选育。

1. 定向选育目标组合。所谓目标组合是指通过人工杂交筛选（包括化学杀雄）出来的强优亚种间杂交稻，它们具有很高的产量潜力，但结实率低或未两系配套。选育方法是将它们的籼型亲本转育成同型的核不育系，粳型亲本导入广亲和基因。如果 F_1 是结实正常的组合，只须将亲本的一方（通常用籼稻）转育成核不育系。目前，大多数目标组合尚处在转育的中世代，但此法的效果已初见端倪。如 9024（籼）／轮回 422（爪）为 F_1 结实正常的目标组合，将核不育基因导入 9024，在 B_2F_2 中选择性状倾向 9024 的不育株与轮回 422 复测，多数杂种的表现已接近原目标组合。

2. 选育优良的粳型广亲和系，然后同各种核不育系测交配组，从中筛选结实正常的强优组合。迄今，一大批熟期不同、农艺性状良好，而又各具特色的矮秆粳型或偏粳型的广亲和系已基本定型，并测配出一些很有希望的苗头组合。如二九青 S 配偏粳型的广亲和系 DT713，F_1 穗大粒多，株叶形态良好，结实正常，小区理论产量比同熟期的威优 6 号高 47%（表 2）。

表 2　二九青 S/DT713 与 CK 考种结果（1989，长沙）

性状	二九青 S/DT713	威优 6 号	对照／%
株高／cm	110.00	103.00	106.30
抽穗日期／天	88.00	88.00	100.00
单株穗数／穗	15.06	17.24	87.35
每亩穗数／万穗	18.83	21.55	87.35

续表

性状	二九青 S/DT713	威优 6 号	对照 / %
每穗颖花数	205.89	112.54	182.94
每亩颖花数 / 万个	3 877.05	2 425.24	159.86
结实率	75.40	83.15	90.68
千粒重 /g	27.80	27.38	101.53
单株粒重 /g	65.00	44.17	147.18
理论产量（kg/ 亩）	812.68	552.14	147.18

注：株行距 6 寸 × 8 寸。

3. 直接利用现有的优良广亲和品种与核不育系测交筛选。此法见效较快，因为亲本双方都是现成的，性状稳定的品种、品系，只要一发现好的苗头，就能很快进入高级试验和生产示范。遗憾的是，大多数的广亲和材料属于古老的高秆地方品种，无直接利用价值，仅有 02428、CPSLO17、Vary Lava 等为数很少的矮秆改良型广亲和品种具有实用意义。尽管配组数量不多，却初步测出几个很有希望的组合，其中以测 64S/Vary Lava 等表现较突出。

4. 将广亲和基因与核不育基因结合于一体，选育粳质、籼核、广亲和的光敏核不育系。粳质对杂种优势有较好的效应；籼稻作母本有利于亚种间制种，解决父母本花时不遇的问题；广亲和基因则能扩大恢复谱，理论上三个亚种的正常品种均可作其恢复系。因此，这是一种较为理想的遗传工具，可暂称为通用型不育系。如基本上育成的培矮 64S 就具有这种"四合一"的性质，它的细胞质和光敏核不育基因来自粳稻农垦 58S，广亲和基因来自爪哇稻培迪，而细胞核则是籼稻测 64 的。实测结果，表明它具有通用性（表 3）。

表 3　培矮 64S 与三个亚种中不同品种测交 F_1 结实率（1989，长沙）

亚种类型	品种名称	测交 F_1 的结实率 / %
籼稻	南京 11 号	73.2
	IR36	81.1
	湘早籼 1 号	76.4
	密阳 46	75.0

续表

亚种类型	品种名称	测交 F_1 的结实率 / %
粳稻	秋光	78.0
	巴利拉	69.8
	城特 232	68.7
	农虎 26	66.0
爪哇稻	培迪	70.5
	轮回	74.8
	CPSLO 17	71.0
	CP 231	70.0

上述结果表明，在两系法亚种间杂交水稻的选育研究方面，已取得可喜的进展。初步认为，籼粳杂种结实率低的问题，应用广亲和基因可基本解决，使之接近正常水平；用等位矮秆基因能把株高降到半矮秆水平；通过不同熟期（主要是早、中熟）的双亲配组，从中可以选到熟期适中的甚至比较早熟的组合。现在还剩下一个籽粒充实度差的问题，尚待解决，这是亚种间杂种优势利用育种所面临的最后一道难关。

籼粳杂种一代籽粒充实度差较为普遍，特别是优势强的大穗型组合，受精率很高，饱满粒少、秕粒多，以致难以高产。例如，籼粳组合 W6154S/ 培 C311，1989 年在广东韶关的晚稻品比中，表现优势强，穗大粒多，乳熟期测产为 850 公斤 / 亩，但成熟时的籽粒仅 50%，以致亩产仅 621 公斤。籽粒充实度差的原因可能与库源不协调，输导组织不畅以及早衰（包括根系、叶片和枝梗）有关，需从形态解剖、生理生化、栽培技术等方面进行研究，找出其症结，以便对症下药加以解决。在育种上的对策，则是要分两步走，首先利用偏籼、偏粳的中间类型进行配组，选育中穗中粒型、籽粒充实度好，比现有杂交水稻增产 10%~20% 的强优组合，待原因弄清后，再进一步扩大亲缘和遗传差异，以选育优势更强的大穗型和特大穗型组合。

作者：袁隆平

注：本文发表于《中国农业科学》1990 年第 23 卷第 3 期。

Prospects and Suggestions for Development of Hybrid Rice Technology in India

Introduction

At FAO RAPA'S request, I was invited as a consultant to visit India from Nov. 2nd to Nov.20, 1990 to take a careful view of hybrid rice research program in the country. This report gives my basic opinions on such affairs.

After several years of probationary research, India has already formulated a network for hybrid rice research programs with 10 centers all over the country. I visited eight network centers during my 18 days' stay in India. At each center, we went to the experimental plots to observe the tested materials, holding talks with Indian scientists to exchange ideas and experiences and to discuss problems encountered in the studies.

My general feeling is that India has established a good newtork and has the capability to achieve success in development of hybrid rice technology at commercial level in the very near future.

Ⅰ) India has already formulated a good plan project proposal specifying its objectives very clearly. The year-wise activity framework spells out clear research strategies to be followed by each centre.

Ⅱ) At almost all the centres, I observed that the scientists concerned have a good working knowledge to tackle the program scientifically.

Ⅲ) India has already identified a very good National Programme Leader, Dr. I. Kumar. I had thorough discussions with him during my trip and my personal impression is that he is not only very responsible to his duty but also has deep knowledge of the subject and is capable of leading this program effectively.

Ⅳ) A satisfactory progress has been made in developing several A & B and many R lines. Some promising hybrdis possessing high yield potential have also been identified in the initial trials, which could outyield the best CK by 15%–25%.

Problems

Ⅰ) Because of lack of experience, some CMS lines developed by India and introduced from IRRI belonged to functional male sterility group. Such lines do not show complete male sterility under different environmental conditions and are thus not suitable for developing rice hybrids.

Ⅱ) CMS line multiplication and hybrid seed production plots have shown low yield level.

Suggestions

1) A line breeding

More emphasis is needed to be put on development of stable CMS lines adaptable to local conditions. Following suggestions may be kept in mind while developing new CMS lines：

Ⅰ) Primarily WA cytoplasm should be used to convert locally adopted elite lines into stable CMS lines by observing pollen sterilitty instead of anther dehiscence.

Ⅱ) As per our experience, only the lines showing complete male sterility in BC_1 should be carried forward. If the line shows segregation even after BC_3, its use should be discontinued.

Ⅲ) In the process of developing A line, besides disease and insect pest resistance emphasis should be given on developing CMS lines having higher out crossing potential. The most important characteristic favorable for out crossing is good flowering behaviour, next to it is high rate of exserted stigma and big size of stigma.

Ⅳ) The introduced and newly developed CMS lines which do not show complete pollen sterility should be discarded, lest it would be wastage of time and resources.

Ⅴ) If some B lines show multiple resistance and good combining ability but do not show complete maintainer reaction, they can be improved by crossing with some other B lines which have very good maintaining ability.

2) Seed production

Ⅰ) The basic requirement in obtaining high seed yield in seed production plots is good field management.

Ⅱ) As per our experience in China, high dosage of GA3 application（100-150 ppm[①]）is the key factor to obtain high yield in seed production plots.

Ⅲ) For better synchronization of A/B and A/R lines it is better to use the leaf number index rather than differences in days of heading of A & B or R line.

Ⅳ) I personally feel that it would be greatly advantageous if scientists from India visit China next season（late June-early July）to practically learn the seed production techniques. In addition it would be better to invite 1-2 Chinese technicians who have rich experience in seed production to India for one season to work hand in hand with Indian scientists.

① 现在出版物中不能使用 "ppm" 这一缩写，此处宜为 mg/L。

3) Others

Ⅰ) In the network there must be a very strong leading research centre. For this purpose the Directorate of Rice Research, Hyderabad is the most suitable centre. Strengthening this research centre with more manpower and resources in the form of land, facilities and budget would be necessary. I fell that the Program Leader must have 4-5 scientists to assist in CMS line development, test crosses and R-line development, seed production and developing two line breeding program.

Ⅱ) To set up a plan to develop hybrid combinations suitable to rain fed land and drought area is necessary. According to our experiences, the hybrids can outyield pure line varieties significantly in less water supply conditions due to their more vigorous root system hence more tolerant to drought.

Ⅲ) It is better to initiate efforts to develop two line breeding system from the next year for which about 20 per cent of effort should be put in. TGMS line developed by Japanese and available at IRRI may be introduced and transferred in indica varieties. Alternatively screening of local breeding material may be done to identify PGMS or TGMS genes as the frequency of such genes it not very low in many rice cultivars. If you pay much attention to look for it, surely you can find it.

In conclusion, I am quite optimistic that India can achieve success in making hybrid rice a commercial feasibility within next 3 years if the programme is effectively implemented.

作者：Yuan Longping

注：本文于 1990 年向 FAO RAPA（联合国粮农组织总部和粮农组织驻亚太区办事处）提交。

Hybrid Rice in China-Techniques and Production

1 Introduction

The phenomenon of heterosis in cereal crops is expressed mainly in the first hybrid generation (F_1). In order to utilize rice hybrids in commercial production, it is first necessary to produce a large amount of F_1 hybrid seeds. However, rice is a strictly self-pollinated crop with tiny florets, so that it is impossible to produce bulk quantities of F_1 hybrid seeds by hand emasculation. This is the primary reason why heterosis in rice has not been utilized in commercial production. To solve this problem, the most effective way, as proved by other self-pollinated crops, is to exploit the phenomenon of cytoplasmic male sterility.

Since Jones (1926) first observed heterosis in rice, suggestions for exploiting heterosis commericially by developing F_1 rice hybrids have been made from time to time (Stansel and Craigmiles, 1966; Shinjyo and Omura, 1966; Yuan, 1966; Athwat and Virmani, 1972; Swaminathan et al., 1972). However, difficulties in hybrid seed production discouraged most of the researchers from continuing their efforts, the notable exception being Chinese scientists (Yuan, 1972).

In China hybrid rice research work was started by L.P.Yuan in 1964, and later a male sterile plant in wild rice was found by his assistant, B.F.Li, in 1970. This male sterile plant was named wild abortive or WA type. This discovery was break-through in hybird rice breeding. Through wide test crosses and successive back-crossing, the first set of WA type cytoplasmic male sterile (CMS) lines and their maintainers was soon developed in 1972. In the next year, the first restorer lines were identified by screening existing varieties introduced from southeast Asian countries. In 1974, some rice hybrids with strong heterosis were developed and a complete procedure of hybrid seed production technology was established in 1975. In 1976, hybrid rice in China was released for commercial use on a large scale (Lin and Yuan, 1980). From 1976 to 1988, the additional growing area of hybrid rice in China was about 84 million ha, and an increase of more than 125 million tons of grain was achieved. In 1989, the hybrid rice growing area in China (total growing area=15.27 million ha) was enlarged by 14 million hectares (Fig.1).

118

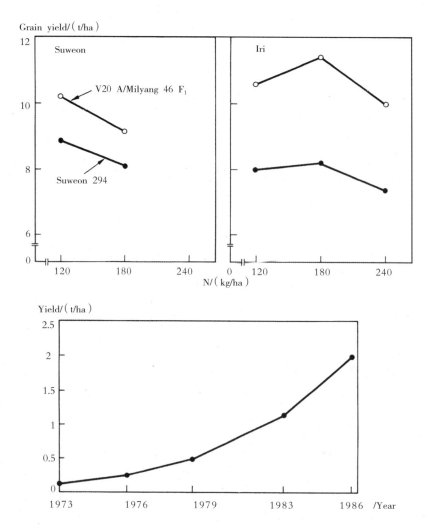

Fig.1 The growing areas of hybrid rice in China (total growing area=15.27 million ha) from 1976 to 1989.

Outside China, hybrid rice research work has been initiated in India, Indonesia, the United States, the Philippines, South Korea, Malaysia, Thailand, Vietnam, Brazil, Mexico, and the International Rice Research Institute (IRRI). Experimental data show the possibility of a yield increase in rice hybrids by 15%–20% over the best semi-dwarfinbred rice varieties. Promising CMS lines and R lines have been developed in some countries, and hybrid seed production has been successful on a small scale in the United States, Philippines, Japan, Indonesia, and South Korea in recent years. More and more countries have recognized the importance of hybrid rice in food production strategy, and are initiating national hybrid rice programs.

2　Heterosis in Rice

The term heterosis in rice refers to the phenomenon in which the F_1 populationobtained by the crossing of two genetically dissimilar parents show superiority to both parents in growth vigor, vitality, reproductive capacity, stress resistance, adaptability, grain yields, and other characters. Jones (1926) observed that some F_1 rice hybrids had more culms and a higher yield than their parents. The extent of heterosis can be estimated in terms of certain paremeters. The following three formulas are usual for the estimation of heterosis in rice, as well as in othe crops.

Mid-parent heterosis or heterosis over the mean parental (MP) value

$$= \frac{F_1 - MP}{MP} \times 100\%$$

High-parent heterosis or heterosis over the better parental (BP) value

$$= \frac{F_1 - BP}{BP} \times 100\%$$

Standard heterosis or heterosis over the check variety

$$= \frac{F_1 - CK}{CK} \times 100\%$$

To utilize heterosis in production, it is essential that the F_1 hybrids show superiority not only to their parents, but also the check variety, i.e., the best commercial variety or hybrids in current use. Thus, heterosis over the check variety is more useful for practical purposes.

2.1　Heterosis for Yield

Experiments on heterosis in rice conducted at Davis. California of U.S., indicated significant yield superiority of 11 of 153 rice hybrids over the best check variety. The standard heterosis ranged from 16% to 63% and averaged 41% (Virmani and Edwards, 1983). A number of experimental rice hybrids were evaluated in comparison to the best available inbred rice varieties at IRRI from 1981 to 1986. The average yield of these hybrids is 16% higher than the best inbred check varieties (Table 1). Dozens of experimental rice hybrids were also evaluated in replicated yield trials in the Philippines, Indonesia, India, South Korea, Malaysia, and Vietnam from 1980 to 1986. The increased average yield of these rice hybrids compared to best check invred varieties ranged from 8% to 17% (Table 2). The yield of most comprehensive commericial hybrid rice under large-scale production in China has exceeded the best inbred varieties by about 20%, and on a country-wide average, the grain yield per hectare is 6.4⁻6.6 t.

2.2　Heterosis for Other Characters

The root quantity and quality of hybrid rice is markedly superior to inbred rice varieties. Nearly all the Chinese hybrid rice showed heterosis for root penetration rate, root depth, width of the rhizosphere, number of adventious roots per plant, and number of root fibrils (Lin and Yuan, 1980). Studies conducted at IRRI have also indicated that some rice hybrids are superior to their parents at comparable growth stages with regard to total root dry weight and root number, length, diameter, and pulling force (O´Toole and Soemartono, 1981). In China, the seeds used per unit area in hybrid rice is only 1/10 to 1/7 of conventional rice varieties, because hybrid rice has higher tillering ability than inbred rice. The superiority of F_1 rice hybrids in physiological traits such as nutrient absorption and transportation ability, respiration intensity, photosynthetic area, chlorophy II content per unit area of leaf, photosynthetic efficiency, and mitochondrial activity has also been reported (Lin and Yuan, 1980; Yuan, 1985).

Table 1　Comparative yield performance of the best experimental F_1 rice hybrids and the best check varieties in IRRI trials, 1981—1986.

Season	No. of Trials	No. of hybrids evaluated	Yield/ (t/ha)		(%) of check	
			Range	Mean	Range	Mean
Dry	14	207	5.4–9.6	7.8	86–141	116
Wet	16	202	2.6–5.6	4.2	100–140	116

Table 2　Comparative yield performance of the best experimental F_1 rice hybrids and the best check varieties in trials conducted in various countries, 1980—1986.

Country / (t/ha)	No. of Trials	Yield of best hybrids/ (t/ha)		(%) of check	
		Range	Mean	Range	Mean
Indonesia	15	4.1–8.9	6.2	102–143	117
South Korea	11	8.1–11.5	9.1	97–142	113
India	21	3.3–9.8	6.2	91–143	116
Malaysia	2	4.2–5.0	4.7	89–127	108
Philippines	8	4.8–7.4	5.4	92–133	114
Vietnam	4	5.3–6.6	6.0	91–122	108

3　Concept of Three lines

The so-called three lines consist of cyto-genetic male sterile line (CMS line or A line), maintainer line (B line), and restorer line (R line), respectively.

3.1　CMS Line

The CMS line refers to a special kind of breeding line whose anthers are abnormal. No pollen or only abortive pollen exists within them, so that no seed is set on this line by selfing, but its pistils are normal and can produce seeds when pollinated by any normal rice variety. As a desirable CMS line, besides good agronomic characters and normal pistils, it should have the following characters:

1. *Stable Male Sterility*. The male sterility should be inherited from generation to generation without any change in pollen sterility, and it should not be influenced by the environment, especially temperature fluctuations.

2. *Being Easily Restored*. This refers to two aspects, first, a proposed male sterile system should have a wider restoration spectrum, so that the possibility of selecting superior hybrid combinations will be higher. Secondly, the seed set of the retored hybrids should be stable and less influenced by adverse environments.

3. *Good Floral Structure and Flowering Habits*. The CMS line should flower normally and the flowering time should synchronize with that of the male parent. Its stigma should be well developed and exserted. The glume opening time should be longer and with a larger angle. The panicle should be fully extended from the sheath of the flag leaf.

3.2　Maintainer Line

The maintainer line is a pollinator variety used to pollinate the relative CMS line and produce progenies which still retain male sterility. If there is no maintainer line, the CMS line cannot be maintained and multiplied for generation after generation. Except for male sterility, the major characters of a CMS line are determined by its corresponding maintainer line, so that great care should be taken initially in the selection of a B line possessing these characters. Besides these, a B line shoud be pure, with uniform population, and have abundant pollen grains which help in multiplying the CMS line.

3.3　Restorer Line

The restorer line is a pollinator variety used to pollinate CMS line to produce F_1 hybrids which are male fertile and thus produce seeds on selfing. As an elite R line, it should have ① strong restoring ability, that is the seed set of its hybrids should be equal to that of a normal variety, or above 80%; ② better agronomic characteristics, good combining ability, and significant heterosis in its hybrids; ③ plant height taller than the CMS line, growth duration close to or a little longer than that of the CMS line; and ④ well-developed anthers with a large number of pollen grains, good flowering habit, and normal dehiscence.

3.4　Interrelations Among the Three Lines

When the three lines have been bred, it is possible to start the production of F_1 hybrid seeds on a large scale; heterosis in rice may thus e commercially utilized.

The CMS line is planted in alternate rows with the B-line in an isolation plot to multiply CMS line seeds as well as maintainer seeds each year. In another isolation plot, the CMS line and the R line are planted in alternate rows to produce hybrid seeds. The genetic relation between three lines and the interrelation among three lines are shown diagrammatically in Fig. 2 and Fig. 3, respectively.

4 Principles and Procedures of Hybrid Rice Breeding

The procedure for hybrid rice breeding can be divided into ① the thre-line breeding phase and ② the hybrid heterosis evaluation phase.

Hybrid Rice in China

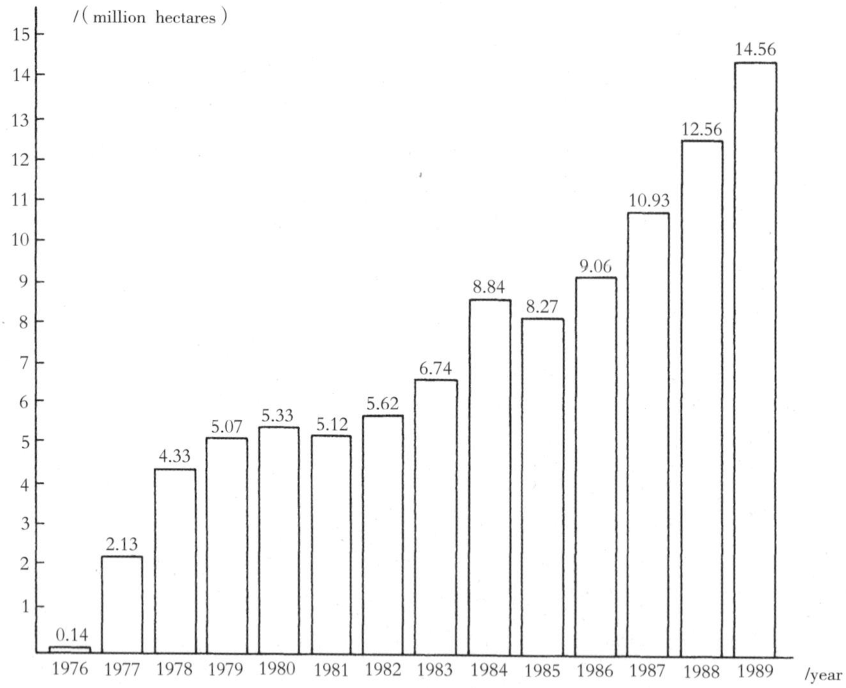

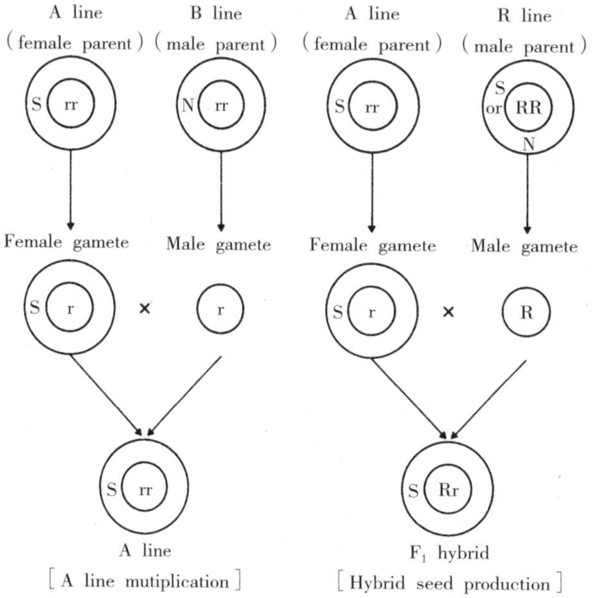

Fig.2 Genetic relation between three lines. S cytoplasmic sterile gene; N cytoplasmic fertile gene; R nuclear dominant gene; r nuclear reccessive gene.

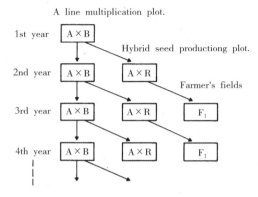

Fig.3　Interrelation between the three lines in production

4.1　Three-Line Breeding

According to breeding objectives, various three-line materials and germplasm resources with different agronomic and biological characters must be collected and planted for crossing, test crossing, and three-line breeding. The CMS lines and their maintainer lines should be grown in isolated plots. The other germplasm material is planted each with 10–20 plants and one seedling per hill. (i.e., the spot in which the rice seedlings are transplanted) in the plot or pots. In order to attain synchronization of flowering, certain materials are needed to be seeded at different times or treated with short day length.

The objective of test crossing is to identify the fertility of F_1 hybrids and screen the R lines and B lines. Usually 10–20 plants are grown in one row for each crossing. A normal check variety is planted after every 10–20 hybrid combinations. If the F_1 of a certain combination proves to be male sterile and its male parent acceptable to the breeding objective with respect to its agronomic traits, that particular F_1 will be developed as a CMS line by successive backcrosses. When the F_1 of a certain combination is normally restored in fertility and has better characters, its male parent could be developed as an R line through re-test crossing with the original CMS line. The F_1s whose male parents have poor restoring or maintaining abilities are generally discarded.

To identify the restoring ability of the male parent again, about 100 plants are grown for each combination with R line or standard commercial variety as check in re-test cross nursery. The heterosis in F_1 is preliminarily observed in this nursery. If the given F_1 is normal in seed set, its male parent is confirmed to be an R line. If the F_1 simultaneously exhibits heterosis, this hybrid can be put into the next stage of the test.

To develop an excellent CMS line and its corresponding B line, the sterile F_1's and the recurrent plants should be grown in pairs in the backcross nursery. Successive backcrosses should be carried out for four to six generations. When their backcross progeny are stable in sterility and apparently conform to the characteristics of the male parent, and the population consists of over 1000 plants, then theseprogenies are designated a CMS lines. The corresponding male parent used as recurrent parent is designated a B line.

124

4.2　Heterosis Evaluation

Hybrid combinations made from crosses between various CMS and R lines should be planted in combining ability evaluation nursery to select the best CMS line and R line with good combining ability as well as better hybrid combination. Each combination is planted in three replication plots and about 100-200 single seedling plants per plot. A standard commercial variety or a leading hybrids is used as control. One or two consecutive replicated trials and regional trials are conducted until release for commercial use. During the regional trial period, farmer's field evaluation can be carried out simultaneously, and regular patterns in cultivation and seed production techniques can be studied.

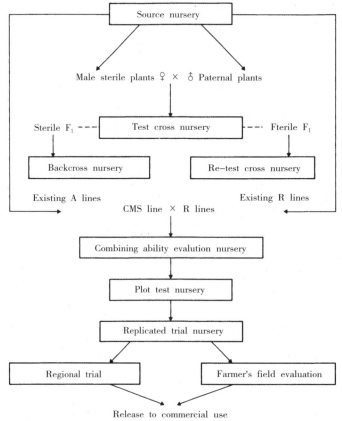

Fig. 4　Hybrid rice breeding procedures

The general course for hybrid rice breeding is shown in Fig.4. It should be practiced in a flexible way. If a hybrid combination is found to be excellent, it may skip any of the regular steps, so that the breeding cycle can be shortened, and the excellent rice hybrids can be put into commercial production relatively earlier.

5　Breeding for CMS Lines and Their Maintainers

Cytoplasmic male sterility (CMS) was visualized as a genetic tool to develop F_1 hybrids in self-pollinating crops. Shinjyo and Omura (1966) created the first CMS.

Line in cultivated rice by substituting nuclear genes of a japonica variety, Taichung65, into the cytoplasm of an indica variety, Chinsurah Boro II . Chinese scientists developed a series of CMS lines and for the first time used CMS lines to develop commercial F1 rice hybrids in 1973 (Lin and Yuan, 1980) . There are three basic approaches to breeding CMS lines and their maintainers, viz., creation of CMS lines with new sterile cytoplasms; transfer of sterile cytoplasms from existing CMS lines into new ones; and direct utilization of the existing CMS lines.

5.1　Breeding for CMS Lines with New Sterile Cytoplasms

It has been proved in practice that nucleus substitution, i.e., using distant crosses followed by successive backcrosses, is a potential method for breeding CMS lines with a new source of cytoplasm. From such nucleus substitution crosses, a nuclear-cytoplasmic hybrid has been bred, whose nucleus and cytoplasm differ in origin (Fig. 5) . The male and female parents are so distantly related that the cytoplasm from the female and the nucleus from the male do not harmonize, which sometimes results in male sterility.

Three kinds of crosses are usually used for breeding CMS lines with new sterile cytoplasm.

1. *Interspecific Cross*. This includes wild rice species crossed with cultivars, and varieties of O. glaberrima crossed with varieties of *O.s ativa* L. The probability of the occurrence of male sterile plants is rather high in the progeny of these crosses, particularly in wild rice crossed with japonica varieties, in which progeny with 100% sterility can be obtained.

2. *Intersubspecific Cross*. Mainly indica crossed with japonica belongs to this category. As indica and japonica are relatively distantly related, male sterility can be expected from certain cross combinations, but the male and female parents should be properly selected.

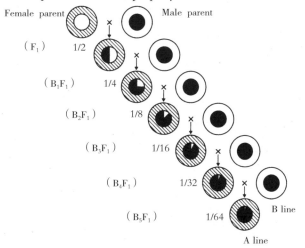

Fig. 5　Nucleus substitution

3. *Intervarietal Cross*. This is effectively used only between distantly related indica rice varieties. If the parents are properly selected, i.e., using the primitive varieties as female parents and adanced varieties as male parents, it is possible to obtain male sterile plants in the progeny.

5.2 Transferring

This is an effective, convenient, and widely used method to develop new CMS lines. In China, nearly all the WA type lines were obtained by this method. The newly developed CMS lines have iso-cytoplasm but allo-nucleus of an existing CMS line.

Select desired varieties or breeding lines to make single-plant crosses in pairs with a selected existing CMS line. Normally, 10−20 F_1 plants are grown side by side with the male parent for each combination. Carefully observe and evaluate the frequency of male sterile plants and their degree of male sterility in the F_1's. Then choose the better ones, which should all be completely male sterile and possess good flowering habits, to make successive backcrosses. Segregation begins in B_1F_1, 30−50 B_1F_1 plants are required to be grown with the recurrent male parent in adjacent rows. Selections should be made in B_1F_1, B_2F_1, and late generations in order of combinations, families, and individual plants. For a new CMS line to be successfully developed and named, the percentage of male sterile plants in a population of about 1 000 plants should be 100% and the degree of male sterility should also reach 100%. The recurrent parent is logically designated its maintainer (B line) .

5.3 Introduction of Existing CMS Lines

Recently, dozens of rice CMS lines of various types have been developed in China and other countries. When introducing existing CMS lines, select the elite ones suitable for local conditions for direct use in the most economic and effective way.

The introduced CMS lines and their B lines are seeded in a normal season, and transplanted side by side with a single seedling per hill. Twenty to 30 plants are needed per line. The criteria for adaptability observation include growth vigor, uniformity, agronomic characters, resistance to diseases and insects, and flowering habits in particular. After heading, the male sterility is identified by using visual inspection, bagging panicles, and microscopic observation methods. During the flowering stage, slightly shake the panicles to see whether the anthers are dehiscent and shed pollen grains or not. Special attention should be paid to examine if any pore dehiscence occurs at the basal part of the anthers. When the male sterile plants just start heading but their florets are still not blooming, cover the panicles with parchment bags to effect selfing, then after about 2 weeks check the bagged panicles. Another method is to sample some anthers from different parts of selected panicles, crush them on a slide with a drop of I-KI solution and examine under the microscope. Normal fertile pollen grains are spherical in shape and stain dark blue with I-KI solution. If the number of fertile pollen is over 1%, male sterility is considered not to be reliable. The CMS lines thus introduced can grow well and show good characters and stable sterility under local conditions; they can thus be directly used in breeding programs.

6　Breeding for Restorer Lines

The practical use of cytoplasmic-genetic male sterility in breeding hybrids in cereal crops became possible only when the effective restorer lines were identified or develped. In rice, a great number of effective restorer lines for BT, WA, and Gam type cytosterility systems have been identified among rice cultivars and elite breeding lines (Shinjyo, 1969, 1975; Lin and Yuan, 1980).

The restoring genes exist in the nucleus of the original female parent, which provides the sterile cytoplasm. A normal rice plant which has sterile cytoplasm certainly contains dominant restoring genes its in its nucleus, otherwise it would be abnormalin seed setting, according to the genetic theory (Yuan ,1985). The restoring ability of rice varieties have been found to be to some extent related to their origin.The frequency of the R line is higher and restoration ability is stronger among varieties closely related to wild rice. Among *indica* rice varieties, the frequency of R line, late-maturing varieties is greater than early-maturing ones. Almost no restorer line has been found in typical *japonica* rice varieties. The commercially used *japonica* R lines in China received the restorer gene (s) from the indica variety IR8 (Mao, 1986). Nearly all breeding methods using conventional varieties can also be used for restorer line breeding, such as screening by test cross; hybridization, rediation, etc.

6.1　Screening by Test Cross

Select typical individual plants from the desired varieties or lines in the source nursery, then test cross them with a representative CMS line in pairs. More than 30 seeds should be collected from each cross. More than ten F_1 plants should be grown for each cross. At the heading stage examine their fertility based on percentage of dehiscent anthers and normal fertile pollen. If the dehiscent anthers are over 99%, the normal fertile pollen is over 80%, and the seed set is normal, test cross them once again. The F_1 population of the re-test cross should be over 100 plants. If anther dehiscence and spikelet fertility are still normal, this indicates that the male parent possesses restoring ability and may be used as an R line. According to the degree of fertility restoration effected by the Rline in the CMS line, the R lines can be classified into strong and weak restorers. The effects of environmental factors on male fertility restoration in hybrid rice would appear to be genotype-specific. The seed set of F_1 hybrids from strong R lines is quit high and is less influenced by environmental fluctuations. The lines confirmed through test cross as possessing strong restoring ability can be used as the R lines to make cross combinations with various iso-cytoplasmic CMS lines. Through regular procedures various F_1 hybrids are then evaluated for yield potential, disease and insect resistance, grain quality, adapt-ability, and other characters, in order to determine which of these lines should be used to obtain a promising hybrid combination. This method is a simple, convenient, and effective way to obtain commercial R lines, for example, the excellent R lines IR 24, IR 26, IR 30, and IR 9761-19-1 etc., which have been widely used in large areas for commercial production in China.

6.2　Cross Breeding

Due to the narrow restoration spectrum of the existing CMS lines, the requirement of obtaining better R lines cannot be met by depending solely on screening existing varieties or lines. New restorers have to be developed through cross breeding, which can enlarge the source of R lines, and new R lines can be developed according to the breeding objective; however, one of the parents should be a restorer source variety. A numer of new R lines have been developed by the cross breeding method in China (Table 3) .

Table 3　R lines obtained by cross-breeding in China.

R line		Major parental varieties
26 Zhai-Zao	(indica)	IR 26, IR 661, Zhai-Ye-Qing 8, Italy B
Ming-Hui 63	(indica)	IR 30, Gui 630
6161-8	(indica)	IR 661, IR 2061
3024-1	(indica)	IR 30, IR 24
3624-33	(indica)	IR 36, IR 24
C 57	(japonica)	IR 8, Ke-Qing 3, Nong-Lin 131
Pei C 115	(japonica)	IR 8, Pei-Di
C Bao	(japonica)	IR 8, Cheng-Bao 1

6.2.1　Single-Cross Method

This includes R×R, A×R, and B×R or R×B. For R×R, two restorer varieties are crossed. The desirble individual F_2 plants are selected for test cross with a CMS line. Based on fertility and other desirable characters of the hybrids, the corresponding male parents with strong restoring ability are selected. Such a process of selection along with test cross continues for several generations until the male parent is stable in both restoring ability and other agronomic characters. In some cases, to improve characters of R lines, they can be crossed with nonresttorer varieties of B lines, but in the progeny, the frequency of nonrestorers is relatively higher than that of restorers. Sometimes, backcrosses are made in order to enrich the traits of the male parent in A×R. Because the restorer line of this kind is identical with the CMS line in cytoplasm, it is called the iso-cytoplasmic R line.

6.2.2　Multiple Crosses Method

Through multiple crosses, the favorable genes (including R genes) in different varieties can be combined in a new R line. For example, the successful early maturing R line 26 Zhai-Zao was developed by such multiple crosses.

6.2.3　Successive Backcross Method (Directional Transfer)

It is necessary to make successive backcrossing in order to transfer a variety into an R line when this variety is evaluated by testcross as having good combining ability and inducing strong heterosis, but being without restoring ability (or with weak restoring ability) .

6.3　Radiation Breeding

This is also an effective way to improve existing R lines. Chinese scientists have developed new R lines by radiation treatment of some existing R lines, which had some shortcomings. For instance, IR 36 is a weak R line, but after radiation treatment, a strong new R line, Radiate 36, was developed. The other example is MH 63, which is an excellent R line widely used in China, but which has the disadvantage of long growth duration. Through radiation breeding, some strong new R lines have been obtained. Most of the characters of these new R lines are similar to MH 63, but they are early maturing, and have been used to develop early-maturing hybrid combinations for double-cropping rice systems in the Yangtze River area.

7　Selection of Parents for Superior Hybrid Combination

Selection of parents for making hybrid combinations plays a very important role in the exploitation of heterosis in rice. Five major factors should be considered.

7.1　Genetic Diversity

Genetic diversity is the basis for including heterosis. Within a certain range, the greater the genetic diversity between both parents, the stronger the heterosis. The genetic diversity mainly refers to both parents being from a distant geographic area or of different ecotype. All the *indica* hybrid rice combinations now used in commercial production in China are made by using parents of different ecotype or different geographic origin, while some distant nuclear components are incorporated into the parents for *japonica* hybrid combinations.

7.2　Complementary Characters

Complementation of the good characters of both parents makes the combined characters of their hybrids superior to the parents and they show obvious heterobeltiosis. For instances, V20A and Zhen-Shan 97A, the female parents of Wei-You 6 and Shan-You 6, are blast-resistant, with short growth duration, weak in tillering ability, and heavier in 1 000-grain weight; whereas their male parent IR 26 is resistant to bacterial leaf blight, longer in growth duration, strong in tillering, and lighter in 1 000-grain weight. The F_1's (Wei-You 6 and Shan-You 6) are, however, resistant to blast and bacterial leafblight, have abover average growth duration, good tillering ability, and above average 1 000-grain weight.

7.3　Combining Ability

As a new CMS or R line is developed, evaluation of their combining ability is needed so as to avoid blindness in selection of parents for crossing. The general combining ability (GCA) and specific combining ability (SCA) of parents has greater effects on yield potential and certain economic characters of hybrid rcie. It is necessary to select parents with high effective value in tillering ability, panicle number per plant, filled grains per panicle, 1 000-grain weight, and grain weight per plant.

7.4　Heritability

The parents for rice hybrids should be selected mainly depending on the parameter of their generalized heritability. A character with high heritability indicates that it is inherited in a major proportion by the parents and less influenced by the environment. It has been practically proved that productive panicles, grain weight per plant, and tillering ability are of weak heritability, while growth duration, brown rice percentage, and 1 000-grain weight are of stronger heritability, and filled grains per panicle and percentage of empty grains are of intermediate heritability. It is very important to understand the heritability and dominance-recessiveness relationship of the major characters of CMS lines and R lines prior to their selection for producing hybrid combinations with significant heterosis.

7.5　High Yielding Ability

A hybrid rice variety can produce higher absolute yield only when both parents or one of them is a high yielding variety. The superior hybrid combinations now used in China, such as the Nan-You system (Nan-You 2,3,6), the Wei-You system (Wei-You 6,30,35,49,64,etc.), and the Shan-You system (Shan-You 2,6,63,30 selection, Gui 8, Gui 33, Gui 34) are all made by using CMS lines and R lines which not only have greater genetic diversity but also have developed from elite varieties (Mao, 1988).

8　Hybrid Seed Production

The procedure of hybrid rice seed production differs from that of pure line varieties. It involves three steps: (1) multiplication of Alines; (2) multiplication of B and R lines; and (3) production of hybrid seeds. Multiplication of B and R lines is performed in the same manner as with conventional varieties. However, multiplication of A lines and production of hybrid F_1 seeds require specific methods. In China, the hybrid rice seed production techniques were fundamentally established in 1975. Before the 1980's, the average yield of hybrid seed production was only about 0.75 t/ha, but in recent years, the average yiedld achieved is 2 t/ha throughout the whole country. The area ratio between A line multiplication, hybrid seed production, and commercial production is determined by the yield obtained in both A line multiplication and F_1 seed production field, and the seed use rate of commercial production field per unit area. The rapid increase in seed yield per unit area raised the field area ratio between A line multiplication, hybrid seed production, and commercial production from about 1 : 30 : 1000 in the 1970's to about 1 : 50 : 5000 between 1985 and 1986. This means that the economic benefit was increased very fast by improvement of hybrid rice seed production (Mao, 1987).

8.1　Techniques of F_1 Seed Production

8.1.1 Choice of Field

The choice of field for hybrid seed production is very important. In the isolated area, the paddy field with fertile soil, the desired irrigation and drainage system, sufficient sunshine, and no serious disease and insect problems, especially those covered by quarantine regulations, is essentially needed. Rice pollen grains are very small and light, and can travel very far with the wind. In order to ensure the purity of hybrid seed and avoid pollination by unwanted varieties, the hybrid seed production field should be strictly isolated.

1. *Space Isolation*. An isolation distance of over 100 m is found to be satisfatory. Within this range, no other rice varieties should be grown except for the pollen parent.

2. *Time Isolation*. Generally, a time of over 20 days is practiced. In other words, the heading stage of varieties grown within 100 m around the seed production field should be over days earlier or later than that of the CMS line.

3. *Barrier Isolation*. In some places topography surface features and artificial obstacles may also be used as the means of isolation.

8.1.2　Synchronization of Flowering

As the seed set on CMS line depends on cross-pollination, it is most important to synchronize the heading date of the male and female parents, especially for the hybrid combinations having parents with quite different growth duration. In addition, in order to extend the pollen supply time, the male parent is usually seeded twice or three times at an interval of 5—7 days and planted alternately.

8.1.3　Row Ratio, Row Direction, and Planting Patterns

Row ratio regers to the ratio of number of rows of the male parent to that of the female parent in the hybrid seed production field. Practically, a row ratio of 1 : (8—10) or 2 : (10—12) is currently used in *indica* hybrid rice seed production, and 1 : 6 or 2 : 8 in *japonica* rice in China. If the R line has more pollen, the row ratio may even be appropriately increased. Generally, the R line is transplanted with a single seedling per hill and the CMS line is transplanted with two seedlings per hill. It is better to make the row direction nearly perpendicular to the direction of the winds prevailing at the heading stage to facilitate cross-pollination.

8.1.4　Prediction and Adjustment of Heading Date

Even if the seeding interval between both parents is accurately determined, the synchronization of flowering might still not be attained because of variation in temperature and difference in field management. Hence, it is necessary to predict their heading date in order to take meansures as early as possible to make the necessary adjustments. Based on the morphological features, the young panicles are classified into eight developmental stages (Fig.6). The synchronization in flowering can be predicted by examining the development of young panicles. About 30 days before heading, the male and female parents are sampled and their young panicles within the main culms and tillers are carefully observed by the naked eye or with a nagnifying glass every 3 days. The criteria for synchronization of flowering include:

1.The male parent should be one stage earlier than the female during the first three stages of young panicle differentiation;

2.Both parents should be in same stages during the fifth, and sixth stages;

3.The female parents should be slightly earlier than the male parent during the last two stages.

If its is found that the synchronization of flowering will not be attained, the earlier developing parent should be applied with quick releasing nitrogen ferlilizer, and the later should be sparyed with 1% solution of phosphate fertilizer, or adjusted by drainage or irrigation,as the R lines are more sensitive to water than the CMS lines. If the difference in flowering period between the two parents reaches 10 days or more, it is necessary to remove the bracts or panicles from the early developing parent and apply nitrogen fertilizer subsequently, thus making its late emerging tillers or unproductive tillers bear panicles and subsequently achieve synchronization of flowering.

8.1.5 Gibberellion Application and Supplementary Pollination

Gibberelin is an effective plant growth hormone which stimulates the elongation of cells. Application of gibberellin has some advantages: (1) To make the panicle base on the CMS line less enclosed in the leaf sheath or fully emerge out of the sheath. (2) To incrcase the rateof stigma exsertion. (3) To adjust plant height. (4) to cause the small tillers to grow faster so that they can catch up with large tillers. (5) By using a large dosage the whole panicle can grow taller than the flag leaf.

Supplementary pollination is generally carried out in the morning, when the CMS line flower. If only the R line flowers, and CMS line does not, supplementary pollination should not be undertaken. In the afternoon when the R line is still blooming, supplementary pollination should be continued even if the CMS line has closed its glumes because the stigmata of some CMS lines always remain outside the glumes. The application is made by shaking the R line's panicles by rope-pulling or rod-driving during anthesis at 30−min intervals, five or more times daily, until no pollen remain in the R line.

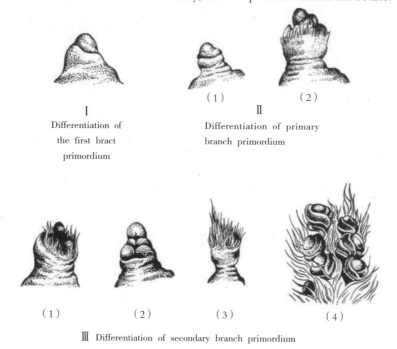

I
Differentiation of
the first bract
primordium

(1) II (2)

Differentiation of primary
branch primordium

(1) (2) (3) (4)

III Differentiation of secondary branch primordium

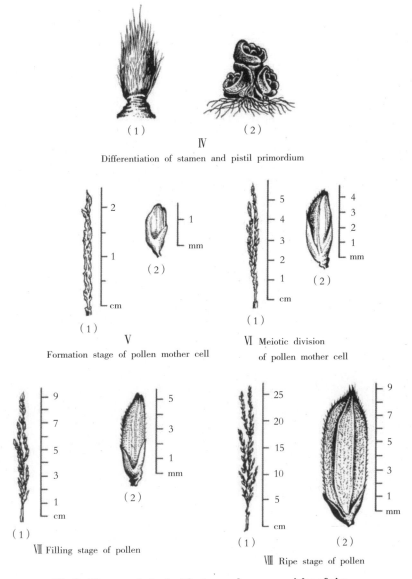

（1）　　　　　　　　（2）

Ⅳ

Differentiation of stamen and pistil primordium

（1）　　　　　　（2）

Ⅴ

Formation stage of pollen mother cell

（1）

Ⅵ Meiotic division
of pollen mother cell

（1）　　　　　　（2）

Ⅶ Filling stage of pollen

（1）　　　　（2）

Ⅷ Ripe stage of pollen

Fig.6　The morphological features of young panicles of rice.

8.1.6　Roguing

The purity of hybrid rice seed used in commercial production must be over 98%. To meet this requirement, the purity of the R lines and CMS lines must be over 99%. Therefore, in addition to strict isolation, it is necessary to make thorough roguing in the seed production field. Roguing should be done two or three times before heading, at the initial heading stage and before harvest. The maintainer plants and semi-sterile plants that appeared in the CMS line rows and other off-type plants mixed in both male and female rows should be removed completely.

8.2 Techniques of CMS Line Multiplication

The techniques of CMS line multiplication are basically similar to those of F_1 seed production except for sowing intervals and row ratio. The CMS line and its B line are like twins, and so do not differ greatly in their growth duration. Since there is also no great difference in plant height between male and female parents, and the male parent is inferior to the female parent in tillering capacity and growth vigor due to delayed seeding, their row ratio is smaller. At present, the row ratio widely adopted in multiplication fields in China is 1 : 3 or 2 : 5.

9 Future Outlook

Research into and utilization of heterosis in rice have made tremendous advances during the past 20 years, but from a stragegic point of view, they are still in their infancy.

9.1 Breeding Methodology

With regard to breeding methodology used to develop heterotic rice hybrids, the classical three-line method involving CMS, maintainer, and restorer lines is expected to remain effective and useful for the future, although it has some short-comings and problems to be overcome. However, it is more complicated than necessary and may be replaced by simpler systems, such as a two-line or one-line method.

The two-line system may involve photosensitive genic male sterile (PGMS) and thermosensitive genic male sterile (TGMS) lines or chemical emasculation. The PGMS and TGMS were developed in China in 1988. Their male sterility is controlled by nuclear gene (s), and has no relation to the cytoplasm. Under long-day or higher temperature conditions, they show complete male sterility, which under short day or lower temperature turn to lower temperature turn to fertility. There are two advantages for using these genic-male sterile lines to develop hybrid rice. Firstly, it can be used for a dual purpose, i.e., seed production and male parent line multiplication under long day and higher temperature (in summer) and short day and lower temperature (in autumn), respectively. Therefore, the A and B are the same variety, and the produres for seed production can be simplified. Secondly, it is more efficient to obtain superior hybrids because nearly all normal varieties have retoring ability to these lines, regardless of whether the male parent has restoring genes or not.

The one-line method is to breed a "true breeding" heterotic F_1 hybrid, i.e., the nonsegregating F_1 hybrids. It would be the best way to use heterosis of rice, because it needs neither seed production nor parent multiplication. The one-line method, involving apomixis, is considered the most worthwhile goal, and may enable true-breeding hybrids with peremanently fixed heterosis. The development of apomictic rice will require such research, probably involving biotechnology.

9.2　Heterosis Levels

From the view point of increasing the level of heterosis, i.e., increasing yield potential, the exploitation of heterosis in rice could be also divided into three levels.

1. *Intervarietal Hybrids.* In this kind of hybrids, the heterosis is greatly limited. Generally, the yield is about 15%−20% over conventional inbred varieties. It seems impossible to further increase the yield of the intervarietal hybrids greatly, unless new materials and methods are found.

2. *Intersubspecific Hybrids.* Indica−japonica hybrids exhibit very strong heterosis and theoretically may yield at least 20% more than the existing intervarietal hybrid rices. However, spikelet sterility in such hybrids prevents the exploitation of this heterosis commercially. The recent discovery of a wide compatibility gene (Araki et al., 1986; Ikehashi and Araki, 1986) should make *indica−japonica* hybrids set normal seed. In order to make this approach practical, wide compatibility gene (s) would have to be transferred to various CMS and restoere lines.

3. *Distant Hybrids* (*interspecific or intergeneraic hybrids*). Utilization of hybrid vigor from wide crosses is hard to imagine today, but with the help of apomixis, biotechnology and genetic engineering tools (see Bajaj, 1989), it may be possible to develop elite lines from interspecific crosses which may have unique gene blocks resulting in wide heterosis in F_1 hybrids.

This strategic plan for hybrid rice development in the future is illustrated in Fig. 7. It is being carried out in China and some other countries. In some steps of this strategic plan, scientists have achieved progress in breeding methods or in breeding malterials. Undoubtedly, after a great amount of work, hybrid rice development should have a brilliant future.

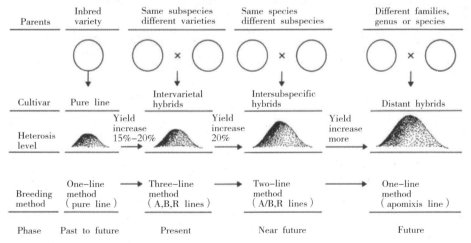

Fig.7　The comparison of diff erent breeding methods and heterosis levels of rice.

136

References

Araki H, Toya K, Ikehashi H (1986) Role of wide compatibility gene (s) in hybrid rice breeding. In: Int Symp Hybrid rice. Changsha, Hunan, China, Oct. 6−10.

Athwat DS, Virmani SS (1972) Cytoplasmic male sterility and hybrid breeding in rice. In: Rice breeding. IRRI, Los Baños, Phil, pp 615−620.

Bajaj YPS (ed) (1989) Biotechnology in agriculture and forestry. Vols 8.9. Plants proteplasts and genetic engineering Ⅰ and Ⅱ. Springer, Berlin Heidelberg New York.

Ikehashi H. Araki H (1986) Genetics of F1 sterility in remote crosses of rice. In: Rice genetics. IRRI. Los Baños, Phil, pp 119−130.

Jones JW (1926) Hybrid vigor in rice. J Am Soc Agron 18:423−428.

Lin SC, Yuan LP (1980) Hybrid rice breeding in China. In:Innovative approaches to rice breeding. IRRI. Los Baños, Phil, pp 35−51.

Mao CX (1986) Parental source analysis of present comercial combinations of hybrid rice in China. In: Int Symp Hybrid Rice. Changsha, Hunan, China, Oct 6−10.

Mao CX (1987) Hybrid rice seed production in China. In: Worksh Rice seed health. Manila. Phil, March 16−20.

Mao CX (1988) Parental source of commercial combinations of hybrid rice in China. In: Hybrid Rice. IRRI, Los Ba?os, Phil, pp 266−267.

O'Toole JC, Soemartono (1981) Evaluation of a simple technique for characterizing rice root systems in relation to drought resistance. Euphytica 30:283−290.

Shinjyo C (1969) Cytoplasmic-genetic male sterility in cultivated rice (*Oryza sativa L*) Ⅱ. The inheritance of male sterility. Jpn J Genet 44:149−156.

Shinjyo C (1975) Outcrossing rate in cytoplasmic male sterile plants of rice in the first and second crop season in Okinawa. Jpn J Breed 25 (Suppl 1):143−144.

Shinjyo C, Omura T (1966) Cytoplasmic male sterility in cultivated rice (*Oryza sativa L*) I. Fertility of F1, F2 and offsprings obtained from their mutual reciprocal backcrosses; and segregation of completely male sterile plants. Jpn J Breed (Suppl 1):179−180.

Stansel JW, Craigmiles JP (1966) Hybrid rice, problems and potentials. Rice J 69:14−15,46.

Swaminathan MS, Sidding EA. Sharma SD (1972) Outllok for hybrid rice in India. In: Rice breeding. IRRI, Los Baños, Phil. pp 609−613.

Virmani SS, Edwards E (1983) Current status and future prospects for breeding hybrid rice and wheat. Adv Agron 36:145−214.

Yuan LP (1966) A preliminary report on the male sterility in rice. Sci Bull 4:32−34.

Yuan LP (1972) An introduction to the breeding of male sterile lines in rice. In: Proc 2nd Worksh Genetics, Hainan, Guandong, China, March, 1972.

Yuan LP (1985) A concise course in hybrid rice. Science and Technology Publ House. Hunan, China.

作者：Yuan Longping　Mao Changxiang

注：本文发表于 *Biotechnology in Agriculture and Forestry* 1991 年。

选育水稻光、温敏核不育系的技术策略

[摘 要] 选育实用的水稻光、温敏核不育系，首先要考虑的是育性对温度的反应而不是光长。导致雄性不育的起点温度要低，这是最关键的技术指标，在这个前提下再追求具有光敏不育特性的材料，同时，温敏不育也是可资实用的好材料。文章对其具体的选育技术和方法进行了介绍和讨论。

培育两系法杂交水稻是作物杂种优势利用育种领域中的一项重大革新，其成败关键在于能否育成适合大面积生产上应用的光、温敏核雄性不育系。近几年来，我国在这方面的研究已取得很大的进展和成绩，选育了一批籼、粳型的光、温敏不育系，配制了不少高产苗头组合，在较大面积的试种、示范中表现良好，个别组合非常突出，初步显示了两系法杂交水稻具有更高的产量潜力和鼓舞人心的前景。但是，必须严肃指出，在前一阶段的研究中，由于受到材料和条件的局限性，研究的深度和广度不够，特别是对光、温条件导致育性转换的规律认识不全面，因而在实际选育不育系的过程中没有正确的指导思想和技术路线，以致现有育成的光、温敏不育系绝大多数并无实用价值。这些不育系，尤其温敏的，在长江流域夏季长日下遇上几天异常低温，就出现育性反复现象；在海南春季短日下，更是随着温度的起伏而"打摆子"；在云贵高原的夏季，不仅温敏不育系，甚至像农垦58S和7001S等典型的光敏粳稻不育系也不表现出明显的稳定不育期；在北方，如北京和辽宁，夏季日长虽长达16小时左右，但遇上偏常低温，许多光敏粳稻不育系的育性也随之波动。以上种种现象表明，要用这样的核不育系在生产上制种是不可靠的，存在着较大的风险。事实上，近三年来，因现有核不育系的育性受有时气温无规律的变化影响，致使制种失败的例子已为数不少。由此可见，现有光、温敏核不育系的育性不够稳定，已成为当前两系法杂交水稻最突出的问题。

探索、揭露和解决问题是科学研究的基本特色，在研究过程中，问题的暴露是好事，有利于认识的深化和推动新对策的产生。本文试就如何选育实用的水稻光、温敏不育系，提出一些新的认识和相应的技术策略，希望能起到拨正航向的作用，促进这项选育工作沿着正确的技术路线开展。

对实用光、温敏不育系的新认识

我国选育水稻光、温敏核不育系的工作始于 20 世纪 80 年代，到 1988 年就育成 17 个不育系并通过了省级以上的技术鉴定，此后又陆续育成许多类似的不育系。从认识上看，过来的选育工作大体可分为两个阶段。

1. 初期摸索阶段

在 1989 年以前，由于早期的某些基础研究所得结论的片面性，加上选育者缺乏经验，因而在工作中存在着相当的盲目性，认为只要是用农垦 58S 作供体所转育成的雄性不育材料，如 W6154S 等，都属于光敏不育性质，新发现的育性可转换的核雄性不育材料，如安农 S-1 等也一律被看作是光敏不育的。对这些核不育系作技术鉴定时，一般都在盛夏长日条件下进行，因不育特性能充分表达，所以较易获得通过。

2. 追求光敏不育阶段

随着基础研究的深入，逐渐认识到温度对育性的表达和转换起着重要作用，特别是 1989 年 7 月下旬出现持续数日的异常低温，导致许多所谓的"光敏"不育系，如 W6154S、安农 S-1 等，发生育性恢复现象，更促使人们明确地认识到，水稻两用核不育材料两大基本类型，即光敏型和温敏型，并且光、温还有互补作用。由于气温的变化有时异常，不像光周期有严格的规律，在生产上应用时很难准确掌握，因此，大多数育种家从 1989 年开始摈弃温敏材料而把重点转到选育光敏的不育系上。这叫吃一堑，长一智，不仅在认识上提高了一步，而且在工作上亦颇见成效，一些较典型的光敏不育系，特别是籼稻，如武汉大学培育的 8902S 等，在短短的两年内就宣告育成并通过了技术鉴定。但是，进一步的观察和研究又表明，即使是典型的光敏不育系，在夏季长日下，若遇上低温，其育性也会受影响，变成部分可育；反之，在秋季短日下，若遇上高温，又会由可育转为不育。也就是说，光敏不育系的育性同样要随异常的温度变化而波动，因而在生产上应用时照样有风险。育种家面临着非常令人困惑的难题，出路何在？

综合现有的研究资料，新近得出一个如下的光敏核不育水稻育性转换模式[※]，对光敏不育系在长日条件下，但温度低时也要"打摆子"的现象作了较好的理论解释。

这个模式表明，光敏不育系只能在一定的温度范围内，才具有光敏特性，即长光下表现不育，短光下可育，超出这个范围，光照长短对育性转换并不起作用。当温度高于临界高温值时，高温会掩盖光长的作用，在任何光长下均表现不育，当温度低于临界低温值时，较低的平温也会掩盖光长的作用，在任何光长下均表现可育。同时，在光敏温度范围内，光长与温度还有互补作用，即温度升高，导致不育的临界光长缩短，反之，温度下降，导致不育的临界光长变长。

品系不同，光温临界指标不同。根据张自国等的研究，诱导农垦 58S 不育的最短日长与最低温度分别为 13.45 小时和 26 ℃左右。周广洽等的研究表明，农垦 58S 在 14 小时长日条件下，只要敏感期的温度低于 24 ℃就出现可育，染色花粉率高达 50%，结实率为 3.3% ~ 10.1%。显然，这类不育起点温度偏高的光敏不育系，在生产上不宜使用。

对温度起主要作用的温敏不育系，也提出了一个育性转换模式：

短日不育温度范围

```
        （可育）    （不育或可育）      （不育）
  ···  ┤      ├──┤            ├───────┤      ├  ···
        生         临           长         生
        物         界           日         物
        学         低           不         学
        下         温           育         上
        限                      温         限
        温                      度         温
        度                      范         度
                   临           围
                   界
                   低
                   温
```

同样，品系不同，导致不育的起点温度不同，如 W6154S 和安农 S-1 为 26 ℃ ~ 27 ℃、5460S 高达 28 ℃以上，长光下略低，短光下略高，但光长的互补作用不如对光敏不育系那

[※] 这是作者根据张自国等和刘宜柏等所提出的相类似的模式加以综合和精简而形成的。

样显著。临界低温值的高低还因低温持续时间、日最低温强度和由不育转可育或由可育转不育而有差异，由于这类温敏不育系的不育起点温度高，加之光长效应小，因而它们的育性较易波动。例如，温敏不育系衡农 S-1，在敏感期内，当一天中出现连续 12 小时的高温（27 ℃）或平温（24 ℃）时，就会引起部分花粉育性发生逆转。很明显，像这类高温敏不育系，在生产上的应用价值更小。

上述两个模式，不仅理顺了水稻光、温敏不育系育性转换与光、温变化的关系，而且还为选育实用的两用不育系指明了方向，主要有以下三点。

第一，可育临界低温的高低是选育两用不育系最重要、最关键的技术指标。无论光敏还是温敏，唯有导致不育的起点温度低，并且在低于临界温度时还需较长的时日才能恢复可育的不育系，才具有实用价值。也就是说，与以前的观点相反，选育光、温敏不育系时，首先要考虑的是育性对温度的反应而不是光长。导致不育的起点温度的具体指标，因地区不同而异，根据湖南长沙的气象记录，7—8 月两月连续三天日均温低于 23 ℃ 的概率是 39 年一遇，因此，在湖南可把临界低温指标定为 23 ℃，时间为三天；广西定的指标是 24 ℃。

第二，符合上述前提且光敏温度范围宽（即长日下不育起点温度低、短日下不育起点温度高）的光敏不育材料为最佳选择对象。这种材料，在夏季长日制种时，即使遇上低温，不至于引起育性恢复；在秋季短日繁殖时，即使遇上异常高温，也不会导致不育。

第三，不育起点温度低、制种无风险的温敏不育系也是可被利用的好材料。这类不育系的主要问题是可育的温度范围窄，因而繁殖种子困难，但并不难解决（这个问题以后将专文讨论）。

选育的技术路线

大量研究表明，同一种光、温敏核不育基因置于不同的遗传背景下，甚至在同一杂交组合的分离后代个体中，其育性的表达对温、光的反应不同，光敏的可变为温敏的，高温敏的可变为低温敏的，等等，虽然，其中的规律现在尚不清楚，但这是一种十分可喜的变异现象，为育种家选择实用的光、温敏材料提供了机会和可能。具体的选育方法因地因材料不同而异，这里只扼要介绍本人推荐的技术路线。

一、材料选择

用现有遗传行为简单的光、温敏不育系或材料为基因供体，以多种不同遗传背景的优良品种、品系作受体，进行杂交。应该强调的是，供体的光、温敏不育基因要越简单越好，以便提高选育效率和避免以后可能带来的麻烦。根据现有资料，安农 S-1 的育性遗传最简单，为一对隐性基因所控制，其次是农垦58S，有两对隐性基因。育性受两对以上基因控制的材料以及遗传特性尚未弄清的不育材料，最好不要使用。

二、选择步骤

1. 初选　将各杂交组合的 F_2 群体，种植在夏季长日高温条件下，促使雄性不育特性能充分表达出来。首先选组合，即重点选可育株与不育株成 3∶1 分离的组合，其次是 15∶1 的组合；然后选单株，即选择不育度为 100% 的优良单株。淘汰育性分离比例过大，无规律可循，并出现很多程度不同的半育株的组合，即使这些组合中有个别完全雄性不育单株也不必选。

当选株让其再生，在秋季短日平温条件下，选收再生稻结实率高的单株（可能属光敏）以及直到晚秋才开始转为可育的单株（可能属低温敏）。

2. 复选　将当选的 F_3 和以后世代的株系，种植在海南春季短日变温条件下。海南南部3—4月，日照较短，温度起伏大（图1），是选育光、温敏不育系较理想的天然筛选场。根据育性转换与温度变化的关系，一般按以下原则进行取舍。

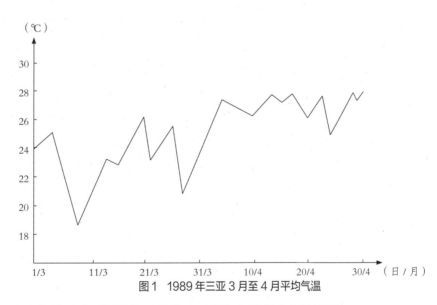

图1　1989年三亚3月至4月平均气温

①选5月上旬前后才开始转为不育的（可能为光敏型或高温敏型）。

②选3月中下旬以后一直表现稳定不育的（低温敏型）。

③淘汰育性随温度起伏而波动的材料（一般温敏型）。

3. 决选　以一分为四法（一个单株掰成四份）在人控或各种生态条件下进行如下四种光温组合处理，从中鉴定出所需类型（表1）。

142

表1　1989年三亚3月至4月的日平均气温

长光高温	长光低温	短光高温	短光低温
S	S	F	F（光敏型）
S	F	S	F（一般温敏型）
S	S	S	F（互作或低温敏型）
S	S	S	S（低温敏型）

注：处理的光照长短和温度高低指标要因地制宜。S为不育，F为可育。

为了能更准确地作出判断，在上述每一个选择程序中，都应设典型的温敏和光敏不育系作对照。决选当选的材料，并不意味着就是实用的两用核不育系，必须在扩大群体后作进一步的鉴定，才能下结论。

References

参考文献

---

［1］张自国等.1990.湖北光敏感核不育水稻在元江不同海拔条件下育性转换初步观察.水稻光（温）敏核不育及亚种间杂种优势利用研究论文选编.农业部科技司.

［2］申岳正等.1990.粳稻光敏感雄性不育"温度效应"初析.论文选编.农业部科技司.

［3］周广洽等.1991.两系杂交水稻基础理论研究进展.湖南师大生物系.湖南省生物研究所.

［4］孙宗修等.1990.光敏核不育水稻的光温反应研究.论文选编.农业部科技司.

［5］张自国等.1991.光敏核不育水稻育性转换光温作用模式研究.（未发表）.华中农业大学.

［6］刘宜柏等.1991.光温条件对水稻两用核不育系育性的作用机理研究.江西农业大学学报，第13卷1期.

［7］龙国炳.1991.两系杂交稻在长沙制种繁殖气象分析.（未发表）.湖南省长沙农业气象试验站.

作者：袁隆平

注：本文发表于《杂交水稻》1992年第1期。

Hybrid Rice Breeding Consultancy for 1992

ACKNOWLEDGEMENTS

I wish to express my deep sense of gratitude and appreciation to Dr. E. A. Siddiq, National Project Director, Directorate of Rice Research, Rajendranagar, Hyderabad, for providing all the necessary facilities and support to successfully carry out this consultancy mission. His active participation in all the group discussions, meeting and field visits, despite his extremely busy schedule, was immensely helpful.

Special thanks are also due to the Hybrid Rice Breeder at the Co-ordinating Centre, Drs. B.C. Viraktamath and M. Ilyas Ahme, who made all the arrangements regarding this consultancy mission and for lively participation in the numerous field visits, their critical and detailed discussions on various facets of hybrid rice research etc.

I express my thanks to all the scientists of hybrid rice research network in India, who provided valuable information and views and had discussion about their experimental material.

I am highly indebted to Dr. M.S. Swaminathan, Dr. Ish Kuman and Dr. K. Govinda Raj for the invitation to their institutions and for the discussion and their hospitality. Dr. V.L. Chopra, Director General, ICAR, New Delhi, Mr. S. I. Zakhariev, Representative in India, FAO, New Delhi, Dr. M. Scaillet, Director, Agricultural Operations Division, FAO, Rome, Mr. V. Kumar and Mrs. Kochar, Programme Officers, FAO, New Delhi, Dr. M. V. Rao, Vice Chancellor, APAU, Hyderabad, Dr. J. G. Ryan, Director General, ICRISAT, Hyderabad, Dr. Y. L. Nene, Deputy Director General, ICRISAT Hyderabad provided the opportunity to meet them and discuss important issues pertaining to hybrid rice research and development in India. I am indeed, very thankful to them.

Dr. C. X. Mao and Mr. X. L. Deng, FAO Consultants for seed production and breeding respectively have been immensely helpful to me in various ways in this consultancy mission. Their help is thankfully acknowledged. Finally I wish to express my thanks for all those who have directly or indirectly helped me during my memorable stay in India for completion of this consultancy mission.

INTRODUCTION

Hybrid rice which has been successfully developed and released in China in 1976 has proved practically that hybrids have 20% - 30% yield advantage over the improved inbred rice varieties. Indian Scientists have started hybrid rice research since 1980s. To accelerate the ongoing hybrid rice program, the Indian Council of Agricultural Research (ICAR) has sponsored a special project through the "Promotion of Research and Development Efforts on Hybrid in Selected Crops" which includes rice crop since December, 1989. The program on hybrid rice has further been strengthened since September, 1991 with the UNDP funded project on "Development and Use of Hybrid Rice Technology (IND/91/008)". The Project has following main objectives: 1) Development of hybrids yielding 15%−30% over the best check varieties; 2) Development of efficient and economical hybrid seed production techniques; 3) Development of productive package of cultivation practices for hybrid rice and 4) Basic and strategic researches on parental line improvement and two-line hybrid rice breeding.

FAO invited me as a consultant and I visited India from 2−20 November, 1990. Now as a Chief consultant for the project IND/91/008, I come to India again. The major mission for the consultancy is:

1. To review the current hybrid rice breeding program at the network centres and suggest measures to develop/improve A, B and R lines (Visit and Discussion).

2. To prepare a detailed technical program for developing two-line breeding using environment sensitive genic male sterility system (EGMS, including TGMS and PGMS) appropriate to India.

Through frequent field visits, observation of experimental materials in the field and discussion with concerned scientists, four major findings come to my mind.

1. A lot of new hybrid combinations have been developed by different research centers or private sectors during last two years and some of them show very high yield potential over the best check inbred varieties.

2. A dozen of CMS lines with good agronomic characters and quite stable sterility have also been developed by Indian hybrid rice breeders or introduced from IRRI. At least there are 2−3 elite CMS lines which could be used in commercial production.

3. Recently, some sterile plants have been found in the experimental fields of DRR's research farm by Chinese and Indian Scientists, which might be TGMS materials. If under relative low temperature condition, say 22 ℃ - 24 ℃, their true TGMS nature is confirmed then they can directly or indirectly be used in the two-line hybrid rice breeding program.

4. One of the hybrids developed by Pioneer Seeds Limited showed the yield potential as high as twice that of conventional inbred rice varieties in farmers' field, that means in farmer's field condition, hybrid rice, no doubt, can bring enough benefit to the farmers.

In conclusion now I have the confidence that the hybrid rice is on the threshold of release to farmers for commercial production (within two or three years) and finally the hybrid rice will play an important role to increase the yield level of rice in India in coming years.

RECOMMENDATION

1.From the technical point of view, the hybrid rice technology developed by Chinese scientists has been basically grasped by Indian Scientists. Now they have reached the point that the hybrid rice is on the threshold of release to farmers for commercial production. This conclusion is based on the following facts:

i) There are several CMS lines that can be used practically which show 1) stable sterility, 2) higher outcrossing rate, 3) good combining ability, 4) better grain quality and 5) Adaptable to local conditions.

ii) A number of hybrids with high yield potential have been successfully developed and their yield advantage over the best check is 1-1.5 t/ha in experimental trials and some of them are being tried in farmers' field, such as IR 58025 A/IR 54742 R, IR 58025 A/IR 40750, PMS-8A/PAU 1106-6-2, PMS-8A/IR 31432-6-2-1 and Hybrid-1 developed by Pioneer Seeds Ltd.

iii) Hybrid rice seed production techniques have been much improved compared to the last two years. The yield of 1.5-2 t/ha has been obtained in smaller plots in many locations. Now it is possible to produce F_1 hybrid seeds commercially from the economic point of view.

2.Many elite restorer lines have been evaluated and identified, and some promising CMS lines (breeding materials) showing good adaptability to local conditions are being developed.

3.Recommendations:

i) A nation wide effective and applicable extension plan must be formulated for commercial use of hybrid rice and this plan can be taken as the key link under which the related developmental programs, such as the scale of A lines multiplicating and of F_1 seed production etc. are worked out step by step and year by year.

ii) The development of new CMS lines should be further strengthened. The most convenient way is to use the "WA" cytoplasmic type for conversion of CMS lines rather than others. It is unnecessary to spend more effort on finding new CMS sources at this moment.

iii) For R line breeding, in addition to identification of the existing varieties by test crosses, it is important to introduce japonica and/or javanica blood into indica R lines to enhance the level of heterosis (The detailed R line improvement action plan is to be prepared by my assistants Dr. Mao and Mr. Deng).

iv) Though the major effort must be focussed on development of three-line system of hybrid rice, two-line hybrid rice breeding program should also be started immediately because two-line system hybrids possess a lot of superiority over the CMS system and especially as my two assistants Dr. Mao and Mr. Deng, FAO consultants for the same project have found some male sterile rice plants in the experimental field of DRR which seem to be TGMS materials.

(For successful carrying out of this research program, the concerned facilities and experts are needed. For the action plan of two-line system hybrid rice program for India, please see the attached papers).

v) Development of inter-subspecific hybrid rice is very important due to their higher yield potential compared to intervarietal hybrids. This programme should be initiated next year (See the attached action plan).

vi) There are still some problems in practices of seed production process in India. Further studies of hybrid seed production and intensive training for seed producers are badly needed, I suggest that some younger scientists who are working or will be working on hybrid rice seed production from public and private sectors could be sent to Hunan Hybrid Rice Research Center, Changsha, China for 3–4 months training course in 1993 (the best time is July to October).

vii) It is necessary to establish factories immediately to produce large amount of GA_3 so as to reduce the price of GA_3 to a reasonable level in India. Since GA_3 application is the key to win high seed yield.

viii) The other two Chinese consultants, Dr. C.X. Mao and Mr. X. L. Deng have rich experiences in hybrid rice breeding and seed production, and have done a lot of effective work appreciated by their Indian Co-workers. Therefore, I would like to recommend them to continue their consultancy for the next whole Rabi season in 1993 (Four months) to share their experience for handling all kinds of hybrid rice research activities including three line and two-line hybrid rice breeding and seed production.

ix) It is necessary to establish an effective system and select some good bases for hybrid rice seed production including F_1 hybrid seed production and foundation seed production.

The ideal conditions for such kind of seed production system are:

(a) Good irrigation and drainage facilities;

(b) Less incidence of diseases and insect pests;

(c) Good isolation condition; and

(d) Seed growers having rich experience in rice cultivation and management.

x) The National Standards for hybrid rice seed production and certification should be also worked out as soon as possible before commercial seed production is started in this country.

xi) After certification and before releasing a new hybrid, high yield demonstration fields and on-the-spot meetings should be arranged so as to convince farmers as well as extension workers at various levels. Meanwhile the correspondents from the media should be also invited to attend these meetings.

APPENDIX 1: Itinerary

Date	Activity
22 Oct., 1992 (THU)	Dep: Changsha 16:20 Hrs by CZ 3037 passed by Hong Kong 17:30 Hrs. −Arrival New Delhi: 23:15 Hrs by BA 36.
23 Oct. (FRI) AM	Visited FAO Office in New Delhi 9:15−10:00 Hrs. Participated the first Tripartite Review Meeting of the Project IND/91/008−Hybrid Rice at Krishi Bhavan, ICAR, New Delhi, 10:30−12:30 Hrs.
PM	Dep. Delhi; Arrival: Hyderabad 19:30 Hrs by ICAR 439.

24 Oct. (SAT) AM

Field visit to experimental farms of Hybrid Rice Research at DRR, Rajendranagar.

PM

Discussion with National Project Director and the Hybrid Rice Scientists of DRR, Hyderabad.

25 Oct. (SUN)

Field visit with counter-part Scientists at DRR and detailed discussion about the on-going programme.

26 Oct. (MON) AM

Detailed review of the work done in Hybrid Rice Research Network since commencement of UNDP-assisted project in September, 1991.

PM

Visited hybrid rice breeding materials in DRR farm with other two Chinese Scientists and discussed the working plan of consultancy with them.

27 Oct. (TUE) AM

Discussion about two-line hybrid rice breeding program with the other two Chinese consultants and counter-part hybrid rice scientists of DRR.

PM

Visited Farmer's a field demonstration of Hybrid Rice organised by PHI biogene Ltd.

28 Oct. (WED) AM

Discussed the detailed schedules for the consultancy activities with Dr. E. A. Siddiq, National Project Director, and other concerned Scientists.

PM

Discussion about two-line hybrid rice breeding programme.

29 Oct. (THU) AM

Discussed the breeding program for Indica/Japonica hybrids with Dr. E. A. Siddiq etc.

PM

Made the action plan for Indica/Japonica hybrid rice breeding program in India.

30 Oct. (FRI)

Continuation of discussion abut Indica/Japonica hybrid rice breeding and visited DRR farm to see the test cross materials in the test cross nursery.

31 Oct. (SAT) AM

Visited National Academy of Agricultural Research and Management and obtained some weather data at different locations in India, especially in Hyderabad for determining the right time to produce or multiplicate F_1 seeds of TGMS lines.

Nov. 1, 1992 (SUN)	Visited another farmer's a field demonstration of Hybrid Rice accompanied by Dr. K. Govida Raj, the hybrid rice breeder of PHI biogene Ltd and the hybrid rice scientists of DRR.
2nd Nov. (MON)	Visited Ramachandrapuram farm of DRR in ICRISAT area and found some sterile plants in the field which seems like TGMS materials.
3rd Nov. (TUE)	Discussed the recommendations with other two Chinese consultants Dr. C. X. Mao and Mr. X. L. Deng and hybrid rice scientists of DRR.
4th Nov. (WED)	Finalized the action plans for two-line hybrid rice breeding and Indica/Japonica hybrid rice program, and discussed the draft of recommendations with Drs. M. Ilyas, B. C. V. Math and other two Chinese consultants.
5th Nov. (THU)	Visited the Ramachandrapuram experimental farm of DRR in ICRISAT area and searched some sterile plants again in the field. Visited PROAGRO Seeds Ltd, during the afternoon and had discussion with production team.
6th Nov. (FRI)	Flew to Madras and Coimbatore to visit the Center for Research on Sustainable Agriculture and Rural Development and TNAU. Met Dr. M. S. Swaminathan and Mr. M. Rangaswamy, Dr. S. Jayaraj, Vice Chancellor of TNAU. Had some discussion with hybrid rice breeders in TamiNadu Agricultural University (TNAU).
7th Nov. (SAT)	Visited TNAU and met Dr. S. Jayaraj, Vice Chancellor. Discussion with hybrid rice scientists of TNAU. Dr. S. Jayaraj, Dr. M. Rangaswamy Dr. T. P. Ranganathan, Dr. Vaidynathan etc. Observed so-called TGMS materials and hybrid rice seed production plots.
8th Nov. (SUN)	Met Dr. M. Scaillet, Director, Agricultural Operation, FAO, Rome and Mr. V. Kumar Programme Officer, FAO.

9th Nov. (MON)	Attended the review meeting of hybrid rice research network —Scientists from all the Centre participated in the meeting there was discussion of on-going programme and formulation of technical programme for next season.
(AM)	Attending the meeting with FAO officials on review of UNDP-assisted project.
10th Nov. (TUE)	Gave a seminar at ICRISAT entitled "Strategic Research for H.R"— Two-line and inter-subspecific hybrid rice breeding programmes.
11th Nov. (WED) (AM)	Preparation of consultancy Report. Attended a press conference. Discussion with Dr. U. R. Murthy, Director, National Research Centre on Sorghum, Hyderabad.
(PM)	Departure from New Delhi.
12th Nov. (THU)	Meeting UNDP/FAO/ICAR officials to appraise about the consultancy mission.
13th Nov. (FRI)	03:15 Hrs Departure for China.

Appendix 2: List of consultancy concerned persons met.

1.Dr. V. L. Chopra	Director General, Indian Council of Agricultural Research, New Delhi.
2.Mr. J. Poulisse	Deputy FAO, Resident Representative New Delhi.
3.Dr. E. A. Siddiq	Director, Directorate of Rice Research, cum National Project Director, ICAR-UNDP Project on Hybrid Rice, Hyderabad.
4.Dr. M. Scaillet	Director, Agri. Operations, FAO, Rome.
5.Dr. James G. Ryam	Director General, ICRISAT, Hyderabad.
6.Dr. Y. L. Nene	Deputy Director General, ICRISAT, Hyderabad.
7.Mr. V. Kumar	Programme Officer, FAO, New Delhi.
8.Dr. M. S. Swaminathan	Director, Center for Research on Sustainable Agriculture and Rural Development, Madras.
9.Dr. M. Ilyas Ahmed	Hybrid Rice Breeder, Coordinating Center, DRR, Hyderabad.

10.Dr. B. C. Viraktamath	Hybrid Rice Breeder, Coordinating Center, DRR, Hyderabad.
11.Dr. C. X. Mao,	Hybrid Rice Breeder, FAO Consultant, cum Deputy Director of HHRRC, China.
12.Mr. X. L. Deng	Hybrid Rice Breeder cum FAO Consultant, HHRRC, China.
13.Dr. M. Govinda Raj	Hybrid Rice Breeder, Pioneer Seed Ltd., Hyderabad.
14.Dr. Ish Kumar	Plant Breeder, Technical Manager, PROAGRO Seed Company Ltd., Hyderabad.
15.Dr. M. Rangaswamy	Professor (Genetics), Tamil Nadu Agricultural University (TNAU), Coimbatore.
16.Dr. S. Jayaraj	Vice Chancellor of TNAU, Coimbatore.
17.Dr. G. S. Sidhu	Senior Rice Breeder cum Director Regional Research Station,Kapurthala, Punjab.
18.Dr. F. U. Zaman	Senior Scientist (Rice), Division of Genetics, IARI, New Delhi.
19.Dr. D. M. Maurya	Senior Rice Breeder, Dean, College of Agriculture, N. D. University of Agril. and Technology, Narendranagar, Faizabad.
20.Dr. R. Vijaya Kumar	Senior Rice Breeder, Agricultural Research Station, Maruteru (A.P.).
21.Dr. A. Singh	Rice Breeder, Rice Research Station, Karnal, Haryana.
22.Dr. R. N. Rao	Senior Scientist, Genetics Division CRRI, Cuttack, Orissa.
23.Dr. V. N. Deshpande	Senior Rice Breeder, Agril. Research Station, Karjat, Maharashtra.
24.Dr. K. Thiyagarajan	Associate Professor, Paddy Breeding Station, TNAU, Coimbatore.

ACTION PLAN I

Title: Development of Rice for hybrids with 2-line system
Objective:

ⅰ) To overcome the limitations of restorer-maintainer system operative in 3-line system.

ⅱ) To enhance the chances of obtaining the heterotic hybrids in rice.

Material:

1.Introduction of existing TGMS lines from China/Japan/IRRI, Philippines, etc.

2.Identification of TGMS lines from the germplasm.

Methods:

-Planning of germplasm should be done in such a way that beading time should coincide with higher temperature and longer day length.

-Once TGMS lines are identified, remove the plants, put them and keep them in glasshouse under 22 ℃-24 ℃.

-Ratooning to be done under two different conditions: 1) low temperature and short daylength conditions and 2) high temperature and long daylength conditions.

-To start with, 500 lines can be screened with a population of a minimum of 100 plants/line.

Confirmation of TGMS Nature.

-Re-evaluation of male-sterility under appropriate conditions next year

　　if TGMS……same behaviour

　　if genetic……No seed set under low temperature

　　if physiological …… phenomenon is not repeated

-Standardization of sensitive stage and temperature range which induces male sterility/fertility under artificially controlled conditions.

-Identifiction conditions of lines carrying TGMS genes.

Transer of TGMS gene (s) to desirable plant type.

-Cross line carrying TGMS gene (s) with desirable plant types.

-Make selection of desirable genotypes carrying TGMS gene (s) in F_2 generation under appropriate conditions.

-Preference may be given for the lines having monogenic TGMS line (3 : 1 segregation) .

-Under low temperature i. e. when fertile, TGMS line should have a seed set of more than 30% for economic and efficient seed production.

-Identification of promising hybrids (by crossing the TGMS line with promising genotypes) .

-Follow the other procedures, as in routine heterosis breeding programmes, for identification of promising hybrids, their evaluation, multi-location testing and on-farm verification trials etc.

-Maintenance of purity of TGMS lines is very important. Isolation distance should be more than 200 m and time isolation i.e. more than 21 days. Other rice should not have been grown in the same field where seed multiplication is being taken up.

-The ratio between S line multiplication, F_1 hybrid seed production and commercial F_1 cultivation in two line system is 1:100:10 000 as against the ratio of 1:50:5 000 in 3-line system.

ACTION PLAN II

Title: Development of Indica×Japonica F₁ hybrids in rice.

Main Objective: To raise the yield ceiling of hybrid rice by 15–20% over the existing intervarietal hybrids.

Materials:

Ⅰ. Source of WC genes:

1.CPSLO17 ······ to be obtained from IRRI, Philippines.

2.Ketan-Nangka (Javanica group) ······ to be obtained from Japan.

3.Dular ······ available from Indian germplasm collection.

Ⅱ. Source of Japonica lines:

1.Europian Type ······ USSR (Azerbaijan, Kazbekistan) Italy, Spain.

2.Taiwan Type······ Taiwan.

3.South-china Type ······ China.

4.Kashmir and North Eastern Indian Type ······ available from Indian germplasm collection.

Ⅲ. Source of Indica lines:

Well adapted genotypes with good flowering behaviour to be included.

BREEDING METHODOLOGY:

Ⅰ.Breeding of Japonica CMS lines with WC genes.

−Cross selected tropical *japonica* lines with lines possessing WC genes such as ① CPSLO ② KetanNangka and ③ Dular.

−In F₂ generation select typical tropic *Japonica* plants possessing WC gene, by using genetic markers [there is a genetic linkage between WC gene and ① purple spikelet tip. ② waxy endosperm].

Waxy endosperm can be checked by staining the pollen with KI-I₂ solution-waxy endosperm is indicated by slightly yellow stained pollen grains.

−Selected F₂ plants can be carried forward to F₄-F₅ generation following the same selection criteria and selecting the desirable plant type based on the objection.

−Test cross the selected *japonica* plants having WC gene with Indica type CMS lines to test for their maintaining ability.

−If found to be maintainer, continue backcrossing programme to BC₄-BC₅ to convert *japonica* type having WC gene into a CMS line.

−In cross progenies, selection must be based on high stigma exsertion and stigma remaining outside for a longer time.

−Make testcrosses extensively with all the known indica restorers to identify the promising hybrids.

−Follow other procedures, as in routine heterosis breeding programmes for identification of promising hybrids, their evaluation, multilocation testing and on-farm verification trials etc.

II . Breeding of *japonica* restorer lines with WC gene.

—Cross tropical *japonicas* with the genotype having WC genes such as CPSLO, KetanNangka and Dular.

—Grow F_1 and F_2 generation.

—Select plants of tropical *japonica* type having purple spikelet tip (there is a linkage between purple tip of spikelet and WC gene) .

—Selected plants should be testcrossed with *indica* restorer lines (after emasculation) in F_3 generation onwards.

—Best *indica* female to be tested would be TGMS line.

—The test for identification of WC gene is that the spikelet fertility is above 75%.

—The tropical *japonica* plants selected with WC gene should be crossed with *indica* CMS line to find out the restorability.

—Select the desirable tropical *japonica* plants having WC gene and good restorability.

作者：Yuan Longping

注：本文为 IND/91/008 项目 1992 年 11 月杂交水稻育种咨询任务报告。

Advantages of Constraints to Use of Hybrid Rice Variety

Based on our studies, the development of hybrid rice breeding may involve the following phases.

First, in terms of breeding methodology there could be three approaches:

1. Three-line method or CMS system
2. Two-line method or PGMS and TGMS system
3. One-line method or apomixis system

Second, from the viewpoint of increasing the degree of heterosis, i.e. to increase yield potential, the forms of heterosis that could be exploited in rice may also be divided into three groups:

1. Intervarietal hybrids
2. Intersubspecific hybrids
3. Distant hybrids (interspecific or intergeneric hybrids)

Each phase marks a new milestone in rice breeding and will result in great increases in rice yield if it is attained.

The existing hybrid rice varieties now used in commercial production in China belong to the category of intervarietal hybrids created using the CMS system. This kind of rice hybrid has created a huge amount of food for the Chinese people. It has been proven practically on a large scale for 16 years that hybrid rice has more than a 30% yield advantage over conventional pure-line varieties (Table 1).

Tab.1 The yield of hybrid rice compared with conventional rice from 1981 to 1990.

Year	Yield			
	Coventional / (kg/ha)	Hybrid / (kg/ha)	Hybrid over Coventional / (kg/ha)	/ %
1981	4 113.0	5 317.5	1 204.5	29.3
1982	4 447.5	5 865.0	1 417.5	31.9
1983	4 774.5	6 375.0	1 600.5	33.5
1984	4 992.0	6 405.0	1 413.0	28.8
1985	4 816.5	6 472.5	1 656.0	34.4
1986	4 857.0	6 600.0	1 743.0	35.9
1987	4 779.0	6 615.0	1 836.0	38.4
1988	4 539.0	6 600.0	2 061.0	45.4
1989	4 534.5	6 615.0	2 080.5	45.9
1990	4 642.5	6 555.0	1 912.5	41.4

The percentage of planting area and total production of hybrid rice in comparison with conventional rice in recent years is shown in the following figure（Fig.1）.

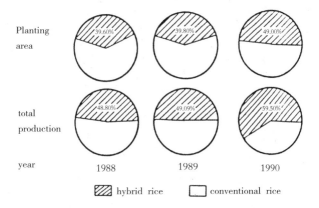

Fig. 1　**Percentages of planting area and total production of hybrid rice compared with conventional rice in recent years.**

The cumulative yield increase of rice grain due to cultivation of hybrid rice from 1976 to 1990 is nearly 169 million tons. If the total rice area is covered by hybrid rice in China, the annual rice production would be 210 million tons, i.e. 40million tons more than the present. Thus it can be seen that the further development of hybrid rice is of strategic significance in increasing food production.

In spite of the great success in the development of hybrid rice through the three-line method in China, some constraints and problems remain to be solved.

1.The yield of the existing hybrid rice varieties has stagnated for years（Fig.2）. This means that they have already reached their yield plateau and it would seem to be very difficult to further increase their yield potential if no new methods and materials are invented and adopted.

2.The sources of male-sterile cytoplasm which can be used for developing better CMS rice lines are poor. Currently, about 95% of the A lines used in commercial production still belong to the "WA" type. This uniform cyto-sterility holds a latent crisis in the long run, in which the hybrid rice could be destroyed by a single disease.

3.The heterosis level in Japonica hybrids is not as good as in Indica hybrids. In addition, the sterility of Japonica CMS lines（BT type）now used is not stable enough to produce very pure F_1 seeds. Therefore, the panting area of Japonica hybrids has been limited to around 0.1 million ha for many years, and, what is worse, the acreage of Japonica hybrids is declining.

4.The yield of hybrid seed production is not consistently stable which sometimes confronts the seed companies with a dilemma.

156

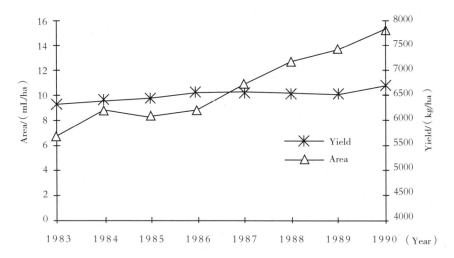

Fig. 2 Area and Yield of Hybrid Rice in China from 1983 to 1990.

In recent year, the average yield in hybrid rice seed production has been about 2.3 t/ha. The field area ratio between A-line multiplication, hybrid seed production and F_1 commercial cultivation is 1 : 50 : 5000.However, because the yield level is much influenced by weather conditions during flowering time, in certain years it is very high and in other years very low. In both cases the consequences are serious. For example, the yield of hybrid seed production decreased sharply due to extremely bad weather for out-crossing in the autumn of 1988 in Hunan Province. In order to fulfill the food plan, many seed companies had to go to Hainan Island that winter to produce 30 000 tons of hybrid seed at very high cost. By contrast, the weather was very favorale for outcrossing in 1990, and, as a result, nearly 25 000 tons of excess hybrid rice seed were produced. Because of the oversupply of seed, a number of seed companies sustained big losses.

Taking the long-range strategy of rice heterosis breeding into account, many Chinese rice scientists have been making attempts to explore new technological approaches as mentioned above so as to replace the CMS system. So far, the most successful one is the development of two-line method hybrids.

Recently two new kinds of rice genetic tools viz., photoperiod-sensitive genetic male-sterile (PGMS) lines and thermo-sensitive genetic male-sterile (TGMS) lines have been successfully developed in China. Their male sterility is mianly controlled by a pair of recessive nuclear genes, and has no relation to the cytoplasmic genomes. Exploitation of these MS lines to develop hybrid rice has the following advantages over the classical three-line or CMS system.

1.The B line is not needed. Under longer daylength the PGMS lines, or under higher temperature the TGMS lines, show complete pollen sterility; thus they can be used for hybrid seed production in these conditions. Under shorter daylength or temperate conditions they show almost normal fertility; thus they can multiply themselves by selfing.

2.The choice of parents in developing heterotic hybrids is greatly broadened. Studies show that over

97.6% of varieties tested（within subspecies）can restore such MS lines. In addition, PGMS and TGMS genes can be easily transferred to any rice cultivar.

3.There are no negative effects caused by sterile cytoplasm and the unitary cytoplasm situation of "WA" will be avoided.

Several better inter-varietal hybrid combinations produced by the two-line system were under regional trial and farmer's field trial in the past three years. The results were promising; some of them outyielded the best existing hybrids by about 5% – 10%. It is estimated that the two-line system of hybrid rice will be released for commercial production by 1993.

The development of intersubspecies, especially Indica/Japonica hybrids with very high yield potential through the two-line system, is still underway. The major problems presented in such F_1 hybrids, namely, semi-sterility, excessive plant height and very long growth duration, have now been solved. However, the plumpness of a large proportion of filled grains is not good in most combinations. Now, efforts are being focused on solving this problem.

Furthermore, tentative research on the one-line method to develop true-breeding rice hybrids by trying to use the mechanism of apomixis has also been started in China. No doubt, this is the best way to use the heterosis of crops but it is a very arduous task to realize. Now, the challenge is facing us.

作者：Yuan Longping

注：本文发表于 *Prceeding of the International Workshop on Apomixis in Rice* 1993 年。

Development and Use of Hybrid Rice Technology in India

SUMMARY

In the past three years, tremendous progress has been made in the development and use of hybrid rice technology in India. More than 35 hybrid combinations which outyielded the best check by one tonne or more have been identified through rigorous testing in national trials and some of them are performing very well in farmer's field also. Now India is on the eve of releasing her rice hybrids to farmers for commercial production.

Three practically usable CMS lines and many good restorers have been developed and identified. More emphasis should be given for the development of new stable CMS lines with high outcrossing potential and good combining ability.

The hybrid seed yield in seed production plots is relatively low. For the time being maximization of seed yield on large scale should be given top priority.

Some promising TGMS rice materials have been found and confirmed. The programme on two-line system of heterosis breeding should be strengthened and intensified.

INTRODUCTION

The successful commercialization of hybrid rice technology in China has practically proved the distinct yield advantage of hybrids over the best rice varieties. Since from the inception of an ICAR project "Promotion of Research and Development Effort on Hybrids in Selected Crops" in 1989, a serious attempt has been made to make the hybrid rice technology a reality in India also. The program has been further strengthened by UNDP through a project on "Development and Use of Hybrid Rice Technology" (IND/91/008) with the following main objectives.

1. To evolve and evaluate heterotic hybrid combinations.
2. To develop efficient and economical hybrid seed production.
3. To develop productive package of cultivation practices for rice

hybrids.

4.To conduct basic and strategic research relevant to the project goals.

The fulfillment of many of these objectives require profound practical knowledge and thorough guidance regarding various aspects of hybrid rice technology. In this context consultancy service play a very important role.

As a chief consultant of FAO, I visited India from 18th November to 8th December 1993.The terms of references and conditions for my third consultancy mission are:

1.To review the on-going programme and to suggest improvement, if any.

2.To visit on-farm verification and experimental trials laid out at all the centers of the research network and suggest measures for further augmentation/correction of the programme.

3.To assist in planning the detailed technical programme centre wise for Rabi 93 and Kharif 1994 seasons.

4.To prepare a detailed plan for large scale seed production in collaboration with private sector seed industries.

5.To deliver special lectures to seed production trainees.

6.To submit mission report according to FAO's attached guidelines.

MAIN FINDINGS

Based on the carefully reviewing experimental data, field visits, observation of experimental materilas at DRR, Hyderabad, visit to hybrid seed production plots of private seed companies (Hindustan Lever and Pro Agro), visit to on-farm verification trials at Nandyal and discussion with concerned scientists, the following major findings were identified:

1.India has developed more than 35 hybrids which out yielded the best check by one tonne or more per hectare.

2.The hybrids viz. IR 58025A/IR 9761, PMS 3A/IR 31802, IR 62829A/IR10198-66 and IR 62829A/MTU 9992 have not only outyielded the best check in national yield trials but also performed remarkably well in farmers' field. Particularly the hybrid IR 58025A/IR 9761 has outyielded best check by 1.2 to 1.4 t/ha in the farmers' field at ten locations.

3.I am very much excited to know that there are two hybrids viz. IR 62829A/IET 8585 and a Pro Agro Company's hybrid which yielded more than two tonnes over the best check in replicated trials. This indicates the great yield potential of India's new rice hybrids.

4.The seed yield of CMS lines in experimental multiplication plots is quite high (1.5 to 2.0 t/ha) in many locations which is quite nearer to the average CMS seed yield in China (2.2 t/ha). If the same packages of cultivation are practised on a large area, the yield levels realized in China could also be achieved in India in coming years.

5.Many new promising CMS lines with stable male sterility are in the advanced stages of conversion program in many research centers.

6.Several newly identified restorers like Swarna, MTU 9992, IET 8585, IR31802, IR 40750, PAU 1106-5-4, IR 35365 etc are being used in hybrid rice breeding. Especially a new R line (Pro Agro R.1) with a strong restoration ability has been identified by Pro Agro Company. The use of such strong restorers can play a very important role in realising the maximum yield potential of the hybrids with large number of spikelets per panicle.

7.Some of the male sterile materials collected from the DRR experimental field last year by FAO consultants show a peculiar stability to fertility transformation. Under high temperature condition during summer there was good seed set but in winter they became completely sterile which indicates that this material may belong to reverse TGMS type. Besides, several male sterile mutants obtained by irradiation mutation were found to be typical TGMS rice plants.

In conclusion, I can say that India has achieved tremendous progress in hybrid rice technology in recent years and is now on the eve of releasing hybrids to farmers for commercial production. In the near future hybrid rice is going to play an important role in breaking the yield plateau in India.

RECOMMENDATIONS

Based on the critical review of various research programmes related to the development and dissemination of hybrid rice technology the following recommendations are made under four headings viz. seed production, F_1 cultural management, parental line improvement and two-line approach.

I . Seed production

1.Since India is on the eve of releasing hybrids for commercial cultivation, production of good quality hybrid seeds on large scale (by involving both public and private sector agencies) should be given priority attention.

2.Some unstable CMS lines viz. IR62829A, PMS 8A, PMS 9A and PMS 10A are being used in developing commercial hybrids, even though, in the short run (1993—1994) some of these lines can still be used but in the long run it is not wise to use such unstable lines for large scale seed production. Wherever such lines being used for seed production, partially fertile types should be removed by involving persons who have rich practical experience in seed production. And the field workers should be trained in identification of partially sterile types in seed production plots.

3.The most important prerequisite for getting good synchronization is the accuracy of recording leaf number in the parental lines.

4.The row ratio for hybrid seed production should be increased from 2 : 4 or 2 : 6 now to 2 : 10 or 2 : 12.The high concentration (200 ppm[a]) and heavy dosage (200 g/ha) of GA_3 should be used in the right stage of panicle emergence. Especially for PMS lines more quantity of GA_3 should be applied because

---

① ppm=mg/L.

their flowering behavior is not good enough for outcrossing.

5.Supplementary pollination should be carried out at the right time, i.e., at the peak pollen shedding stage, therefore the peak period of pollen shedding should be carefully determined.

6.The hybrid rice seed production programme for Rabi 1993—1994 should be taken in an area of 150-200 hectares. In order to ensure a seed yield of 1.5 t/ha and to pursue a high seed yield of 2 t/ha, it is suggested that:

(a) Skillful use of GA_3 is the key for obtaining super high seed yield, so at least 20 to 30 kg of GA_3 are needed for the above mentioned area of seed production. It is better to obtain such amount of GA_3 from China because the price of GA_3 in China is the cheapest in the world market.

(b) Two Chinese hybrid rice seed production experts had better be invited to India from 15th January to 15th April 1994, to work hand in hand with the Indian seed producers to ensure high seed yield.

(c) A video cassette on "Super high yields in Hybrid Rice Seed Production" has been prepared and patented by a seed company in China. The patent right in India may be transfered to Directorate of Rice Research, Hyderabad, so that the multiplied copies can be distributed widely and used extensively during Rabi season among all the Indian hybrid rice seed producers, which will help to learn and adopt all the important techniques to obtain the super high yields in hybrid rice seed production. About the three recommendations mentioned above, I can help to contact the concerned and get the needful done at the earliest.

II . Crop management

1.Much attention should be given at nursery stage. Sparse seeding in the nursery at the rate of 150 to 200 kg of seed per hectare of seed nursery should be adopted in order to exploit the early vigor of hybrids and to get strong and healthy seedlings with more tillers.

2.Transplanting of two seedlings per hill instead of one is recommended because if the purity of the hybrid is not very high, this practice can minimize the yield losses especially when some unstable CMS lines are used.

3.In order to get good seed set and better grain filling, generally 20 percent more potassium (k) is needed for hybrids as compared to conventional rice varieties.

III . Parental line improvement

India has achieved a remarkable progress in the development of superior combination of hybrids. Further jump in yield is possible only by a strong parental line improvement programme.

1.CMS line development:

(a) So far "WA" cytosterile source is still the most effective and convenient way for developing new CMS lines. In order to avoid the use of single source of male sterility which may cause genetic vulnerability of the hybrids to the pests, two line method of heterosis breeding should be given more emphasis rather than looking for diversification of cytoplasmic male sterile sources.

(b) While developing new CMS lines, besides stable male sterility, more attention should be paid to outcrossing traits . In this context the flowering behavior and stigma exertion rate are very important.

(c) In BC_1 generation of new CMS line development, if there is segregation for sterility, select 3 to 5 male plants for back crossing with completely sterile plants. In BC_2, if all the families still show segregation for male sterility, the material should be discarded. On the other hand if some families are showing 100 percent male sterility, it is better to select a few best plants from a best family for further back crossing.

2.Restorer line improvement:

(a) Enhancement of heterosis level is possible only by the use of genetically diverse parents. In this context, the utilization of inter-subspecific crosses involving indicas, javanicas, tropical japonicas and European japonicas for the development of intermediary types which can later be used as parents should be given due emphasis. Inter subspecific heterosis cannot be exploited just by using major WC gene alone but some minor genes need to be incorporated. The details of the program is furnished at the end of the report (appendix) .

(b) Development of Iso-cytoplasmic restorers by using good hybrid combinations needs to be undertaken.

IV. Development of two-line system

1.Two line system of heterosis breeding is superior to three line system in many ways. It is time to concentrate more and to put more research efforts for this programme in India.

2.Theoretically there are 16 possible types of EGMS materials. Out of these, four types viz. TGMS, PGMS, reverse TGMS and reverse PGMS can be used practically. The existing EGMS rice materials in India should be critically examined and classified into different groups.

3.The temperature and day length standards (critical male sterility inducing temperature as well as day length) must be set up for different locations based on meteorological observations. For example, in Hunan Hybrid Rice Research Center the standards of daily mean temperature for screening practically usable TGMS lines is 24 ℃ (19–28 ℃) for three consecutive days. According to meterorological data of Hunan Province in China only once in 30 years the daily mean temperature goes below 24 ℃ for three consecutive days in the summer season.

4.The most critical and sensitive stage for transformation of sterility to fertility and vice versa should be identified for all the EGMS materials.

5.A couple of scientist from India may be asked to visit China for a month during August–September 1994 to acquaint themselves with two-line method of rice heterosis breeding.

ACKNOWLEDGEMENT

I express my deep sense of gratitude and appreciation to Dr. E. A. Siddiq, National Project Director, Directorate of Rice Research, Hyderabad for providing all the necessary facilities and support to carry out this consultancy mission. I am very much thankful to Dr. K. G. Pillai for his kind help.

I am very much thankful to Dr. M. Ilyas Ahmed for making all necessary arrangements for various programmes. I thank Dr. N.Sree Rama Reddy, Dr. K.Vijaya Kumar and Dr. Raju, Associate Director, RAPS, Nandyal for making necessary arrangements for visit to on-farm verification trial in Kurnool district.

My thanks are due to Dr. Ish Kumar, Dr. I. J. Anand and Dr. K. Govinda Raj for their lively discussions and hospitality.

I am highly indebted to Mr. S. I. Zakhariev, FAO Representative in India and Mr. V. Kumar, Programme Officer FAO, New Delhi for making necessary arrangements for consultancy mission.

I am thankful to Dr. M. S. Ramesha and to Research Associates i.e. Mr. Sukhpal Singh, Patil and Sridar for helping me in writing the consultancy mission report.

Finally I wish to express my thanks for all those who have directly or indirectly helped me during my memorable stay in India for the completion of this consultancy mission.

APPENDIX

Exploitation of Inter-subspecific Heterosis by Using WC Genes

The hybrids between *indica* and *japonica* show higher heterosis than *indica×indica* or *japonica×japonica* hybrids. But the exploitation of *indica×japonica* heterosis has been hampered due to hybrid sterility. At least two dominant WC genes are known to restore the fertility of these crosses, of which one is a major gene and other is a minor. Theoretically in both *japonica* and *indica* types, four combinations are possible. Let us assume that A is the major gene and B is a minor gene, so in each subspecies the genotypes of AABB, aaBB, AAbb and aabb types are possible. More than 95 percent of both *indica* and *japonica* rice are of aabb type. Remaining less than 5 percent mostly belong to either aaBB or AAbb types, as for the AABB type which is extremely rare. Identification of aaBB and AAbb types and the development of AABB types which can complement each other to give good seed set is the most important step in exploiting inter subspecific heterosis. Large number of test crosses involving *indica* rice and *japonica* rice should be undertaken to identify the desirable combinations, and in addition, the AABB type which has broader spectrum of wide compatibility can be obtained.

作者: Yuan Longping

注: 本文为 IND/91/008 项目 1993 年 12 月印发杂交水稻发展任务报告。

Increasing Yield Potential in Rice by Exploitation of Heterosis

【Abatract】The degree of heterosis in different hybrid rice varieties has the following general trend: indica/japonica>indica/javanica>japonica/javanica>indica/indica>japonica/japonica. The yield potential of the best existing indica/indica hybrids developed by the CMS system, in terms of per day per unit area, is around 75 kg/ha (from seed to seed), about 15% higher than pureline varieties. The best two-line system indica/indica hybrids can outyield CMS indica/indica hybrids by 5%–10%. Heterosis of japonica/javanica hybrids could be used to increase japonica hybrid yields. Their grain quality retains japonica characteristics. Indica/japonica hybrids possess the highest yield potential in both sink and source. Theoretically, they may have a 30% yield advantage over the best existing indica/indica hybrids. However, there are some problems in such intersubspecific hybrids-semisterility, plants too tall, very long growth duration, and many poorly filled grains. So far, most problems have been overcome through two-line breeding methods, except for the poor filling of a number of fertilized grains. The strategy of developing very high-yielding indica/japonica hybrids for commercial production is discussed.

Hybrid rice has helped China to increase rice production by nearly 200 million t from 1976 to 1991.Hybrid rice has a yield advantage of more than 30% over conventional pureline varieties (Table 1). In 1991, the area under hybrid rice was 17.6 million ha, 55% of the total rice area in China, and production of hybrid rice was 66% of the total rice output.

Although research on the commercial utilization of heterosis in rice has made tremendous gains during the last 20 years, it is, from a strategic point of view, still in its infancy because the high yield potential of hybrid rice has not been fully tapped yet.

Strategies for developing hybrid rice will involve the following:
· Breeding methodology
  −Three-line method or CMS system
  −Two-line method or PGMS and TGMS systems
  −One-line method or apomixis system
· Exploitation of heterosis
  −Intervarietal hybrids

−Intersubspecific hybrids

−Distant hybrids (interspecific or intergeneric hybrids)

Table 1　Yieds of hybrid and conventional rices from 1981 to 1990.

Year	Conventional / (t/ha)	Hybrid / (t/ha)	Hybrid over conventional / %
1981	4.1	5.3	29.3
1982	4.4	5.9	31.9
1983	4.8	6.4	33.5
1984	5.0	6.4	28.8
1985	4.8	6.5	34.4
1986	4.9	6.6	35.9
1987	4.8	6.6	38.4
1988	4.5	6.6	45.5
1989	4.5	6.6	45.9
1990	4.6	6.6	41.4

Breeding methodology

Three−line method or CMS system

Hybrid rice varieties used commercially are intervarietal hybrids produced by the CMS system. Many years of experience have proved that the CMS system, or three-line method, is an effective way of developing hybrid varieties and will continue to play an important role in the next decade. However, there are some constraints and problems in such a system. The most serious is that yields of existing hybrid rice varieties, including newly developed ones, have stagnated (Fig. 1). They have already reached their yield plateau, and further increase in yield potential is unlikely if no new methods and materials are invented and adopted.

Two−line method or PGMS and TGMS systems

Chinese rice scientists have been exploring new technological approaches to replace the CMS system and raise the yield ceiling in hybrid rice. So far, the most successful is the development of two-line method hybrids. This method is based on two new kinds of rice genetic tools, photosensitive (PGMS) and thermosensitive (TGMS) genic male sterile lines, which have been developed successfully in China. Male sterility is mainly controlled by one or two pairs of recessive nuclear genes and has no relation to cytoplasm.

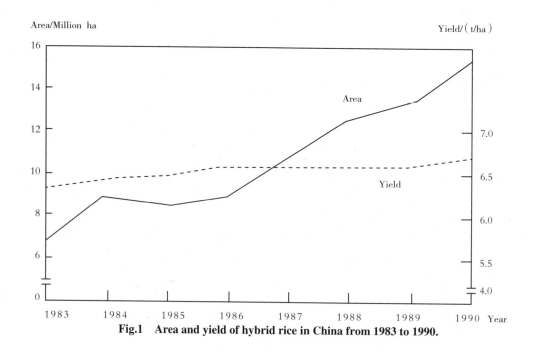

Area/Million ha

Yield/(t/ha)

Fig.1   Area and yield of hybrid rice in China from 1983 to 1990.

Developing hybrid rice varieties with these systems has the following advantages over the classical three-line or CMS system:

－Maintainer lines are not needed. The PGMS lines (under longer daylength) and the TGMS lines (under higher temperature) show complete pollen sterility and can thus be used for hybrid seed production. Under shorter daylength or temperate conditions, they show almost normal fertility and can be multiplied by selfing.

－The choice of parents for developing heterotic hybrids is greatly broadened. Studies showed that more than 97% of varieties tested (within subspecies) can restore such MS lines. In addition, PGMS and TGMS genes can be transferred easily to almost any rice lines with desirable characteristics.

－There are no negative effects due to sterile cytoplasm, and the unitary cytoplasm situation of WA will be avoided.

Exploitation of heterosis

Intervarietal F_1 hybrids

Breeding F_1 rice hybrids through two-line systems may be classified into two categories: intervarietal hybrids and intersubspecific hybrids. There are no genetic barriers or fertility problems in intervarietal hybrids, and the male sterility of PGMS or TGMS lines is controlled by simple recessive gene (s) which do not need a special R gene for fertility restoration; almost all normal varieties and lines within the same subspecies can restore fertility. Consequently, the probability of obtaining excellent intervarietal hybrid combinations through the two-line method is much greater than if the three-line method were used. A number of improved two-line system intervarietal rice hybrids have been successfully developed recently, and replicated tests, regional trials, and farmers' field trials on a large scale have been carried out. The total planting area was 15 300 ha in 1991.The results were promising; some of them outyielded the best existing hybrids by 10%−20%. Several examples are given in Table 2.

Table 2　Yield of intervariental rice hybrids produced by the two-line method in large demonstration fields, 1991

Location	Season	Hybrid	Area /ha	Yield / (t/ha)	Percent of check
Tian−Dong County Guangxi Province	First crop	KS-9/03	470	9.5	+21
Sha County Fujian Province	Middle crop	W6111 S/Vary Lava	20	9.8	+18
Hunan Province	Second crop	W6154 S/Teqing	270	9.4	+14
Hunan Province	Second crop	Pei−Ai 64 S/Teqing	10	9.2	+12

The highest yield of W6154S/Teqing in an experimental plot was 15.7 t/ha during a single crop season in Yunnan Province and 11.3 t/ha during the second crop season in Hunan Province. This hybrid variety has created new record yields in rice.

Intervarietal two-line system hybrids will be released to farmers in 1993; area to be planted will be 2 million ha in 1995.

Intersubspecific F_1 hybrids

Our studies indicated that the degree of heterosis in different kinds of hybrid rice varieties has the following general trend: indica/japonica>indica/javanica>japonica/javanica>indica/indica>japonica/ japonica (Fig. 2, 3). The first three kinds are intersubspecific hybrids, the latter two are intervarietal.

There are fewer fertility problems in japonica/javanica hybrids, and the grain quality of most javanica varieties is similar to that of japonica varieties in many aspects. It is therefore reasonable to assume that heterosis of japonica/javanica hybrids would be an effective approach for increasing japonica yields while retaining japonica-type grain quality. The development of indica/javanica hybrids may also improve grain quality as well as yield.

Indica/japonica hybrids possess the highest yield potential in both sink and source. Their theoretical yield may be 30% more than the existing highest yield of intervarietal hybrid varieties. Exploiting the strong heterosis in indica/japonica hybrids has been the major goal of our two-line system hybrid breeding program. However, in order to achieve it, four barriers normally found in such F_1 hybrids must be overcome: low seed setting rate, plants too tall, very long growth duration, and many poorly filled grains. So far, encouraging progress has been made.

168

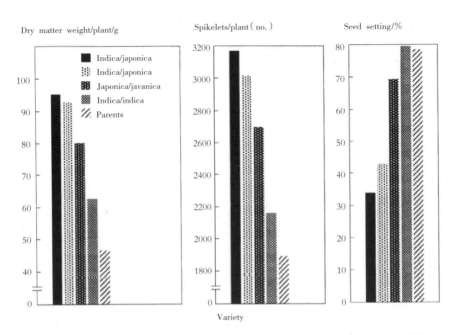

Fig.2   Heterosis in different F1 rice hybrids. Hunan Hybrid Rice Research Center, Changsha, 1988.

By using wide compatibility (WC) genes, the low seed-setting rate caused by semisterility due to incompatibility between indica and japonica can be raised to nearly normal levels (Tables 3 and 4). A large number of japonica lines and several indica TGMS lines possessing WC genes have been developed.

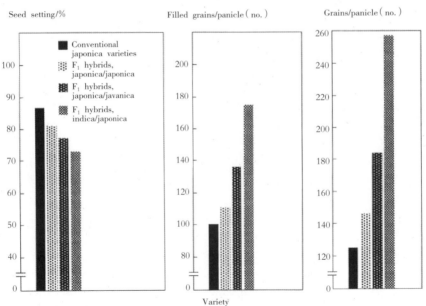

Fig.3   Comparison of economic characters in F₁ hybrids derived from types of indica and *japonica* varieties. Shenyang, 1990.

Table 3　The wide compatibility test of Pei-Ai 64 S（an indica TGMS line with WC gene）, Changsha, China, 1989.

Female parent	Tester		F$_1$ seed setting/%
	Nagjing 11	（Indica）	73.2
	IR36	（Indica）	81.1
	Xiang-Zao-Xian 1	（Indica）	76.4
	Milyang 46	（Indica）	75.0
	Akihikari	（Japonica）	78.0
	Banila	（Japonica）	69.8
Pei-Ai 64 S	Cheng-Te 232	（Japonica）	68.7
	Nong-Hu 26	（Japonica）	66.0
	Pei-Ti	（Japonica）	70.5
	Lun-Hui 422	（Japonica）	74.8
	CP SLO	（Japonica）	71.1
	CP 231	（Japonica）	70.0

Table 4　The wide compatibility test of Linggul 66（a japonica line with WC gene）, Hainan Island, 1992.

Male parent（japonica）	Tester（indica）	F$_1$ seed setting/%
	Pei-Ai 64 S	95.53
	1356 S	88.29
	8902 S	85.23
Linggui 66	An-nong S	83.82
	645 S	83.15
	8526 S	65.28
	735 S	50.10
Shan You 55（check）		91.93

Transferring an allelic dwarf gene（*Sd1*）into male as well as female parents can lower the plant height of indica/japonica hybrids to a semidwarf level, yet the hybrids still express very strong heterosis.

By crossing parental lines of different growth duration,（except for photosensitive late varieties）, indica/japonica hybrids with medium and even shorter growth duration can be obtained.

Efforts are now focused on solving the last problem-i.e., the poor plumpness of a number of fertilized grains. It is expected that new breeding strategies will overcome this barrier within 2-3 years.

作者：Yuan Longping

注：本文发表于 Hybrid Rice Technology：New Developments and Future prospects 1994 年。

水稻光、温敏不育系的提纯和原种生产

　　导致雄性不育的起点温度必须相对地低是选育实用的水稻光、温敏不育系最重要的技术指标。根据这一标准，我国近几年来已成功地选育出一批达到实用要求的水稻光、温敏不育系，正在或即将投入大面积生产应用。

　　但是，水稻光、温敏不育系的育性具有随温度变化而波动的特点，由于变异，同一不育系不同个体之间存在着差异，因此，在不育系的繁殖过程中，若按一般的常规良种繁育程序和方法选种、留种，不育系的不育起点温度不可避免地会逐代升高，最终将导致该不育系因起点温度过高而失去实用价值。这是因为繁殖过程中，不育起点温度较高的个体，其可育的温度范围较广，在温度经常变化且有时幅度很大的自然条件下，结实率一般较高，因而它们在群体中的比例必然逐代加大，出现遗传漂移现象，繁殖2~3代后，达标的不育系就会降级为不合格的不育系（见表1）。

　　例如，培矮64S的不育起点温度，在1991年通过省级鉴定时为23.3℃，1993年已上升到24℃以上，有的株系高达25℃以上。

　　为了防止光、温敏不育系在繁殖过程中产生高温敏个体的比例逐年增加的遗传漂移现象，特提出下述的水稻光、温敏不育系的提纯方法和原种生产程序：

　　单株选择→低温或长日低温处理→再生留种（核心种子）→原原种→原种→制种

　　我们的具体做法是，根据植株形态，选择若干典型单株，在敏感期内进行为期4~6天的长日低温处理（14小时光照，日均温24℃，温度变幅为19~27℃）。抽穗时镜检花粉育性，凡花粉不育度在99.5%以下的单株一律淘汰，当选的植株立即刈割再生，使再生株在短日低温条件下恢复育性，所结的自交种子就是核心种子。核心种子在严格的条件下，繁殖原原种，然后再繁殖出原种，供制种用。

表1 水稻光、温敏不育系不育起点温度的遗传漂移

项目	个体数	占群体的百分比	个体数	占群体的百分比	个体数	占群体的百分比	个体数	占群体的百分比
低温敏株[*]	1	33.3%	100	16.7%	10 000	7.1%	1 000 000	2.8%
中温敏株[*]	1	33.3%	200	33.3%	40 000	28.6%	8 000 000	22.2%
高温敏株[*]	1	33.3%	300	50.5%	90 000	64.3%	27 000 000	75.0%
群体总数	3		600		140 000		36 000 000	
世代	当代		第一代		第二代		第三代	

注：假定低、中、高温敏株每株分别结 100、200、300 粒种子。

这种提纯方法和原种生产程序，不仅能保证光、温敏不育系的不育起点温度始终保持在同一个水平上，而且简便易行，生产核心种子的工作量很小。按一株不育系平均结 200 粒自交种子计算，一株再生稻可生产出供 100 亩制种田的原种。因此，建议这一程序可作为水稻、光温敏不育系提纯和繁殖的新体系，在生产上推广应用。

作者：袁隆平

注：本文发表于《杂交水稻》1994 年第 6 期。

两系法杂交水稻研究I. 1991—1995年研究概况

在"七五"攻关的基础上，本中心"八五"期间在两系杂交水稻育种以及应用基础研究方面取得了长足的进展，育成的优良组合正在投入大面积生产应用。本文就过去五年所做的工作简结如下。

1 基础研究

1.1 总结和提炼出选育实用光温敏不育系的基本指标

通过"八五"期间的育种实践以及对光温敏核不育系育性转换规律认识的不断深化，在"八五"末期提出了较为全面的选育地域适应性广、风险极小的光温敏核不育系的四项基本指标，即①起点温度低（在长江流域应在23℃以下）；②光敏温度范围宽（宜在23℃~29℃之间）；③临界光长短（短于13小时）；④长光对低温和短光对高温的补偿作用强，从而为选育实用的水稻光温敏核不育系确立了基本指标和指明了方向。

1.2 对广亲和系进行了分类

通过5年的研究，在育种应用上对广亲和性有了进一步的认识。现有广亲和系可以分为四种类型：①广谱广亲和系；②部分广亲和系；③弱广亲和系；④非广亲和系。育种实践证明，这种分类对选育过硬的广亲和系和选配结实率正常的亚种间组合具有十分重要的指导作用。

1.3 提出了以籼爪交和粳爪交为主的水稻亚种间杂种优势利用的育种策略

由于典型的籼粳亚种间杂交稻存在着许多不协调的因素，如籽粒充实度不好；米质不籼不粳，不能适应人民消费的需要等。我们在1994年的海南年会和扬州会议上提出了主要利用带有爪哇稻血缘的品系进行水稻亚种间杂种优势利用的育种策略，即南方籼稻区以籼爪交为主，兼顾籼粳交；粳稻区以粳爪交为主，兼顾籼粳交，并以部分利用而不直接利用籼粳杂种优势为上策。这为顺利利用水稻亚种间杂种优势起了导航作用。

1.4 发明冷水串灌法，解决了低温敏核不育系繁殖难以高产的问题

1991 年，本中心罗孝和发现水稻光温敏核不育系对温度敏感的部位是水稻植株基部，并进一步发明了 18 ℃～22 ℃的冷水串灌法能使低温敏不育系的育性恢复正常和繁殖高产。该技术在湖南、广东等省大面积生产中都得到了成功的应用，1995 年湖南 250 亩 "培矮 64S" 的繁殖亩产量已达 220 kg。

1.5 提出了光温敏核不育系的核心种子和原种生产程序

从培矮 64S 等的繁殖实践中发现，如果按照常规的方法繁殖水稻光温敏不育系，将会发生不育起点温度逐渐升高的遗传漂移现象，繁殖 2～3 代后，原来达标的不育系就会降级为不合格的不育系。因此，我们在 1994 年扬州会议上提出以生产核心种子为中心环节的水稻光温敏核不育系的提纯方法和原种生产程序，即单株选择→低温或长日低温处理→再生留种（核心种子）→原原种→原种→制种，从而解决了水稻两用核不育系起点温度逐步提高的遗传漂移现象，为两系杂交水稻原种生产体系的建立奠定了科学基础。

2 育种实践

2.1 育成 4 个通过省级鉴定的实用水稻光温敏不育系

"八五" 期间育成了培矮 64S、安湘 S、香 125S 和轮回 422S 4 个通过省级以上鉴定的两用核不育系，它们各自的主要特点如下。

培矮 64S：爪籼型，具有广谱广亲和性，配合力好，湖南、广东、辽宁、陕西等省的育种实践表明，易于配制出强优两系杂交水稻组合。

安湘 S：籼型，具有弱广亲和性，异交结实率高，易于制种高产，米质优。

香 125S：籼爪中间型，具有部分广亲和性，熟期早，再生能力强，起点温度较低，米质好，有香味。

轮回 422S：爪粳型，具有广亲和性，起点温度较低。

2.2 广亲和恢复系的选育

在 "八五" 期间，本中心共选育出 5 个广亲和系，即广谱粳型广亲和系零轮和 CB-1，部分广亲和系零培、双光（爪哇型）、510（粳爪型）等，并用这些广亲和恢复系测配出了如测 64S/零轮和安湘 S/510 等一系列强优苗头两系杂交水稻亚种组合。

2.3 两系亚种间组合选育

2.3.1 籼爪交组合：利用本中心的培矮 64S 选配出一系列强优势组合。如本中心选育的 "培矮 64S/特青"，其产量明显超过对照汕优 63，为全国第一个通过省级审定的两系杂交水稻组合，在湖南连续四年多点的大面积试种和示范中均比对照增产显著。另一个苗头组合是

安湘 S/510，其穗粒优势特强，且结实率和籽粒充实度良好，可望在产量上有重大突破，计划明年在湖南省以一定的规模进行中试示范。

广东茂名市和陕西汉中地区用"培矮 64S"作为母本选配的强优两系杂交水稻组合"培矮 64S/山青 11 号"和"培矮 64S/特三矮"，连续 2～3 年表现丰产，并通过了地、市级的品种审定。

2.3.3　爪粳交组合：辽宁省稻作所用本中心的"培矮 64S"和"轮回 422S"所选配的"培矮 64S/C418""N422S/418"连续两年在江苏连云港地区表现比对照汕优 63 增产显著。

2.3.3　籼粳交组合：用低温敏籼型水稻两用核不育系"测 64S"所选配的籼粳交组合"测 64S/零轮"具有较强的产量优势。1995 年利用籼型两用核不育系"安湘 S"和粳型广亲和系"株选"系列选配出一系列强优早熟两系杂交早稻组合，不仅早熟和米质优，而且产量比同熟期的对照增加 15%～20%，极有希望在长江流域作为早中熟杂交早稻进行大面积推广。

2.4　两系杂交水稻制种技术已经成熟

经过近两年的攻关，本中心的两系杂交水稻制种产量大大提高，特别是培矮 64S 系列组合制种产量低的"瓶颈"已被突破。1995 年湖南制种培矮 64S 系列组合 2 600 亩，平均亩产达 160 kg，高产丘块在 350 kg 以上。三系杂交水稻"不育系繁殖∶制种田∶大田栽培"的面积比例约 1∶50∶（6 000～7 000），由于两系杂交水稻不育系繁殖和制种产量提高，使这个比例上升为 1∶100∶（15 000～20 000），因此两系杂交水稻比三系杂交水稻可节约 50% 的繁殖田和 20%～30% 的制种田。

2.5　成果中试开发

在"八五"期间共繁殖培矮 64S 等水稻光温敏不育系近 500 亩（1995 年为 300 亩），两系杂交水稻制种面积共近 6 000 亩（1995 年为 3 000 亩），湖南省两系杂交水稻示范面积为 110 万亩（其中 1994 年为 70 万亩，1995 年为 40 万亩），本中心育成组合的累计示范面积为 30 万亩。目前两系杂交水稻技术已经成熟配套，开始进入大面积推广应用阶段。

"八五"期间，本中心在两系杂交稻研究方面虽取得了很大的成绩和进展，但今后仍任重道远，我们初步规划的"九五"重点任务是：①育成比现在常规早稻增产 20% 左右的早、中熟早稻组合；②育成比三系杂交稻增产 15% 的亚种间组合；③选育特级优质米组合，其产量与三系杂交稻持平；④超高产制种技术和超高产栽培技术的研究；⑤化学复雄的研究。

作者：袁隆平　李继明

注：本文发表于《湖南农业科学》1995 年第 6 期。

选育水稻亚种间杂交组合的策略

选育高产亚种间杂交稻，从根本上说，就是要应用育种艺术，克服各种障碍，将水稻亚种间强大的生物杂种优势协调地转化为经济产量优势，特别是要把解决杂种结实率低而不稳和籽粒充实度不良的问题作为主攻对象。下面论述的八项选育原则，是在已经具有广谱亲和系的基础上提出的，这是我们近 10 年来在这方面的研究工作体会。谬误难免，敬请指教。

1 矮中求高

利用了等位矮秆基因，亚种间杂交稻植株过高的问题已获解决。反过来又要求在不倒伏的前提下，适当增加株高，借以提高生物学产量，使之具有充足的源，为高产奠定基础。

2 远中求近

以部分利用亚种间的杂种优势选配亚亚种组合为上策，克服纯亚种间杂交因遗传差异过大所产生的生理障碍和不利性状。

3 显超兼顾

既注意利用双亲优良性状的显性互补作用，又特别重视保持双亲有较大的遗传距离，避免亲缘重叠，以发挥超显性作用。

4 穗求中大

以选育每穗颖花数 180 粒左右，每公顷 300 万穗左右的中大穗型组合为主，不片面追求大穗和特大穗，以利协调库源关系，使之有较高结实率和较好的籽粒充实度。同时，在提高穗粒数方面，是以增加穗长和一次枝梗数为主，追求过大的着粒密度，不利于灌浆和籽粒的充实。

5 高粒叶比

粒叶比值是衡量一个品种光合效率的重要指标。通过测定，选择粒叶比值高的组合，把凭经验的形态选择与能定性和定量的生理功能选择结合起来，使选择技术建立在更科学的基础上，从而能大大提高选择的准确性和效果。

6　以饱攻饱

根据观察和经验，杂种一代籽粒的饱满度与亲本这方面的性状密切相关，因此选择籽粒充实良好和特好的品种、品系作亲本，是解决亚种间杂交稻籽粒充实不良的途径之一。另一方面，选用千粒重不大但容重大的作亲本，也是一条有效途径。

7　爪中求质

选用爪籼中间型的长粒种优质材料，与籼稻配组，米质优良且倾籼型；选用爪哇型或爪粳中间型的短粒型材料，与粳稻配组，米质优良且倾粳型。

8　生态适应

籼稻区以籼爪交为主，兼顾籼粳交；粳稻区以粳爪交为主，兼顾籼粳交。

作者：袁隆平

注：本文发表于《杂交水稻》1996 年第 2 期。

从育种角度展望我国水稻的增产潜力

【摘　要】农业生产和作物育种的历史表明，凡在育种上有所突破，就会给农业生产带来一次飞跃。如杂交玉米和高粱、矮秆水稻和小麦的育成和应用，都在世界范围内大幅度地提高了这些作物的产量。

我国的水稻育种，已有两次突破，第一次是 20 世纪 60 年代初矮化育种的成功，把水稻产量提高了 20%～30%；第二次是 70 年代中期杂交水稻的研究成功，水稻产量又在矮秆良种的基础上增长了 20% 左右。

当前，我国的水稻育种正在酝酿着第三次突破，即利用水稻亚种间杂种优势的研究已取得重大进展，预计在近期内，具有更强大杂种优势的两系法亚种间杂交水稻就能培育成功，其产量潜力要比现行的杂交水稻高 15%～20%。

水稻育种更高层次的发展是通过生物技术利用远缘杂种优势。1995 年我们在野生稻中发现了两个重要基因位点，每一基因位点可对杂交稻再贡献 20% 的产量，现在正开始进入这方面的研究，将常规育种手段与分子育种技术结合起来，利用水稻的远缘杂种优势，预期在下世纪初将会取得重大突破。

【关键词】水稻；育种；增产潜力

我国现有人口 12 亿，人均耕地面积不足 0.093 hm^2，预计 21 世纪 30 年代，我国人口将增加到 16 亿，人均耕地将不足 0.067 hm^2。未来靠谁养活中国？这是人人关心的头等大事。对这个问题，美国经济学家莱斯特·布朗散布了悲观的论调，认为中国今后粮食不能自给，需要大量进口，从而会引起全球性的粮食短缺和粮价暴涨。布朗的论证虽然有一定的根据，但是在某些重要地方则很片面，其中最主要的一点就是低估或轻视了科技进步对提高生产力的巨大潜力。

我是从事水稻育种研究的，水稻蕴藏着巨大的产量潜力，进一步提高它的产量具有广阔发展前途。这里只想从育种科学的角度，回顾一下我国水稻产量的增长情况并展望今后我国水稻的增产前景。

农业生产和作物育种的历史表明，凡在育种上有所突破，就会给农业生产带来一次飞跃。如杂交玉米和高粱、矮秆水稻和小麦的育成和应用，都在世界范围内大幅度地提高了这些作物的产量。美国的玉米产量，在 20 世纪 30 年代全国平均每公顷只有 2.25 t，由于杂交玉米的推广应用和不断改良以及其他农业技术和条件的改进，到 80 年代，平均产量超过了 7.50 t/hm²，45 年内把玉米产量提高了 2 倍多。在诸多的增产因素中，品种改良的贡献最大，接近 60%。

我国水稻的平均单产在 1950 年只有 2.21 t/hm²，到 1995 年达到 5.93 t/hm²，居世界第 7 位；45 年内把水稻单产提高了 1.8 倍，稻谷总产由 1950 年的 5 687 万 t 增加到 1995 年的 18 250 万 t。水稻产量的增长速率大大超过了人口增长的速率。从技术上讲，提高水稻产量的主要因素是扩大灌溉面积、增施肥料、改进栽培技术和选用优良品种，其中，优良品种的贡献占 1/3 以上。

我国的水稻育种，已有两次突破，第一次是 60 年代初矮化育种的成功，把水稻产量提高了 20%～30%；第二次是 70 年代中期杂交水稻的研究成功，水稻产量又在矮秆良种的基础上增长了 20% 左右。在这两个方面，我国都处于世界领先地位。水稻品种的产量潜力，可用每公顷的日产量来计算，过去的高秆品种，草多谷少，收获指数只有 30% 多（即 1/3 为稻谷，2/3 为稻草），其产量潜力每日仅 45 kg/hm² 左右。中熟品种（全生育期 125 d）的产量很难达到 6.00 t/hm²；现代矮秆良种，不仅耐肥抗倒，而且收获指数提高到 50% 左右，产量潜力每日为 60.0～67.5 kg/hm²，中熟品种的产量，每公顷可超过 7.50 t/hm²。以矮秆良种为亲本的杂交水稻，不仅保持了收获指数高的优点，而且由于有杂种优势，其生物学产量也有很大的增长。因此，杂交水稻的产量潜力每日为 75 kg/hm²，中熟组合每公顷可超过 9.00 t。

近几年，我国杂交水稻的年种植面积为 1 533 万～1 600 万 hm²，占水稻总面积的 50% 左右，产量则占稻谷总产的 57%；单产 6.60 t/hm²，常规矮秆良种为 5.25 t/hm²。从 1976 年到 1995 年，杂交水稻累计种植面积 1.87 亿 hm²，增产粮食 2.8 亿 t。由此可见，进一步发展杂交水稻对解决我国的粮食问题具有十分重要的战略意义。

科学技术的发展无止境。当前，我国的水稻育种正在酝酿着第三次突破，即利用水稻亚种间杂种优势的研究已取得重大进展，预计在 1～2 年内，具有更强大杂种优势的两系法亚种间杂交水稻就能培育成功，其产量潜力要比目前生产上应用的杂交水稻高 15%～20%。

现行的杂交水稻属品种间杂种优势利用的范畴，由于品种间的亲缘关系较近，遗传差异相对较小，杂种优势有较大的局限，增产幅度一直在 20% 左右徘徊多年，很难再上一个台阶。为了进一步提高水稻的产量，从 20 世纪 80 年代中期起，我们就在探索和寻求新的优势利用途径，其中最有希望的首推利用水稻亚种间的杂种优势。亚种间杂交稻的库大源足，理论上的产量要比品种间杂交稻高 30% 左右，因此，利用水稻亚种间的杂种优势，一直是育种家多年梦寐以求的愿望。但是，要选育实用的亚种间杂交稻，在技术上存在许多难关，特别是结实率低而不稳定的问题很难解决，以致有些人对利用水稻亚种间杂种优势有望洋兴叹之感。

我国许多水稻科学家通过近 10 年的努力和协作研究，技术上的难题已基本解决，育出一批很有希望的亚种间苗头组合，在试验田进行对比，产量比品种间杂交稻高 20% 左右，其产量潜力每日为 82.5~90.0 kg/hm²，中熟组合每公顷可产 10.50~11.25 t。预计，亚种间杂交稻可在"九五"后期应用于生产，并将在下世纪初大面积生产中发挥巨大的增产作用。以年种植面积 1 333 万 hm²、每公顷增 1 125 kg 计算，每年能增加 150 亿 kg 粮食，相当于一个中等产粮省的全年总产。

水稻育种更高层次的发展是通过生物技术利用远缘杂种优势，1995 年我们用分子标记的方法，结合田间试验，在野生稻中发现了两个重要的 QTL 基因位点（RZ776 和 RG256），分别位于 1 号和 2 号染色体上，每一基因位点具有比现有高产杂交稻威优 64 增产 20% 的效应。目前正开始进入这方面的研究，将常规育种手段与分子育种技术结合起来，利用水稻的远缘杂种优势，预计在 21 世纪初将会取得又一次重大突破。

综上所述，通过育种科学技术的进步和运用，水稻的产量可跳跃式地不断登上新台阶。举一反三，水稻如此，其他粮食作物同样具有美好的发展前景。提高农作物产量在技术上的因素很多，而每一项技术进步，都能对增产起一定的作用。因此，我深信，随着科学技术特别是高、精、尖技术的向前发展，再加上国人的自我努力和追求，中国能依靠自己解决吃饭问题。

作者：袁隆平

注：本文发表于《杂交水稻》1996 年第 4 期。

野生稻中拥有可以显著提高水稻产量的基因

水稻（*Oriza sativa L.*）是当今世界上 57 亿人口之中一半以上人口的口粮作物，全世界都有栽培。据预测，世界人口将以每年增加九千万的速度递增，到 2030 年，世界水稻年产量必须增长 70% 才能满足人口增长的需要。过去，提高水稻产量大都是通过扩大种植面积、增加肥料投入、应用化学药剂控制病虫害等措施来实现的。但是，这些措施今天已不再起多大作用。未来提高水稻产量必须通过从遗传上提高水稻的产量潜力来实观。

尽管全世界都投入了相当大的力量进行水稻育种，但现代水稻品种的产量潜力在过去几十年中一直停滞不前。国际水稻所最近育成的水稻品种的产量潜力只与 1966 年首次育成的 IR8 相当。在中国这个世界最大的水稻生产国，现代水稻品种的产量潜力自 20 世纪 70 年代中期开始推广杂交水稻以来也没有显著提高（袁隆平，未发表）。

与大多数作物相似，水稻由人类从当地植物的野生群体中驯化而来。在这一驯化过程中，人类仅获得了存在于自然界中的一小部分遗传变异。正是这一小部分遗传变异构成了当今所有栽培品种的遗传基础。DNA 多态性研究结果表明，稻属中大多数有利的遗传变异仍存在于尚未挖掘的野生近缘种中。

尽管从农业生产的观点来看，作物野生近缘种的产量远远低于现代栽培品种，但近来有证据显示，野生近缘种中可能拥有既可以提高作物产量又可以改良作物品质的基因。然而，这些有利基因常常为不良基因的效应所掩盖，如果不利用分子作图技术，则很难检测出来。近来，Tonksley 等提出了应用分子遗传图谱挖掘野生种的遗传潜力以改良现代作物品种产量和品质的技术策略。为了检验这一策略，我们试图从栽培稻的一个杂草型野生近缘种，普通野生稻（*Oryza rufipogon*）中筛选能够提高中国高产组合产量的基因。普通野生稻与其他野生近缘种一样，在中国或其他国家以前都没有用于水稻产量的遗传改良。

　　威优 64 由 V20A 和测 64 配组而成，是在中国种植表现非常突出的 F_1 代杂交种。用来源于马来西亚的普通野生稻（IRGC 105491）作父本与细胞质雄性不育系 V20A 杂交，产生的种间杂种 V20A/ 野生稻表现出非常强的营养优势。用这个 F_1 与 V20B（V20A 的保持系，核基因组与 V20A 相同）杂交，产生 52 个 BC_1 单株，1993 年夏把它们种植在中国长沙。根据植株形态，生育期以及小穗育性，选择 10 株最好的 BC_1 单株再与 V20B 回交，形成了 3 000 多株 BC_2 群体。从该群体中再选择 300 株与测 64 杂交，构成 300 个 BC_2 测交 F_1 家系。除了很小部分 DNA 片断为普通野生稻的 DNA 所替代外，每个家系都应该基本具有威优 64 的遗传组成。平均来讲，每个 BC_2 测交 F_1 家系含有 5% 的普通野生稻 DNA。1994 年夏这些 BC_2 测交 F_1 家系，连同威优 64、V20B 以及普通野生稻种植在中国国家杂交水稻工程技术研究中心。完全随机区组设计，二次重复，每小区 3 行，每行 11 株。收获时按小区测产并取样考种。

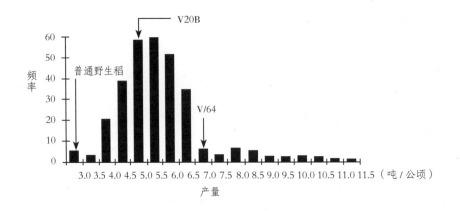

图 1　种间杂种（*O.sativa/O.rufipogon*）BC_2 测交 F_1 群体的产量分布
注：箭头所示为普通野生稻、V20B 和威优 64（V20A/ 测 64）的产量。

　　图 1 表明了 BC_2 测交 F_1 家系及亲本对照的产量分布。普通野生稻表现最差，比对照威优 64 产量低 55%，单株粒数少，千粒重低。这是意料之中的事，因为普通野生稻不是栽培种。大多数 BC_2 测交 F_1 家系的产量也低于威优 64，但一小部分小区的产量较高，比威优 64 高 50%。整体来讲，有 15% 的 BC_2 测交 F_1 家系产量高于威优 64；有 14% 的 BC_2 测交家系单株粒数高于威优 64；有 56% 的 BC_2 测交 F_1 家系千粒重高于威优 64。13 个（4.3%）BC_2 测交家系的产量至少比威优 64 高 30%。这些结果表明，尽管普通野生稻本身在产量及产量

性状方面远远不及栽培水稻，但是有些普通野生稻基因可以提高优良水稻品种的产量。

如果普通野生稻的基因确实可以提高某些 BC_2 测交 F_1 家系的产量，那么就应该可以检测到携带这些基因（常称为数量性状位点或 QTL）的野生稻染色体片段的存在，而且这些基因与产量之间应该存在显著相关。高密度水稻分子遗传图谱已用来对决定杂种优势、生物或非生物胁迫的耐性和各种农艺性状的 QTL 位点进行遗传作图和特征标识。为了寻找普通野生稻基因组中可以提高产量的区域，我们选择覆盖整个水稻基因组，间隔约为 12 cM（厘摩）的 100 个信息性很强的 RFLP 和 20 个微卫星 DNA 标记对 300 个 BC_2 测交 F_1 家系进行了分析（每个家系取 20 株混合进行 DNA 分析）。

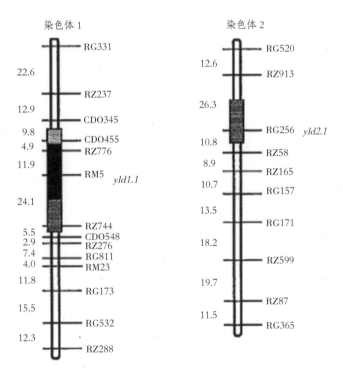

普通野生稻中与增产密切相关的等位基因位点
（P 分别为：黑区 $P < 0.005$，灰区 $P < 0.050$，白区 $P < 0.050$）

图 2　来自野生稻的增产基因 *yld1.1* 和 *yld2.1* 在分子图谱上的位置

QTL 作图根据 BC_2 测交 F_1 家系数据进行，作图时采用标准的方差分析（ANOVA）程序将 F_1 家系的田间表现对标记基因型进行回归分析，并假定测交 F_1 家系内的野生和栽培稻的等位基因位点正常分离。大多数情况下，普通野生稻基因位点的渗入没有增产效应或具有减产效应。但是，普通野生稻中处在第一染色体标记位点 RM5 和第二染色体标记位点 RG256 上的等位基因位点具有增产效益（$P<0.006$），见表 1。用 yld1.1 和 yld2.1 来命名这两个 QTL 位点。携带这两个普通野生稻 QTL 位点的测交 F_1 家系分别具有每公顷增产 1.2 t 和 1.1 t 的表型效应，相当于比对照威优 64 增产 18% 和 17%（表 1，其中实验数据转换成田间产量进行计算，种植密度和一般大田种植密度相同）。yld1.1 和 yld2.1 都能非常明显地增加单株粒数（$P<0.005$），但对千粒重，株高和生育期没有多大影响。就产量和所测定的其他性状而言，没有发现 yld1.1 和 yld2.1 之间彼此存在上位性效应。现正在建立近等基因系来分离这两个增产基因以进一步确定这两个基因的特性。

表 1　显著杂交水稻产量的野生稻 QTL 等位基因的特征

QTL	标记	染色体	P	"V/C"		（1/2 "V/C" +1/2 "R/C"）		"R/C"		基因效应	比 V64 增产的百分比 %
				PM	N	PM	N	PM			
yld1.1	RM5	1	0.003 6	5.77	251	6.38	45	6.99		1.22	18.26
yld2.1	RG256	2	0.005 8	5.76	247	6.33	49	6.90		1 014	17.07

注：V/C 和 R/C 是单个标记的表型类型：V/C 是 V20A/ 测 64 杂合体；R/C 是 BC$_2$ 测交群体中的种间杂合体 O.rufipogon/ 测 64；PM 为表型平均数；R/C 的 PM 值是在假定家系内基因按正常比例分离时，按下式计算，PM（R/C）=2×PM（1/2 "V/C" +1/2 "R/C"）–PM（V/C），P 为该标记与产量无关的概率。N 代表具有该基因型的家系个数。

栽培作物的许多祖先仍然存在于本地野生群体中或被收集保存在种质库里。尽管这些野生种被视为独一无二的遗传变异资源，但它们的产量很低，一般不认为它们对于提高作物产量有什么用处。本文的研究结果表明，尽管普通野生稻（栽培水稻的近缘野生种之一）本身表现差，但是它携带有大幅度提高水稻产量的基因。这一发现显示，地球上的野生或外来的种质可能在将来提高水稻产量潜力方面起到关键作用。本文所采用的研究策略不仅正在用于在其他水稻野生种中寻找提高产量的基因。而且可以供其他作物的遗传改良参考。

作者：肖金华　Silvana Grandillo　Sang Nan Ahn
Susan R.Mclouch　Stever D.Tanksley　李继明　袁隆平

注：本文是 1996 年发表在 Nature 第 384 卷上 "Genes from wild rice improve yield" 的译文。

我国两系法杂交水稻研究的形势、任务和发展前景

【摘　要】从育种上讲，杂交水稻育种可以分为三系法、两系法和一系法三个发展阶段；从利用杂种优势水平来看，杂交水稻育种可以分为品种间、亚种间和远缘杂种优势利用三个发展阶段。与三系法比较，两系法具有配组自由、种子生产成本低、无不育细胞质负效应、易转育新不育系等优点。迄今为止，我国已育成 20 多个通过省级鉴定的实用光温敏核不育系，7 个通过省级审定的两系法杂交稻组合，解决了低温敏不育系繁殖产量低而不稳的难题，实施了光温敏不育系提纯和原种生产的程序，两系法杂交水稻技术基本成熟配套，先锋组合在大面积示范中表现良好。预计到本世纪末，全国可推广 333.33 万 hm^2 两系杂交稻，增产粮食 25 亿 kg。

【关键词】杂交水稻；两系法；光温敏核不育系

1　杂交水稻育种的战略

农业生产和作物育种的历史表明，凡在育种上有所突破就会给农业生产带来一次飞跃。如杂交玉米和杂交水稻，矮秆水稻和矮秆小麦等的育成和应用，都大幅度地提高了这些作物的产量。如果没有新的突破，作物的产量就很难再上新台阶。

杂交水稻育种的战略，从育种方法上说，可分为三系法、两系法和一系法三个发展阶段，朝着程序由繁到简而效率越来越高的方向发展。从提高杂种优势水平上看，可分为品种间、亚种间和远缘杂种优势利用三个发展阶段，朝着优势越来越强的方向发展。每进入一个新阶段都是育种上的一次突破，从而会把水稻的产量推向一个更高的水平。

2　三系法杂交水稻的现状与面临的挑战

现行的杂交水稻属三系法品种间杂种优势利用的范畴，虽方兴未艾，并且在本世纪末以前仍将在生产上起主导作用，但是却面临如下挑战：

（1）单产多年徘徊不前。从 1986 年到 1995 年的 10 年间，全国杂交稻的平均亩产一直为 440 kg 左右。

（2）缺乏强优早熟组合，特别是适合在长江流域作双季早稻的早、中熟组合。

（3）杂交粳稻的优势不够强、增产幅度不大，同时现行的 BT 型不育系的育性不够稳定，难以保证杂交种子的纯度。因此，杂交粳稻的种植面积近年来每况愈下，仅 6.67 万 hm² 左右。

（4）不育细胞质单一。当前生产上应用的不育系，85% 以上仍然属于野败型，而且大多数新质源不育系的恢、保关系与野败型雷同。

3　两系法的优越性和问题

以光温敏不育系为遗传工具选育两系法杂交水稻的优点是：

（1）不受恢保关系制约，配组自由度大，选到优良组合的概率高。据不完全统计，在同一亚种范围内，90% 以上的现有品种品系能使大多数光温敏不育系的育性恢复正常，而能恢复三系不育系的则不到 5%。

（2）不需保持系，生产程序简化，种子成本降低，特别是不育系繁殖的产量大大高于三系不育系。

（3）不育性与细胞质无关，可避免不育细胞质的负效应，解决了母本细胞质单一的问题。

（4）不育性受一两对隐性基因控制，较易转育。一般通过一次杂交或两三次回交就可选择到基本上符合要求的不育系。

但是，光温敏不育系的育性受光、温条件所左右，其稳定性不如三系不育系，因此，制种和繁殖都存在一定的风险。前几年，广西、福建、湖南在这方面都有过失败的教训。主要原因是当时对光、温敏不育系的育性转换规律尚不甚了解，前期研究所下的结论有很大的片面性，认为育性的表达只受光长所左右而与温度无关。

4　两系法杂交水稻育种研究的进展和成绩

（1）掌握了光温敏不育系育性转换与光、温变化关系的基本规律，从而提炼出选育实用光温敏不育系的四项指标，使制种风险率降到 1% 左右：

·导致不育的起点温度低。这是最重要、最关键的指标。具体的温度指标因地而异，在长江流域应定为 23 ℃（日均温，变幅为 19 ℃～27 ℃），偏北地区须更低，华南亚热带地区可稍高。

·光敏温度范围宽 [（22 ℃～23 ℃）～（29 ℃～30 ℃）]。即光敏不育系在此温度范围内，长日导致不育，短日导致可育。

·临界光长短。以 13 小时为宜，这样才有更广泛的应用范围。

·长日对低温、短日对高温的补偿作用强。即在长日条件下，遇上低于起点温度的低温仍表现不育或在短日条件下遇上高于临界高温时也表现可育。

（2）全国已育成 20 多个通过省级鉴定的实用光温敏不育系。

（3）育成 7 个通过省级审定的两系法杂交稻组合，这些组合在大面积示范中表现良好，其中尤以培矮 64S/ 特青这一先锋组合更加突出，在我国中南部主要稻区多年、多点的试种、示范面积超过 6.67 万 hm^2，处处成功，造造增产。与同熟期三系对照组合比较，增产幅度在 10% 左右。

（4）在敏感期采用冷水串灌法彻底解决了低温敏不育系繁殖产量低而不稳的难题。

（5）大面积制种产量已达三系杂交稻的平均水平。

（6）培矮 64S 系列组合的不育系繁殖、制种和大田生产面积比例为 1∶100∶15 000，高于三系法杂交稻。

（7）实施了以生产核心种子为关键环节的光温敏不育系的提纯和原种生产程序，从而能在一定程度上防止不育起点温度逐代升高的漂移现象，保证不育系的不育起点温度始终保持在同一水平上。

水稻光温敏不育系的提纯和原种生产程序为：

单株选择→低温或长日低温处理→再生留种（核心种子）→原原种→原种→制种。

（8）提出了育性转换受一两对主基因控制，而起点温度的高低为数量性状遗传的假说，不仅较圆满地解释了起点温度发生漂移的原因，而且为根除漂移现象指明了选育对策。

总之，两系法杂交水稻在技术上现已基本成熟配套，完全可以因地制宜地推向大面积生产应用。

5 "九五"期间的任务和目标

（1）选育符合上述 4 个指标的不育系，特别是要在低温、短日双重因子的作用下才能恢复可育的不育系。在生产中应用这种不育系，其风险率几乎为零，华中师范大学在这方面已获得很有希望的苗头。

（2）选育广谱广亲和系。

（3）选育长江流域的品种间早、中熟双季早稻组合，增产 10% 左右，米质 2 级以上。

（4）选育比同熟期三系品种间杂交稻增产 15% 以上，米质 2 级以上的亚种间组合。

（5）到 2000 年，全国推广两系法杂交水稻 333.33 万 hm^2，增产粮食 25 亿 kg。

6 前景展望

通过科研单位的协作努力和有关方面的积极配合，预计到本世纪末全国推广 333.33 万 hm^2 两系法杂交水稻、增产粮食 25 亿 kg 的指标是能够实现甚至会超额完成的。根据是：

（1）已经有一批可比三系杂交稻对照增产 10% 左右的过硬组合；

（2）有不少比三系杂交稻对照增产 15% 左右的苗头组合；

（3）总结出选育亚种间组合的八项原则。在这些原则指导下，亚种间组合选育取得了长足进展，优势更强、米质更好的组合将在近年内陆续育成和投入应用；

（4）国家杂交水稻工程技术研究中心与美国康奈尔大学合作，通过分子标记辅助选择技术，1995 年在野生稻（*Oriza rufipogon*）中发现两个重要数量基因位点，每个位点对三系杂交稻具有增产 18% 的效应。这项研究结果表明，利用远缘基因来提高水稻的产量，已取得突破性进展，预计到下世纪初可以培育出比亚种间优势更强的远缘杂交稻。

作者：袁隆平

注：本文发表于《农业现代化研究》1997 年第 18 卷第 1 期。

杂交水稻超高产育种

1 超高产水稻的概念

什么叫水稻超高产育种，迄今并没有一个统一的标准和严格的定义，因此各家各派提出的产量指标并不相同。

1980年，日本制定的水稻超高产育种计划，要求在15年内育成比原有品种增产50%的超高产品种，即到1995年要在每公顷原产$5.00 \sim 6.50$ t糙米的基础上提高到$7.50 \sim 9.75$ t（折合稻谷为$9.38 \sim 12.19$ t）[1-2]。

1989年，国际水稻研究所提出培育"超级稻"，后又改称"新株型"育种计划[3]，目标是到2005年育成单产潜力比现有纯系品种高$20\% \sim 25\%$的超级稻，即生育期为120 d的新株型超级稻，其产量潜力可达12 t/hm²。

1996年我国农业部立项的"中国超级稻"育种计划[4]，产量指标见表1。

表1 超级稻品种（组合）产量指标

单位: t/hm²

类型阶段	常规品种				杂交稻			增产幅度
	早籼	早中晚兼用籼	南方单季粳	北方粳	早籼	单季籼、粳	晚籼	
现有高产水平	6.75	7.50	7.50	8.25	7.50	8.25	7.50	0
1996—2000年	9.00	9.75	9.75	10.50	9.75	10.50	9.75	15%以上
2001—2005年	10.50	11.25	11.25	12.00	11.25	12.00	11.25	30%以上

注：连续2年在生态区内2个点，每点6.67 hm²面积上表现。

超高产水稻的指标，当然应随时代、生态地区和种植季别的不同而异，但笔者认为，在育种计划中应以单位面积的日产量而不用绝对产量作指标比较合理。这种指标不仅通用而且便于作统一的产量潜力比较，因为生育期的长短与产量的高低密切相关，对生育期相差悬殊的早熟品种和迟熟品种要求具有相同的或相差很小的绝对产量，显然是不科学的。

根据当前我国杂交水稻的产量情况、育种水平，特别是最新的突破性进展，笔者建议在"九五"期间超高产杂交水稻育种的指标是：每公顷每日的稻谷产量为 100 kg。这个指标与国际水稻研究所提出的相同，但是，依据现已获得的试验结果，笔者充满信心地预见，我国新近育成的两系法亚种间杂交组合会比国际水稻研究所至少提前 5 年在较大面积上（6.67 hm^2 和 66.67 hm^2 级）实现这一超高产指标。

2　超高产稻株的形态模式

优良的植株形态是超高产的骨架，自从 Donald 提出理想株形的概念[5]以来，国内外不少水稻育种家便围绕这一育种上的重要主题开展了研究，设想了各种超高产水稻的理想株型模式，如库西的少蘖、大穗模式[3]，黄耀祥的"半矮秆丛生快长超高产株型模式"[6]，杨守仁的"理想株形"和"巨型稻"[7]，周开达的"重穗型"[8]，等等。当然，这些模式是根据一定的理论和实践经验设计出来的，对水稻超高产育种都很有参考价值。但是，必须指出，设想能否成为现实，尚有待实践证明。

近年来，江苏农科院与本中心协作，用培矮 64S 作母本，通过大量测交筛选，从中选到几个具有超高产潜力的苗头组合，其中培矮 64S/E32，1997 年在南京、苏州、高邮 3 个点共试种 0.24 hm^2，平均实收产量 13.26 t/hm^2。培矮 64S/E32 属中熟中稻组合，全生育期 130 d 左右，因此可以说，该组合已在 0.07 hm^2 级水平上达到每公顷日产稻谷 100 kg 的超高产指标。

对该组合做了较详细的观察后，我们得到一些重要启发，悟出超高产杂交水稻在形态上最主要的特点是上部 3 片功能叶要长、直、窄、凹、厚（图 1）。修长而直挺的叶片，不仅叶面积较大，而且可两面受光和互不遮蔽；窄而略凹的叶片，所占的空间面积小但整叶的面积并不因窄而减小；较厚的叶片光合效率高且不易早衰。总之，具有这种形态结构的水稻品种，才能有最大的有效叶面积指数和光合功能，为超高产提供充足的源。

库大源足是高产的前提，可是，包括笔者在内的许多水稻育种工作者，在进行超高产育种设计时，重库轻源，片面追求一定的穗数、每穗粒数和千粒重，其结果往往是库大源不足，不能实现超高产。

下述 2 个组合是库很大且相等但产量相差悬殊的实例。

图1　培矮64S/E32植株形态

培矮64S/E32在江苏农科院试验田的产量结构是260万穗/hm²，260粒/穗，结实率88%，千粒重23.5 g，理论产量13.95 t/hm²，实际产量12.87 t/hm²（江苏农科院邹江石交流材料）。

29S/510在本中心试验田的产量结构是270万穗/hm²，236.7粒/穗，千粒重25 g，受精率90%，如果每个受精粒都能充实的话，理论产量可达14.25 t/hm²，而实际产量只有7.35 t/hm²。

根据典型株的测定，本中心试验田的培矮64S/E32，上部3叶平均每叶的面积为75 cm²、长度为53.2 cm，29S/510分别为41 cm²和39 cm。由于29S/510上部3片功能叶面积较小而薄，再加上倒二叶较披，所制造的光合产物不能装满大库，以致秕粒率高达35%以上，实际产量不高。由此可见，在进行超高产育种时，在扩库的同时，更要特别重视"开源"。就水稻育种的现状而论，增源是实现超高产的关键环节。

参照培矮64S/E32这个已具有超高产潜力组合的植株形态，针对长江中下游生态区的中熟中稻（生育期130 d左右），我们初步提出如下的超高产稻株形态模式，供育种家们参考和指正。

（1）株高100 cm左右，秆长70 cm左右，穗长25 cm左右。

（2）上部三叶的形态特点如下：①修长。剑叶50 cm左右，高出穗尖20 cm以上，倒二叶比剑叶长10%以上，并高过穗尖，三叶尖达到穗中部。②挺直。剑叶、倒二叶和倒三叶的角度分别为5°、10°和20°左右，且直立状态经久不倾斜，直到成熟。③窄凹。叶片向内微卷，表现较窄，但展开的宽度为 2 cm 左右。④较厚。培矮 64S/E32 上部 3 叶 100 cm² 的干重为 0.98 g，而产量为 8.25 t/hm² 的一般高产组合 312S/桂云粘为 0.73 g。

（3）株型。适度紧凑，分蘖力中等，灌浆后稻穗下垂，穗尖离地面 60 cm 左右，冠层只见挺立的稻叶而不见稻穗，即典型的"叶下禾"或"叶里藏金"稻。

（4）穗重和穗数。单穗重 5 g 左右，每公顷有效穗为 270 万穗左右。

（5）叶面积指数和叶粒比。以上部 3 叶为基础计算，叶面积指数 6.5 左右，叶面积和粒重之比为 100∶2.3 左右，即生产 2.3 g 稻谷上部 3 叶的面积要有 100 cm²。

（6）收获指数为 0.55 以上。

3 技术路线

根据杂交水稻育种的特点，原则上超高产育种要从两个方面着手：一是充分利用双亲优良性状的互补作用，在形态上作更臻完善的改良；二是适当扩大双亲的遗传差异，以进一步提高杂种优势的水平。二者密切结合，相辅相成。关于形态改良，已在前面述及，现就第二个问题扼要提出我们正在进行和即将实施的技术路线。

3.1 利用亚种间的杂种优势选育超高产组合

亚种间杂交稻比品种间杂交稻具有更强的杂种优势，因此，这是当前最现实的有效途径。在近期内以打"短平快"为主，即以培矮 64S 为重点，进行更广泛的测交筛选，从中选出超高产组合。培矮 64S 是一个株叶形优良，亲和谱较广，配合力良好的籼粳中间型光温敏不育系，用它不仅选配出一批通过省级审定、已在大面积生产上应用的高产两系法先锋组合，而且还选出如培矮 64S/E32 等几个超高产苗头组合。据此，笔者深信，只要对准前述的形态模式，不论用籼稻还是粳稻测交，选到超高产组合都有较大的可能性。

从长远着眼，为了更充分地发挥亚种间的杂种优势和提高超高产育种的效率，我们计划的重点是放在各种类型广亲和系的选育上，其中包括不同熟期的籼型、粳型和籼粳中间型的恢复系和不育系，特别是广谱的广亲和系，为选育各种熟期和适应不同生态地区的超高产组合打好基础。

3.2 利用野生稻的有利基因选育超高产组合

1995 年，我们与美国康奈尔大学合作，采用分子标记技术，结合田间试验，在野生稻（*O. rufipogon* L.）中发现两个重要的 QTL 基因位点，每一基因位点具有比日产潜力为 80 kg/hm² 的高产杂交稻威优 64 增产 18% 左右的效应[9]。通过分子标记辅助选择技术，正在选育携有该两个 QTL 位点的相应亲本的近等基因系。

3.3　利用新株型超级稻选育超高产组合

库西预言[3]，国际水稻研究所培育的新株型稻将比现有的高产纯系品种增产20%，进一步，新株型稻将用于选育籼粳亚种间杂交稻，其产量又可增加20%～25%，二者相结合，可把热带水稻的产量潜力提高50%。

1995年，在本中心试验田我们对国际水稻研究所初选的21个新株型品系作过观察，发现这些材料的优点是秆粗、穗大、分蘖少，但籽粒不充实，产量很低。它们同样存在着库大源不足的缺点。尽管如此，我们对该所的新株型育种计划仍寄托着很大的希望，因为目前这些品系还只是属于初选材料，缺点难免。一旦新株型水稻品种育成，将其应用于杂种优势育种上，水稻的产量潜力很可能会再跃上一个新台阶。

—————————— R e f e r e n c e s ——————————

参考文献

［1］佐藤尚雄.水稻超高产育种研究.国外农学·水稻，1984，（2）：1-16.

［2］金田忠吉.应用籼粳杂交培育超高产水稻品种.JARQ，1986，19（4）：235-240.

［3］Khush，G.S.，Prospects of and approaches to increasing the genetic yield potential of rice. In "Rice Research in Asia，Progress and Priorities"，edited by R.E. Evenson et. al. CAB International and IRRI，1996，59-71.

［4］中国农业部.中国超级稻育种——背景，现状和展望.见：中国农业部编.新世纪农业曙光计划项目.1996.

［5］Donald，C.M. The breeding of crop ideotypes.

Euphytica，1968，17:385-403.

［6］黄耀祥.水稻丛化育种.广东农业科学，1983，（1）：1-5.

［7］杨守仁，张步龙，王进民等.水稻理想株形育种的理论和方法初论.中国农业科学，1984，（3）：6-13.

［8］周开达，马玉清，刘太清等.杂交水稻亚种间重穗型组合的选育——杂交水稻超高产育种的理论与实践.四川农业大学学报，1995，13（4）：403-407.

［9］Xiao，J.，Grandillo，S.，Ahn，S.N. et al. Genes from wild rice improve yield. NATURE，1996，384:223-224.

作者：袁隆平

注：本文发表于《杂交水稻》1997年第12卷第6期。

农作物两系法杂种优势利用的现状与前景

一、三系法面临的挑战

杂种优势利用是本世纪作物育种上最引人注目的成就之一。20 世纪 30 年代美国玉米产量平均每公顷 2.5 吨左右。到 80 年代平均产量达到 7.5 吨 / 公顷，杂交玉米起了关键性的作用，其贡献占 60%。中国推广杂交水稻 20 年来，已为中国增产粮食 3 亿吨以上，单产约高出常规水稻 30%。据联合国粮农组织统计（1990）杂交水稻种植面积占世界水稻总种植面积的 10%，而产量却占水稻总产的 20%。

利用雄性不育性生产杂交种子是利用作物杂种优势最主要的途径，经典方法是利用核质互作型雄性不育系的三系法，即雄性不育系、雄性不育保持系和雄性不育恢复系。水稻、高粱杂交种几乎全部由这种方法生产，部分玉米杂交种也是采用这种方法生产的。

三系法虽然是利用农作物杂种优势行之有效的方法，并且在当前和今后一段较长的时期内仍将在生产上起主导作用，但是三系法存在某些内在的不利因素：

1. 受严格的恢保关系制约，以致配组不自由，选到优良组合的概率低。以杂交水稻为例，在国际水稻所选育的籼稻品系中，对"野败型"不育性具有完全保持能力的不到 1/1 000，恢复力好的品系低于 5%。

2. 不育细胞质较单一。如杂交水稻的野败型不育系占 85% 以上，细胞质单一化存在着受毁灭性病虫害侵袭的潜在危险，杂交玉米在这方面曾有惨重的教训。

3. 育种程序复杂，效率较低；生产环节多，推广速度慢，种子成本较高。

由于上述原因，杂交稻产量的进一步提高和种植面积的扩大受到很大程度的限制。例如，中国的三系杂交水稻的单位面积产量一直在 440 公斤 / 亩左右徘徊了十多年，种植面积近七八年来也在 50%~53% 之间徘徊（图 1）。

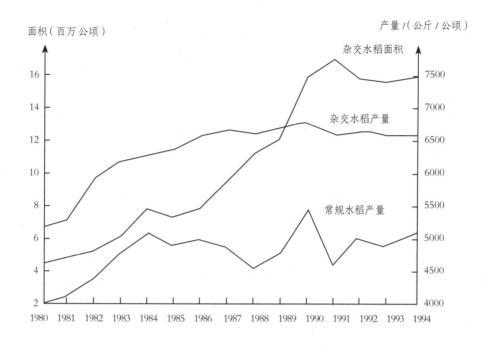

图1　中国杂交稻和常规稻的产量与面积

二、两系法及其优越性

从作物杂种优势利用的长远战略考虑，许多育种家都在探索新的技术路线，以期采用简易而效率更高的方法来取代三系法。从目前的发展水平和优越性来看，首推两系法的利用。所谓两系法是相对三系法而言的，即在育种和制种过程中省掉保持系。

广义的两系法包括：

1. 光温敏核不育系的利用。

2. 隐性核不育系的利用。一般用于繁殖系数较高，人工杂交有效的作物，并与标记性状结合应用，如棉花等作物。

3. 显性核不育系。可在产品为营养体的作物上应用。

4. 雌雄异熟系。如中国湖南利用棉花雌雄异熟系配制棉花杂交种。

5. 其他。如常异花授粉作物标记性状的利用。

目前研究最广泛、最有利用价值的是光温敏不育系的利用。光温敏不育系的育性由外界光温条件控制，具有可育与不育转换机制，解决了隐性核不育利用中不育系繁殖时要发生分离的问题。本文要讨论的基本上属于此法。

1973 年，石明松在水稻品种农垦 58 中发现雄性不育株，后来的研究表明该不育株育性能够随光温的变化而转换。敏感期处在长日高温条件下，表现雄性不育，处在短日低温条件下，则表现为可育，这个发现为在中国开展两系法杂交水稻研究揭开了序幕。利用光温敏不育系配制农作物杂交种具有如下意义。

1. 不受恢保关系制约，配组自由度大，选择到优良组合的概率高。这是两系法的最大优点。例如，95% 以上的水稻品种都能使水稻光温敏不育系的育性恢复正常。

2. 光温敏不育系由 1~2 对隐性基因控制，选育优良不育系较容易。

3. 生产程序简化，种子生产成本降低。不育系的繁殖没有异交过程，只要条件满足就能完全可育。

4. 可避免不育细胞质对杂种优势的负效应，以及消除细胞质单一可能造成的遭受毁灭性病虫害侵袭的潜在危险。

三、已有的光温敏不育类型及获得途径

1. 已发现的光温敏不育类型（表 1）。

表1　不同类型光温敏不育系对光温的反应

类型	长日高温	短日高温	长日低温	短日低温
高温不育型	S	S	F	F
低温不育型	F	F	S	S
长日不育型	S	F	S	F
短日不育型	F	S	F	S

注：F 表示可育，S 表示不育。

纯粹由光照长度控制的两用不育系不育性会更稳定，更易被人们操作，因为日长变化很有规律。

2. 光温敏不育材料的获得途径。目前发现有 3 种方法可以产生光温敏雄性不育系。

（1）自然突变。

（2）亲源关系远的亲本间杂交。

（3）辐射或化学诱变。

表 2 列出了通过上述 3 种方法获得的各种农作物光、温敏不育材料。

四、两系法杂交水稻的研究进展

无论是从提高杂种优势水平，还是从简化杂交水稻的育种程序来看，杂交水稻育种都可以分为三个战略发展阶段。

表 2　光温敏不育材料

材料	物种	不育转换类型	来源	获得途径	发现地点	发现报道年份
农垦 58S	水稻	长日高温	农垦 58	自然突变	湖北仙桃市农科所	1981
5460S	水稻	高温	IR54	辐射诱变	福建农学院	1989
安农 S	水稻	高温	超 40B/285 // 6029-3	自然突变	湖南安江农校	1988
87N123	水稻	高温	长芒野生稻 / R0183 // 测 64	远缘杂交	湖南衡阳农科所	1989
农林 12	水稻	高温	黎明	γ 射线辐射诱变	日农研中心	1989
M201S	水稻	长日高温	M201	EMS 诱变	美国 Rutger	1990
Calrose76S	水稻	长日高温	Calrose76	花培	美国 Rutger	
822	水稻	高温	印度品种	自然突变	湖南农科院	1992
IR32364S	水稻	高温	IR32364- 20-1-3-28	γ 射线辐射诱变	IRRI	
琼 6QMS	玉米	高温	琼 6	自然突变	海南省农科院	1992
湘糯粱 S-1	高粱	短日低温不育	（湘白糯高粱 × TX3197B） × TX3197A	自然突变	湖南省农科院土肥所	1994
ES	小麦	短日低温不育	普通小麦 × 野燕麦	远缘杂交	湖南农业大学	1992
C49S	小麦	短日低温不育	普通小麦 × 栽培二粒小麦	远缘杂交	重庆市作物所	1992
901S	小麦	短日低温不育	皖 901	自然突变	安徽省农科院	1995
88-428 By-3	大豆	短日不育	土梅豆	自然突变	山西省农科院品资所	1994
292	谷子	长日低温		自然突变	河北张家口市坝下农科所	1991

续表

材料	物种	不育转换类型	来源	获得途径	发现地点	发现报道年份
湘油 91S	十字花科油菜	高温	86A/// 中油 821 // 湘油 5 号 / Marnoo	自然突变	湖南省农科院作物所	1994
	大麦		毛-X21-4	辐射	武昌东湖农科所	1991

提高优势水平：品种间杂种优势利用→亚种间杂种优势利用→远缘杂种优势利用。

简化育种程序：三系法→两系法→一系法。

每进入一个新阶段都是育种上的一次突破，能使水稻产量登上一个新台阶。目前中国的两系法杂交水稻已在理论研究和实践上取得了重大进展，技术上已基本成熟配套，正在向大面积生产过渡，两系法杂交水稻面积累计推广 950 多万亩，比三系法杂交稻增产 10% 左右，至1996 年增产粮食 3 亿公斤以上（表 3）。

表3 两系法杂交水稻的推广面积和产量（1991—1997）

年份	1991	1992	1993	1994	1995	1996	1997
面积 / 千公顷	4.30	12.00	27.00	67.00	73.00	250.00	400.0
单产 /（公斤 / 公顷）			7 170	7 005	7 215	7 190	

应用最广泛的光温敏不育系和杂交组合分别是国家杂交水稻工程技术研究中心育成的第一个实用籼粳中间型光温敏不育系培矮 64S 和第一个通过省级审定的两系法杂交稻组合培矮64S/ 特青。培矮 64S/ 特青到 1997 年已累计推广 200 万亩以上，并创造了我国和湖南省水稻史上的 5 个第一（表 4）。

表4 培矮 64S/ 特青创造的高产记录

地点	面积	时间	季别	产量 /（公斤 / 亩）	纪录
湖南省区试		1992	中稻	631.8	湖南省中稻区试
云南省永胜县	1.04	1992	单季稻	1 140.85	中稻单季稻实际验收产量
湖南省黔阳县文丰	1.3	1994	一季中稻	864.8	湖南省中稻
湖南省湘潭泉塘子	1.5	1994	双季晚稻	775.17	长江流域双季晚稻
湖南省汉寿县护城村	2.13	1995	单季 + 再生	1 007.8	湖南省单季加再生

几项创新纪录表明，两系法杂交稻虽然处于初期发展阶段，却打破了三系法单产长期徘徊的局面，这充分显示了两系法杂交水稻的增产潜力。

迄今，中国已育成 10 个通过省级审定的组合（表 5），鉴定了 10 多个实用光温敏核不育系（表 6）。

表5　已育出通过省级审定的组合

组合	审定省份	审定时间
培矮 64S/ 特青	湖南	1994
培矮 64S/288	湖南	1996
培矮 64S/ 余红 1 号	湖南	1997
7001S/ 秀水 04	安徽	1994
7001S/ 皖恢 9 号	安徽	1994
700S/1514	湖北	1995
5088S/R187	湖北	1995
7001S/ 双九	安徽	1997
培矮 64S/ 山青 11	广东	1996
蜀两优 1 号	四川	1996

表6　若干实用水稻光温敏不育系

光温敏不育	亚种	鉴定时间	鉴定省份	不育临界温度/℃	类型	主要利用地区	不育基因来源
培矮 64S	*Javanica*	1991	湖南	23.5	高温不育型	湖南、广东、江西、广西、四川	农垦 58S
7001S	*Japonica*	1989	安徽	24	长日高温不育型	安徽、湖北	农垦 58S
GB028S	*Japonica*	1996	辽宁	23	长日高温不育型	辽宁	农垦 58S
5088S	*Japonica*	1992	湖北	24	长日高温不育型	湖北	农垦 58S
安湘 S	*Indica*	1994	湖南	24	高温不育型	湖南、福建	安农 S
810S	*Indica*	1995	湖南	24	高温不育型	湖南	安农 S
香 125S	*Indica*	1994	湖南	23.5	高温不育型	湖南	安农 S
测 64S	*Indica*	1996	湖南	23.1	高温不育型	湖南	安农 S
蜀光 612S	*Indica*	1996	四川	23	高温不育型	四川	农垦 58
GD2S	*Indica*	1995	广东	23	高温不育型	广东	农垦 58

五、其他作物两系法杂种优势利用的研究情况

1. 两系高粱。湖南省土肥所选育的湘两优糯粱 1 号（湘糯粱 S× 湘 10721）通过了湖南省审定。该组合在湖南、四川、贵州等 10 多个省、市试种、示范 2 500 余公顷，比三系杂交高粱增产 10% 以上。表现早熟高产、优质、抗性好、品质好、适应性广、再生能力强。两系法杂交高粱技术已基本配套。

2. 两系油菜。湖南省作物所用温敏不育系湘 1S 所配组 7 个组合，在 1995 年的品比试验中，比常规对照良种增产幅度达 20%，1996 年小面积试种示范显示出较强的优势。

3. 两系小麦。湖南农业大学选育的两系杂交小麦，如 E 优 137，表现在用种量少（每亩 3 公斤以下），生长势强，分蘖旺盛，增产显著（20% 以上），品质较好（面筋含量高，上白粉率高）。

4. 两系棉花。用雌雄性异熟系配制的两系杂交棉花异优 3 号和异优 6 号，1996 年在品比和示范中比对照常规棉增产 30% 以上，最高亩产皮棉 206 公斤，创湖南省单产最高纪录。

两系法杂种优势利用已经显示出广阔的应用前景。本人深信两系法在未来的杂种优势利用中将占据重要地位。我们对两系法杂交水稻提出了一个超高产育种目标，并争取在下世纪初实现，即每公顷日产量达到 100 公斤稻谷，短生育期品种 100 天左右，产量达到 10 000 公斤 / 公顷，中熟组合 120 天左右，产量达到 12 000 公斤 / 公顷，150 天的迟熟组合，产量达到 15 000 公斤 / 公顷。我们相信各学科相互交流，将有利于促进农作物两系法杂种优势利用的进一步发展，为下一个世纪的作物增产增值作出越来越大的贡献。

作者：袁隆平

注：本文是作者在 1997 年 9 月 6—9 日于湖南长沙"农作物两系法杂种优势利用国际学术讨论会"上所作的报告，原系英文。

水稻广谱广亲和系的选育策略

【摘　要】选育广谱广亲和系对于利用水稻亚种间杂种优势具有极其重要的意义。根据现有的研究结果，初步假设水稻亚种间亲和性基本上可以分为四大类：广谱广亲和性、部分广亲和性、弱亲和性和非亲和性。决定亚种间亲和性的基因可能有两类：一类是广亲和基因 Wc，另一类是辅助亲和性基因 Sc。这种辅助亲和性基因既有可能是几个主基因，也有可能是一系列微效多基因。广谱广亲和系可能同时具备 Wc 和 Sc 基因；部分广亲和系可能只具备 Wc 基因；弱亲和系可能只具备 Sc 基因；非亲和系则两者都不具备。选育广谱广亲和系的途径有两个，一是利用现有的广谱广亲和种质直接转育，二是利用部分广亲和系与弱亲和系杂交，组合广亲和基因与辅助亲和性基因。筛选广谱广亲和系可以分三步进行：初选，即在低世代用两个具有代表性的品系进行测交鉴定；复选，即在中世代增加测交品系个数，进一步作亲和性鉴定；决选，即用更多的非亲和品系与当选株系测交，在进一步确定广谱广亲和性的同时筛选优势组合。

【关键词】广亲和性；选育方法；杂种优势；水稻

　　水稻亚种间杂种具有强大的杂种优势，但亚种间的不亲和性导致杂种一代结实率低，限制了这一强大优势的直接利用。许多研究结果表明，亚种间的不亲和性受遗传基因控制，但由于选用的研究材料不同，采用的研究手段和分析方法有别，故不同研究者提出的控制亚种间不亲和性的遗传模式也不一致[1,6,7]。不过，亚种间的不亲和性并不是绝对的，Ikehashi 等从源于印度尼西亚、孟加拉、菲律宾等国家的 74 个水稻品种中，经测交筛选获得 6 个品种，与普通籼稻和粳稻杂交，F_1 无论是花粉育性还是结实率都表现正常，并将这类品种称为广亲和品种（wide compatibility varieties）。进一步的研究结果表明，广亲和品种的广亲和性受一对主要广亲和基因 S_5^n 控制，位于第 I 连锁群，与色素原基因 C 之间的交换值为 $3.9\% \sim 5.6\%$[5,6]。

　　广亲和品种和广亲和基因的发现，为克服亚种间不亲和性找到了有效的途径，使得直接利用亚种间杂种优势成为可能。1985 年以来，我们在这一方面做了大量的研究，育成一批新的广亲和系，通过测交鉴定，发现不同的广亲和系的亲和性不同，有的广亲和系对所有测交的品种和不育系都表现亲和，有的则部分亲和、部分不亲和。据此初步认为，培育亲和谱广的广亲和系基本上可以解决亚种间杂交稻结实率低的难题，从而能大大提高选择强优势亚种间杂交组合的概率。本文扼要论述选育广谱广亲和系的途径与方法，供育种家指正。

1　水稻品种亚种间亲和性的分类

许多试验结果表明，虽然亚种间存在不亲和现象，但亚种内不同品种或品系间也存在亲和性差异。1993 年我们用 6 个籼稻材料：IR36、南特号、安农 S-1、1356S、545S、735S 和籼爪中间型材料培矮 64S 与 10 个粳稻材料：零培 13、零轮 11、CB-1、Cpslo17、02428、零贵 66、88-13、秋光、巴利拉和 5088S 进行研究。所有杂交 F_1 的小穗育性列于表 1。

表1　72 个不完全双列杂交组合 F_1 的小穗育性（1993，长沙）

♀		♂（粳）								
		零培13	02428	零贵66	Cpslo17	CB-1	零轮11	88-13	秋光	巴利拉
籼	1356S	94.0	89.3	87.5	92.3	91.2	92.3	48.7	44.8	46.7
	安农 S-1	91.2	85.3	88.9	89.3	80.1	90.6	49.1	50.2	48.2
	545S	82.5	80.5	87.2	90.7	92.7	90.6	45.3	51.5	39.8
	IR36	86.0	82.2	87.5	90.2	90.7	93.7	40.1	52.1	43.2
	南特号	51.2	38.7	58.9	87.5	87.2	85.6	3.7	22.2	18.7
	735S	49.2	12.1	52.3	90.6	80.2	85.7	1.2	35.3	3.0
中间型	培矮 64S	87.2	89.2	90.1	91.7	90.8	95.4	89.2	85.2	84.6
粳	5088S	89.5	92.3	88.2	88.8	85.2	95.7	80.2	92.1	95.3

1.1　供试粳稻材料对粳稻具有良好的亲和性

用典型的粳稻不育系 5088S 和其他 9 个正常粳型品种或品系配组，各 F_1 的小穗育性均表现正常。

1.2　不同粳稻品种或品系与籼型不育系或品种的亲和性有差异

对表 1 数据经反正弦转换后作方差分析，结果表明，不同粳型品种或品系对籼型不育系

或品种的亲和性差异达到了极显著水平（表 2），以零轮 11 的亲和性最好，小穗育性平均为
90.8%；其次是 Cpslo17 和 CB-1，其亲和性与零轮 11 没有显著差异（表 2）；亲和性最差
的是 88-13，平均小穗育性仅 34.4%。根据表 2 的比较，可以将供试粳稻品种或品系分成四
类：第一类以零轮 11、Cpslo17 和 CB-1 为代表，对籼型不育系或品种的亲和性均很好，各
F_1 的小穗育性都在 80% 以上，亲和性最好。第二类以 02428 为代表，平均小穗育性 70%
左右，与部分籼型不育系或品种，如 1356S、安农 S-1、545S 和 IR36 杂交的 F_1 小穗育
性在 80% 以上，表现为亲和性良好；而与另外一些籼稻不育系或品种，如南特号、735S 杂
交，F_1 的小穗育性都在 60% 以下，表现为亲和性差。第三类以秋光为代表，仅对培矮 64S
亲和，与其余的籼型不育系或品种概不亲和，但小穗育性尚可以保持在一定水平上，最低的
有 22.2%（南特号 / 秋光）。第四类以 88-13 为代表，也仅仅对培矮 64S 亲和（实际上培矮
64S 为一个中间型的广谱广亲和系），对其余的籼型不育系及品种概不亲和，平均小穗育性较
低，最低的仅 1.2%（735S/88-13）。

零贵 66 和零培 13 所配组合的平均小穗育性虽然与零轮 11、Cpslo17 和 CB-1 所配
的组合无显著差异，但其亲和谱比零轮 11、Cpslo17 和 CB-1 要窄，与南特号和 735S 杂
交，F_1 的小穗育性都在 60% 以下，故应划入第二类。巴利拉配组的小穗育性和 88-13 配组
的无显著差异，与 735S 杂交的 F_1 小穗育性为 3%；而与南特号杂交的 F_1 的小穗育性也较
低，故暂将巴利拉列为第四类。

表 2　不同父本的亲和性比较

♂	小穗育性 $\bar{x}$ /%	a=0.05	a=0.01	亲和性类型
零轮 11	90.8	a	A	
Cpslo17	90.4	a	A	广谱广亲和
CB-1	87.9	a	A	
零贵 66	80.4	ab	AB	
零培 13	79.4	ab	AB	部分广亲和
02428	69.8	b	BC	
秋光	50.6	c	CD	弱亲和
巴利拉	41.4	cd	D	
88-13	34.4	d	D	非亲和性

第一类品种（系）的亲和谱很广，对所有供试的籼稻品种或不育系都亲和，故可以称之为

广谱广亲和系；第二类品种（系）的亲和谱比第一类窄些，对部分籼稻品种（系）亲和，而对另外一些则不亲和，故可以称为部分广亲和系；第三类品种仅对籼爪中间型广谱广亲和不育系培矮64S亲和，但与其余的籼稻不育系或品种杂交，F_1小穗育性也能维持在一定水平上，表现出弱的亲和性，故可以称之为弱亲和系；第四类只对广谱广亲和籼爪中间型材料培矮64S亲和，而与少数籼稻不育系或品种杂交，F_1小穗育性极低，表现为亲和性很差，故可以称之为非亲和系。

1.3 不同籼型不育系及品种与粳型广亲和系或品种的亲和性有差异

分析结果表明，不同籼型不育系及品种对粳型广亲和系或品种的亲和性差异也达到了极显著水平（表3）。最高的为籼爪中间型培矮64S，对所有粳型广亲和系或品种都表现为亲和，说明它是具有广谱广亲和性。其次是1356S、安农S-1、545S和IR36，对所有广亲和系的亲和性都尚好，而对非广亲和系的亲和性差，但小穗育性都能维持在一定水平上，说明它们具有弱亲和性。再次是南特号和735S，与粳型广谱广亲和系杂交，F_1小穗育性正常；与粳型部分广亲和系杂交，F_1小穗育性不正常；与粳型弱亲和系杂交，F_1小穗育性比较低；与粳型非亲和系杂交，F_1小穗育性极低（几乎为零），说明这类籼稻品种对粳稻没有亲和性。因此，供试籼稻或籼爪型不育系或品种可以分为三类：一类是广谱广亲和系，如培矮64S；二类是弱亲和系，如1356S、安农S-1等；三类是非亲和系，如南特号和735S。

表3　不同母本的亲和性比较

♀	小穗育性 $\bar{x}$/%	a = 0.05	a = 0.01	亲和性类型
培矮64S	90.6	A	A	广谱广亲和
1356S	79.2	Ab	A	
安农S-1	76.8	B	A	弱亲和
IR36	76.4	B	A	
545S	75.7	B	A	
南特号	50.1	C	B	非亲和
735S	42.7	C	B	

上述结果很难用单个广亲和基因S_5^n的遗传理论来解释，因为假如只有单个广亲和基因起作用，那么部分广亲和系02428等就应该对所有的籼型不育系及品种亲和，而事实并非如此。因此，我们初步假设水稻中除了前人业已发现的广亲和基因Wc外，还可能存在另一类亲和性基因，暂且称为辅助亲和性基因（supplementary compatibility gene，Sc）。这

种辅助亲和性基因的性质，目前尚不清楚，既有可能是少数几个基因，也有可能是多个微效基因。而且不同来源的材料可能携带不同的辅助亲和性基因。因此，广亲和品种的亲和性强弱还可能与广亲和系所携带的辅助亲和性基因数目有关。只有当亚种间杂种中同时存在 Wc 和足够的 Sc 基因时，F_1 才能正常受精结实。根据上述分析，广谱广亲和系中可能同时存在 Wc 和足够的 Sc 基因，故无论与对应亚种中的哪一类品种杂交，F_1 皆能正常受精结实；部分广亲和系中可能仅存在 Wc 基因，不存在或只存在少数 Sc 基因，故只有与对应亚种中具有所需的 Sc 基因的品种杂交，F_1 才能正常受精结实；弱亲和系可能仅仅具备 Sc 基因，故只有与对应亚种中具有 Wc 基因的品种杂交，F_1 才能正常受精结实；非亲和系可能既不具备 Wc 基因，也没有或仅具有少数 Sc 基因。故只能与同时具备 Wc 基因和足够 Sc 基因的广谱广亲和系杂交，F_1 才能正常受精结实（表 4）。

表4　水稻亚种间亲和性模式

籼	粳			
	广谱广亲和系	部分广亲和系	弱亲和系	非亲和系
广谱广亲和系	F	F	F	F
部分广亲和系	F	?	F	S
弱亲和系	F	F	S	S
非亲和系	F	S	S	S

注：F. 两者杂交，F_1 受精正常；S. 两者杂交，F_1 受精不正常；?. 两者杂交，F_1 受精情况尚不明了。

值得指出的是，我们所选用的籼稻材料仅仅包括了三类，这并不表明籼稻中就不存在广谱广亲和系，很有可能只是试验材料中没有收集到这类品种。

2　广谱广亲和系的选育途径

自然界的广谱广亲和资源基本上都是爪哇稻即热带粳稻。典型的籼稻或温带粳稻品种基本上都是非亲和材料或弱亲和材料。近年来我国人工育成的广亲和系如02428等，大都属于部分广亲和系。因此，培育籼型或粳型广谱广亲和系对于利用籼粳亚种间杂交优势具有极其重要的意义。

选育广谱广亲和系主要有两个途径。

2.1 利用鉴定出来的广谱广亲和资源（WcWcScSc）进行转育

热带粳稻（爪哇稻）中存在一些具有广谱广亲和性的优良种质，如 Cpslo17 等。应用这些广谱广亲和种质与需要转育的籼稻或粳稻品种杂交，于后代中选择具有广谱广亲和性的个体。培矮 64S 的广谱广亲和性可能就源于广谱广亲和材料培迪[2,3]。先用培迪与籼稻品种矮黄米杂交，选育出具有培迪广亲和特性的培矮；然后用培矮与高配合力的恢复系测 64 杂交，育成了具有广谱广亲和特性的培矮 64；再用农垦 58S 与培矮 64 杂交并回交，育成了广谱广亲和不育系培矮 64S（图 1），经测交鉴定，培矮 64S 与粳稻杂交 F_1 结实率比与籼稻杂交的 F_1 结实率还要高（表 1）。

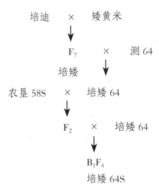

图 1 培矮 64S 系谱

2.2 组合广亲和性基因和辅助亲和性基因

02428（螃蟹谷 / 汲滨稻）和轮回 422［芦苇稻 / 培 422（Paddy/C57）］是两个来源不同的广亲和系，经测交鉴定，它们的广亲和性都受 S_5^n 控制[5]。两者杂交选育的零轮，其亲和性较 02428 和轮回 422 大大拓宽，在 1996 年的对比测试中，对所有测交品种或品系都基本上是亲和的（表 5）。零轮之所以具有广谱广亲和性，很有可能是来源不同的广亲和系 02428 和轮回 422，除都有 S_5^n 外，可能还各携带了不同的辅助亲和性基因，只是在数量上不够，故无广谱广亲和性。两者杂交后，不同来源的辅助亲和性基因累加在一起，使得零轮不仅携带了双亲共同的广亲和基因 S_5^n，而且获得了在数量上超过了双亲的辅助亲和基因，从而表现出广谱广亲和性。

表 5　零轮与 02428（对照）的广亲和性测验（1996, 长沙）

测验系		结实率 / %	
		零轮	02428（CK）
籼稻	香 125S	76.50	64.46
	133S	75.30	55.72
	1356S	72.34	68.73
	810S	77.21	73.25
	2-2S	70.56	54.42
	南京 11 号	78.42	67.59
	IR36	80.30	71.57
	水源 287	81.09	68.51
	密阳 46	73.36	72.80
	密阳 49	75.60	71.72
	IR24	77.56	62.57
	特青	79.17	74.55
	测 64S	77.93	70.53
	1148S	68.04	37.43
		$\bar{x}=75.96$	$\bar{x}=65.28$
粳稻	秋光	82.17	71.12
	巴利拉	82.72	72.17
	5088S	84.01	78.55
	867S	76.49	83.44
		$\bar{x}=81.35$	$\bar{x}=76.32$

　　此外，广谱广亲和性可以与细胞质雄性不育恢复基因结合在一起。用部分广亲和系 AB240 与具有恢复基因的弱亲和系新恢 73 杂交选育出的 CB-1 不仅具有广谱广亲和性，而且能够很好地恢复野败和 BT 型不育细胞质（表 6）。

表 6　CB-1 部分组合的结实率（1992, 长沙）

组合	结实率 / %
7001S（J）/CB-1	71.22
寒丰 A（J）/CB-1	73.37

续表

组合	结实率 / %
299S（Ⅰ）/CB-1	83.55
珍汕 97A（Ⅰ）/CB-1	79.86

注：J. 粳稻；I. 籼稻。

3 广谱广亲和系的选择方法

现以零轮的选育为例予以说明（图 2）。

年份	地点		流程
1990	长沙		02428/ 轮回 422
1990	三亚	F_1	30 株（1）
1991	长沙	F_2	选 45 株与安农 S-1 和 1147S 测交（2）
	初选		
1991	三亚	F_3	零轮 -1 的 2 个组合的结实率都在 80% 以上，从该株系中选 20 个单株与 1356S 和 W6154S 测交（3）
	初选		
1992	长沙	F_4	零轮 -1-4 的 2 个组合的结实率都在 75% 以上，当选 15 个单株，与衡农 S、8902S、735S 和 870S 测交（4）
	复选		
1992	三亚	F_5	零轮 -1-4 的 4 个组合结实正常，性状基本整齐一致，继续选单株与 1356S 等 21 个不育系测交（5）
	确选		
1993	长沙	F_6	零轮 -1-4-11 性状稳定，与 21 个不育系测交结实正常，其中测 64S/ 零轮等组合优势很强（6）

图 2　广谱广亲和系零轮的选育过程

3.1 初选

在选育的早期世代，将当选单株与两个分别具有弱亲和性和非亲和性的品系进行测交，从中选择与两个不育系配组结实都正常的单株。如在选育零轮的过程中，在 F_2 代和 F_3 代就是用具有弱亲和性的安农 S-1 或 1356S 和非亲和性材料 1147S 或 W6154S 进行测交，再选亲和性好的单株。

3.2　复选

在选育的中期世代，增加测验品系的个数，进一步鉴定中选单株的亲和性。如在零轮的选育过程中，我们用了四个不育系进行复测鉴定。其中8902S和衡农S具有弱亲和性，735S和870S为非亲和性核不育系。

3.3　决选

对复测鉴定中选的株系，加大选择压力，即进一步用更多的非亲和品系多次重复测交筛选，一方面鉴定亲和性，另一方面鉴定杂种优势。如在零轮的选育过程中，我们就用了1356S、1147S和测64S等21个不育系与F_5中当选株系进行了测交，不仅进一步鉴定了零轮的广谱广亲和性，而且筛选出了测64S/零轮和香125S/零轮等强优势苗头亚种组合。

─────── R e f e r e n c e s ───────

参考文献

［1］张桂权，卢永根. 栽培稻杂种不育性的遗传研究. Ⅱ .F_1花粉不育性的基因模式. 遗传学报.1993，20（3）:222-228.

［2］罗孝和，袁隆平. 水稻广亲和系的选育. 杂交水稻.1989，（2）:35-38.

［3］李新奇，等. 培迪广亲和基因重组效应的研究. 杂交水稻.1990，（4）:36-38.

［4］顾铭洪，等. 水稻品种广亲和基因等位关系的遗传分析. 两系法杂交水稻研究论文集.北京：农业出版社,1992:259-268.

［5］Ikehashi H, Arak H. Varietal screening of compatibility types revealed in F1 fertility of distant crosses in rice. Japan. J. Breed，1984，34:304-318.

［6］Ikehashi H, Arak H. Genetics of F_1 sterility in remote crosses of rice. Rice Genetics，Procedings of the International Rice Genetic Symposium. IRRI. 1985:119-130.

［7］Oka HI. Analysis of genes controlling F_1 sterility in rice by the use of isogenic lines. Genetics.1974，77:521-534.

作者：袁隆平　武小金　颜应成　罗孝和

注：本文发表于《中国农业科学》1997年第30卷第4期。

Hybrid Rice Breeding in China

【Abstract】From 1976 to 1995, hybrid rice helped China to increase rice production from 129 million t to 200 million t annually. Hybrid rice varieties yield on average 6.6 t/ha compared with 5 t/ha for conventional rice varieties. Hybrid rice seed production technology has been well developed to achieve a nationwide average seed yield of 2.4 t/ha. Most existing commercial rice hybrids belong to the category of intervarietal hybrids based on the CMS system. Future emphasis is on developing two-line rice hybrids using PGMS and TGMS systems. To increase the yield potential of hybrid rice in China, emphasis is given to intersubspecific hybrids (*Indica/Japonica, Indica/Javanica, and Japonica/Javanica*). The discovery of QTLs for yield in wild rice species has opened up a new avenue for raising the heterosis level by using distant genes.

Current status of hybrid rice in China

Since 1976, when hybrid rice was first released commercially, its area has increased consistently (Fig. 1). Farmers growing hybrid rice obtain more than a 30% yield advantage over conventional pure-line varieties (Table 1). From 1976 to 1995, hybrid rice helped China to increase rice production from 129 million t to 200 million t. In recent years, hybrid rice has yielded about 6.6 t/ha versus 5 t/ha for conventional rice. In 1994, hybrid rice covered 15.7 million ha, 50% of the total rice area, and hybrid rice production was 57% of the total rice output in China. The largest hybrid rice—growing province is Sichuan, where 3 million ha (95% of the rice area) is under hybrids and the average yield has remained at 7.5 t/ha for years. Hunan is the second-largest hybrid rice-growing province, where the average yield of the second crop grown on 2 million ha is 6.8 t/ha . The highest yield recorded for hybrid rice from a single crop on a large scale (1,000 ha) was 11.2 t/ha and from a small plot (0.1 ha) 16.8 t/ha. So far, the highest yield recorded for double-cropped hybrid rice was 23.3 t/ha. In Hunan Province, more than 0.2 million ha of double-cropped hybrid rice produce 15 t/ha of grain every year.

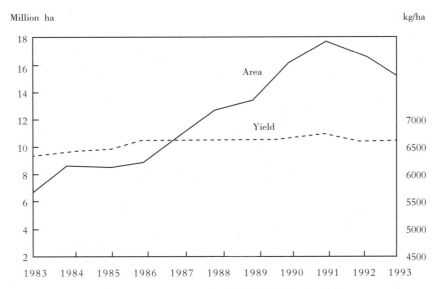

Fig.1　Area and Yield of Hybrid Rice in China

Table 1　Yield of hybrid rice compared with that of conventional rice from 1986 to 1993 in China

Year	Conventional variety / (kg/ha)	Hybrid rice / (kg/ha)	Hybrid over conventional/%
1986	4,857	6,600	36
1987	4,779	6,615	38
1988	4,539	6,600	45
1989	4,787	6,615	38
1990	5,315	6,675	26
1991	4,551	6,565	44
1992	4,986	6,636	33
1993	4,950	6,675	35

The technology of hybrid rice seed production has been well developed. The nationwide average seed yield is 2.4 t/ha (Table 2). Many new cytoplasmic male sterile (CMS) lines with a high outcrossing rate and good grain quality have been developed recently. Many new rice hybrids with good grain quality and multiple resistance have been released to farmers. The field area ratio of A line multiplication, hybrid seed production, and F_1 commercial cultivation was 130 : 1,000 in the late 1970s. This has been increased to about 1 : 50 : 6,000 recently. The highest hybrid seed yield recorded (7.4 t/ha) was obtained in 1993 by the Zixing Seed Company of Hunan Province on a small plot (0.2 ha).

Table 2　Area and yield of hybrid rice seed production from 1986 to 1993 in China

Year	Area/ha	Yield/ (kg/ha)
1986	100,500	1,995
1987	154,100	2,010
1988	135,800	1,628
1989	171,900	1,956
1990	192,000	2,250
1991	124,700	2,252
1992	146,600	2,438
1993	105,900	2,214

Constraints and challenges

Research on the commercial use of heterosis in rice has made tremendous achievements during the past 20 years. From a strategic point of view, however, it is still in the juvenile stage because the high yield potential of hybrid rice has not yet been fully tapped. Hybrid rice breeding still has a bright future. Based on our studies, to derive full benefit from hybrid rice breeding, future developments may involve intensifying research on breeding methods and increasing the degree of heterosis. Three approaches are involved:

1.The three-line method using the CMS system.

2.The two-line method using the photoperiod-sensitive genic male sterility (PGMS) or hermosensitive genic male sterility (TGMS) system.

3.The one-line method using the apomixis system.

The rice hybrids used in commercial production belong to the category of intervarietal hybrids based on the CMS system. Many years of practice and experience have proved that the CMS system or three-line method is an effective way to develop rice hybrids and will continue to play an important role in this century. But this system has some constraints and problems.

· In all the rice hybrids developed so far, the level of yield harvested has stagnated for years (Fig. 1). This means that we have already reached the yield plateau for rice hybrids. It would be difficult to further increase the yield potential in new rice hybrids if no new methods and materials are invented and adopted.

· The sources of male sterility-inducing cytoplasm that can be used to develop better CMS lines are poor. Currently, about 85% of the A lines used in commercial production still belong to the wild abortive (WA) type. The dominant cytosterility situation of the WA type could produce a crisis in the long run, which could make hybrid rice susceptible to destructive pests.

· The heterosis level in japonica hybrids is not as good as in *indica* hybrids. In addition, the currently used CMS lines (BT type) in *japonica* are not stable enough to produce pure F_1 seeds. Therefore, the planting area of *japonica* hybrids has been limited to about 0.1 million ha for many years and, what is worse, the area of *japonica* hybrids is declining.

· We do not have very early maturing combinations with strong heterosis suitable for the first crop in rice double-cropping regions, which is one of the major reasons why the area under hybrid rice cannot increase.

To increase the yield potential of hybrid rice, the magnitude of heterosis must be increased by adopting the following strategies:

1. Intersubspecific hybrids.

2. Distant hybrids (interspecific or intergeneric hybrids) .

In each of these phases, if the objectives are achieved, this will mark a new breakthrough in rice breeding and will result in a large increase in yield.

Strategies for the 21st century

Development of two-line hybrids

Taking the long-range strategy of rice heterosis breeding into account, many Chinese rice scientists have been exploring new technological approaches to replace the CMS system. So far, the most successful outcome is the development of two-line hybrids.

This method is based on two new kinds of rice genetic tools: photoperiod-sensitive genic male sterile (PGMS) lines and thermosensitive genic male sterile (TGMS) lines that have been successfully developed in China recently. Their male sterility is mainly controlled by one or two pairs of recessive nuclear genes, and it has no relation to cytoplasm. Exploitation of these P (T) GMS lines to develop rice hybrids has the following advantages over the classical three-line or CMS system:

· The maintainer line is avoided. The PGMS lines under longer daylength or the TGMS lines under higher temperature show complete pollen sterility; therefore, they can be used for hybrid seed production in these conditions. Under shorter daylength or moderate temperature conditions, they show almost normal fertility and can thus multiply themselves by selfing.

· The choice of parents in developing heterotic hybrids is greatly broadened. Studies showed that more than 95% of varieties tested (within the same subspecies) can restore such GMS lines. In addition, PGMS and TGMS genes can be easily transferred into almost any rice lines with desirable characteristics.

· No negative effects are caused by sterile cytoplasm and the dominant cytoplasm situation of WA will be avoided. Several achievements in this research area have been made.

Table 3　Area and yield of two-line hybrid rice from 1993 to 1997 in China

Year	Area/ha	Yield/ (kg/ha)
1993	17,190	7,170
1994	60,000	7,005

Continued

Year	Area/ha	Yield/ (kg/ha)
1995	75,330	7,215
1996 (estimated)	120,000	—
1997 (estimated)	600,000	—

To develop P (T) GMS lines that can be used commercially, one important criterion is that the male-sterility-inducing temperature (critical temperature) must be relatively low (mean temperature 23 ℃ in the temperate zone and 24 ℃ in the subtropics). If the critical temperature is relatively high (such as 26 ℃) for these male sterile lines, regardless of PGMS or TGMS, the temperature is not safe in hybrid seed production because a temperature below this point, which can induce sterile pollen into fertile pollen, sometimes occurs in the hot season. After nine years' research, considerable progress has been made. Now more than 20 P (T) GMS lines have been registered in China. Among these, two *japonica* lines belong to PGMS and the others are *indica* TGMS lines. Seven combinations have been certified and released for commercial production. The area under the two-line system for hybrid rice has been increasing steadily (Table 3). Experimental tests and commercial practices have proved that the best two-line hybrids outyield three-line hybrids by 5%–10%. Another advantage of the two-line system over the three-line system is that the yield area ratio of P (T) GMS line multiplication, seed production, and commercial use of F_1 is 1 : 100: (12,000–15,000). The expansion of this ratio can reduce seed cost. The area of two-line rice hybrids will be extended to 3.4 million ha by the end of the 20th century. We believe that two-line hybrids will replace 70% of the three-line hybrids in the first decade of the 21 st century.

Development of intersubspecific hybrids

Our studies have indicated that the degree of heterosis in different kinds of rice hybrids has the following general trend: *indica/japonica>indica/javanica >japonica/javanica>indica/indica>japonica/japonica* (Yuan, 1994).

The first three kinds are intersubspecific hybrids and the latter two are intervarietal hybrids. *Indica/japonica* hybrids possess the highest yield potential, considering their sink and source. Their theoretical yield may be 30% higher than that of the existing intervarietal hybrids (Yuan, 1994). Exploiting the strong heterosis in *indica/japonica* hybrids has been the major goal of our two-line system hybrid breeding program. To achieve this, however, five barriers commonly found in such F_1 hybrids must be overcome: low seed set rate, too tall plant height, very long growth duration, many poorly filled grains, and poor grain quality. So far, progress has been encouraging in our attempts to overcome these barriers.

By using wide compatibility (WC) genes, the low seed-setting rate caused by semisterility from incompatibility between *indica* and *japonica* lines can be raised to nearly normal levels. A large number of *japonica* lines and several *indica* TGMS lines possessing WC genes have been developed recently.

Transferring an allelic dwarf gene (Sd_1) into male and female parents can lower the plant height of *indica*/*japonica* hybrids to a semidwarf level, and still allow the hybrids to express strong heterosis.

By crossing parental lines of different growth duration, except photosensitive late varieties, *indica*/ *japonica* hybrids with medium and even shorter growth duration can be obtained.

Efforts are now focused on solving the last two problems, that is, the poor filling of a num ber of fertilized grains and poor grain quality. Within a year or two, revised breeding strategies are expected to help overcome these barriers. The new strategies emphasize developing *indica*/*javanica* hybrids rather than typical *indica*/*japonica* hybrids in the indica rice-growing region and *japonica*/*javanica* hybrids in the *japonica* rice-growing region. The superiority of this strategic change involves:

· Fewer fertility problems.

· Ecological adaptability, which will solve the problem of poor grain filling.

· The *indica*/*javanica* hybrids have similar or improved grain quality compared with that of *indica* rice; the *japonica*/*javanica* hybrids also have similar or better grain quality.

Several combinations that performed well in the experimental field in 1995 under went regional trials in 1996. The intersubspecific hybrids with super high yield potential ($100 \text{ kg} \cdot \text{ha}^{-1} \cdot \text{d}^{-1}$) are expected to be released to farmers by the end of this century. These hybrids will play a major role in increasing rice yield in the 21 st century.

Development of the one-line system of hybrid rice

Theoretically, there could be several approaches to fixing heterosis; among these, the use of apomixis to develop true-breeding F_1 hybrids appears to hold promise. Although this research program began in the late 1980s, it is still tentative. After extensive screening, some apomictic rice lines with a low frequency of apomixis (adventitious embryo and apospory) were found. But their frequency is too low (only 1%– 5%) for practical use. Transferring the obligate apomixis gene from wild grasses (such as *Pennisetum*) via genetic engineering combined with conventional breeding should be an effective way to develop the one-line system of rice hybrids. This could be achieved in the 21st century.

Using distant genes to raise the heterosis level

To use stronger heterosis observed in distant crosses, a marker-assisted advanced breeding strategy can be used to rapidly discover and transfer valuable novel genes for high yield potential and grain quality from wild species into elite combinations of hybrid rice. Based on careful evaluation in the experimental field and with the help of molecular markers, we discovered two important quantitative trait loci (QTL) from a wild rice in 1995 (Xiao et al., 1996). The two genes are located on chromosomes 1 and 2, respectively, each bringing about an increase in grain yield of 20% compared with the control hybrid. We have just started the new strategy to exploit heterosis of distant hybrids. We expect that the skillful combining of conventional breeding methods with molecular techniques could lead to another breakthrough in hybrid rice breeding in the first decade of the 21st century.

216

References

1. Xiao JH, Grandillo SN, Ahn SN, McCouch SR, Tanksley SD, Li J, Yuan LP. 1996.Genes from wild rice improve yield. Nature 384:223–224.

2. Yuan LP. 1994.Increasing yield potential in rice by exploitation of heterosis. In: Virmani SS, editor. Hybrid rice technology: new developments and future prospects. Manila（Philippines）: International Rice Research Institute. p1–6.

作者：Yuan Longping

注：本文发表于 *Advances in Hybrid Rice Technology* 1998 年。

Hybrid Rice Development and Use: Innovative Approach and Challenges

Hybrid rice has about a 30 percent yield advantage over conventional pure line varieties. From 1976 to 1995 hybrid rice technology had helped China to increase rice production by nearly 300 million tonnes. A well-established package of the technology for hybrid rice seed production would greatly boost the expansion of hybrid rice. The technology has been well developed in China to give an average seed yield of 2.3 tonnes/ha nationwide. The field area ratio between A line multiplication, hybrid seed production and F_1 commercial cultivation has been increased to about 1:50:6 000.

Outside China, 16 countries and two international research institutes are endeavouring to develop hybrid rice technology. The International Rice Research Institute (IRRI) has made great progress in breeding CMS lines. The situation for the development and commercialization of hybrid rice in India and Viet Nam is very encouraging. However, the flow of hybrid rice technology to farmers need to be accelerated. FAO is playing an important role in supporting national hybrid rice programmes for fighting world hunger.

The exploitation of photoperiod-sensitive and thermosensitive genic male sterile (PGMS and TGMS) lines to develop two-line system rice hybrids has great advantages over the classical three-line or CMS system. China is again a pioneer in this innovative approach of hybrid rice technology. The area under the two-line system hybrid rice is rapidly expanding in China. Recently, a super hybrid rice breeding programme has been set up in the China National Hybrid Rice Research and Development Centre. The bottleneck caused by maintaining a higher harvest index without lodging while huge photosynthates are produced has been overcome by Chinese scientists. The goal of breeding super hybrids yielding 100 kg/ha per day is expected to be realized by the year 2000.

CHINA'S ACHIEVEMENTS IN HYBRID RICE R&D

China is the first country to produce hybrid rice commercially.

Hybrid rice research was initiated in 1964 (Yuan, 1966) and the genetic tools essential for breeding hybrid rice varieties, such as the male sterile line (A line), the maintainer line (B line) and restorer line (R line), were developed by 1973 (Yuan and Virmani, 1988). Several hybrid combinations with good heterosis and higher yield potential were identified in 1974.Hybrid seed production techniques were basically established in 1975.The first batch of rice hybrid varieties was released commercially in 1976.Since then the area under hybrid rice has been increasing year after year. It has been proved on a large scale for 20 years that hybrid rice has about a 30 percent yield advantage over conventional pure line varieties (Table 1). From 1976 to 1995, hybrid rice had helped China to increase its production by nearly 300 million tonnes. In recent years the yield of hybrid rice has been around 6.7 tonnes/ha while conventional rice varieties yield 5.2 tonnes/ha on average. In 1994 the area under hybrid rice was 15.7 million ha, which covered 50 percent of the total rice area, and the production of hybrid rice was 57 percent of the total rice output in China. The largest hybrid rice-growing province is Sichuan, which has 95 percent of its rice area (3 million ha) under hybrids and has had an average yield of 7.5 tonnes/ha for several years. Hunan is the second largest hybrid rice-growing province, where the average yield from 2 million ha of second crop is 6.8 tonnes/ha. The highest yield record from a single crop on a large scale (1 000 ha) is 11.2 and 16.8 tonnes/ha on a small plot (0.1 ha). The highest yield record from double cropping hybrid rice is 23.3 tonnes/ha so far. In Hunan Province more than 0.2 million ha of double cropping hybrid rice produces 15 tonnes/ha for years.

Table 1　Comparison of average hybrid and conventional rice yields in China, 1986—1995

Year	Conventional rice/ (kg/ha)	Hybrid rice/ (kg/ha)	Hybrid advantage over conventional/%
1986	4 857.0	6 600.0	35.9
1987	4 779.0	6 615.0	38.4
1988	4 539.0	6 600.0	45.4
1989	4 786.5	6 615.0	38.2
1990	5 314.5	6 675.0	25.6
1991	4 551.0	6 565.5	44.3
1992	4 986.0	6 636.0	33.1
1993	4 950.0	6 675.0	34.8
1994	5 149.0	6 670.0	29.5
1995	5 098.0	6 678.0	31

Sources: Yuan (1996); Fu and Gong (1994).

In the meantime, hybrid seed production technology has been well developed, giving an average seed yield of 2.3 tonnes/ha nationwide (Table 2). The field area ratio between A line multiplication, hybrid seed production and F_1 commercial cultivation was 1:30:1000 in the late 1970s and has increased to about 1:50:6 000 recently. The highest yield record is 7.4 tonnes/ha created by the Zixing Seed Company of Hunan Province on a small plot (0.2 ha) in 1993.

Table 2　Area and yield of hybrid rice seed production in China, 1981—1993

Year	Area/×1 000 ha	Yield/ (kg/ha)
1981	110.4	669
1982	154.6	909
1983	138.8	1 290
1984	104.7	1 415
1985	87.7	1 655
1986	100.5	1 995
1987	154.1	2 010
1988	135.8	1 628
1989	171.9	1 956
1990	192	2 250
1991	124.7	2 252
1992	139.4	2 396
1993	105.9	2 214

Sources: Adapted from Yuan (1994a) ; Fu and Gong (1994) .

HYBRID RICE DEVELOPMENT OUTSIDE CHINA

Following China's success in the commercialization of hybrid rice technology in the late 1970s, IRRI revived its hybrid rice research in 1980, and much progress has been made by programmes undertaken in many other countries-Japan, the United States, India, Viet Nam, the Philippines, the Republic of Korea, Thailand, Indonesia, Myanmar, Brazil, Egypt, Colombia, Malaysia and the Islamic Republic of Iran-with support from FAO (Trinh, 1992, 1993, 1994; McWilliam, Ikehashi and Sinha, 1995) as well as from public agencies or the private sector. Technical support in hybrid technology has been provided by IRRI and China: with financial support from FAO, the China National Hybrid Rice Research and Development Centre (CNHRRDC-the former Hunan Hybrid Rice Research Centre) has held six international courses on hybrid rice production and trained more than 70 rice scientists from India, Viet Nam and Colombia.

India

India was one of the first countries to start academic studies on hybrid rice. The India Council of Agricultural Research established a goal-oriented network project on hybrid rice in 1989, and this has received further support from the United Nations Development Programme and FAO since 1991. The project is now being operated as a well-organized national research network with 12 centres across the country and the Directorate of Rice Research in Hyderabad acting as coordinator (Paroda and Siddiq, 1996).

The situation for the development and commercialization of hybrid rice in India is very encouraging. Between 1990 and 1996, more than 700 experimental hybrids were developed and evaluated, and the yields of over 100 combinations exceeded that of the best traditional variety by more than 1 tonne/ha. Several hybrid varieties have been released for commercial cultivation. The yield advantage of these hybrids over their check varieties ranged from 16.2 to 44.2 percent (Table 3). Another two hybrids proposed for release are CRH-1 (Coordinated Rice Hybrid-1) from the Directorate of Rice Research and PBH 71 from the Pioneer Overseas Corporation. Besides these, six to eight rice hybrids are being marketed by private seed companies.

Table 3   Public hybrids released in India

| Hybrid | Parentage | Duration /days | Yield in on-farm trials | | Advantage over check /% |
			Hybrid/ (tonnes/ha)	Check	
APRH−1	IR58025A/ Vajram	130−135	7.14	5.27	35.4
APRH−2	IR62829A/ MTU9992	120−125	7.52	5.21	44.2
MGR−1	IR62829A/ IR10198−66−2 R	110−115	6.08	5.23	16.2
KRH−1	IR58025A/ IR9761	120−125	6.02	4.58	31.4
CNRH−3	IR58025A/ Ajaya R	125−130	7.49	5.45	37.4

Source: Adapted from Paroda and Siddiq (1996).

After years of extensive trials, a generalized optimum package of hybrid seed production technology has been established in India, with a hybrid seed yield of about 1.5 to 2 tonnes/ha.

In 1995 India planted 10 000 ha of hybrid rice which outyielded inbred varieties by approximately 1 tonne/ha. The area under hybrid rice was 60 000 ha in 1996, and India aims to increase this to 2 million ha by the start of next century (Paroda and Siddiq, 1996) —there are indications, in fact, that this target may be reached earlier than expected. Thus India has emerged as the second largest hybrid rice-growing country in the world.

Viet Nam

Viet Nam started its hybrid rice research in 1983 by introducing breeding lines from China and IRRI to Haugiang in the Mekong River Delta and Hanoi in the Red River Delta. Several experimental rice hybrids from IRRI have shown a 18 to 45 percent yield advantage over Viet Nam's best local inbred varieties at Cuu Long Delta Rice Research Institute (Table 4). Some rice hybrids, such as Shanyou 63, Shanyou Gui 99, Shanyou Quang 12 and Boyou 64, were introduced from China to northern Viet Nam where they have yielded 6.5 to 8.5 tonnes/ha, 18 to 21 percent more than the conventional inbred varieties (Nguyen, Nguyen and Virmani, 1995).

Table 4　Yield performance of some experimental rice hybrids from IRRI in Viet Nam

Year/season	Hybrid	Yield	Advantage over check/%	Check variety
		/ (tonnes/ha)		
1989/90 DS	IR54752A/IR64R	7.5	1 311	OM80
	IR54752A/IR64R	7.2	1 251	OM80
	IR54752A/OM80R	6.7	1 181	OM80
1990 WS	IR58025A/IR29723R	7.6	1 431	MTL 58
	IR62829A/IR29723R	6.7	1 261	MTL 58
1990/91 DS	IR62829A/IR29723R	6.1	1 231	MTL 61
	IR58025A/IR29723R	6	1 221	MTL 61
1992 WS	IR58025A/IR52287R	6.7	1 311	IR64
1992/93 DS	IR58025A/IR32358R	6.8	1 451	IR64

Note：① DS=dry season; WS = wet season.

② Significantly higher than check variety at 5 percent level using LSD test.

Source: Nguyen, Nguyen and Virmani (1995).

In 1992, Vietnamese Government officials called on FAO to help them intensify development of their hybrid rice technology. With FAO's assistance provided through the project TCP/VIE/2251 (A), considerable progress has been made since then in hybrid rice cultivation and seed production. The area under hybrid rice increased rapidly from 20 ha in 1990 to' 102 800 ha in 1996 (Table 5), with a yield advantage of 20 to 30 percent over the improved semidwarf varieties. The Chinese hybrids are highly adaptable in the northern mountainous area of Viet Nam, where growing conditions are similar to those of southern China, and farmers harvest up to 10 tonnes/ha in Dien Chau (Nghe An Province) and Phu Xuyen (Ha Tay Province). Some Chinese hybrids yield up to 14 tonnes/ha in Dien Bien (Lai Chau Province), 12 tonnes/ha in Hoa An (Cao Bang Province) and 12.6 tonnes/ha in Van Quan (Lang Son Province). The widespread success of hybrid rice in Viet Nam has depended on large amounts of imported seed that has been developed and produced by the Chinese. However the rice hybrids introduced from China are not suited to the tropical conditions of the Mekong River Delta in the south, where IRRI-bred rice hybrids and parental lines instead grow very well.

Table 5　Hybrid rice area and yield in Viet Nam

Year	Area/ha	Yield/ (tonnes/ha)
1992	11 137	6.66
1993	34 828	6.71
1994	60 007	5.84
1995	73 503	6.14
1996	102 800	6.58

Source: Yin (1997).

The yield of hybrid seed production was only 200−680 kg/ha in 1992, but the technology has been improved by years of research in the country. In 1996, F_1 seed yield of 2.1 tonnes/ha was obtained on a large area (Table 6). The yield of F_1 seed production with Boyou 64 was as high as 3 tonnes/ha. However, there is still a big gap between the total seed production and the need for rapidly expanding cultivation area. Viet Nam's dependence on seed imports from China and the resulting scarcity of seed adapted to its southern regions are still major constraints to increased use of rice hybrids in this country.

Table 6　Hybrid seed production in Viet Nam

Year	Dry season		Wet season	
	Area /ha	Yield / (kg/ha)	Area /ha	Yield / (kg/ha)
1993	141.4	550	13.2	550
1994	52	630	71	400
1995	46	760	55	1 150
1996	169	2 100	98	1 150

Source: Adapted from Yin (1997).

Japan

Japan has been researching hybrid rice since the 1950s. Sinjyo developed the A, B and R lines of japonica rice (Taichong 65) towards the end of that decade but it has not been used commercially (IRRI, 1980). The Ministry of Agriculture, Forestry and Fisheries started the natio-nal hybrid rice breeding programme in 1983, and the first three-line rice hybrid, Hokuriku-ko 1, was developed in 1985, outyielding the inbred check by about 20 percent (Uehara et al., 1997). The National Federation of Agricultural Cooperative Associations (Zen-Noh) and several private companies, such as RAMM Hybrid International Cooperation, Kirin Brewer Co. Ltd and Sumitomo Chemical Co., are also involved in developing and testing rice hybrids using different approaches (Kato, Maruyama and Uchiyamada, 1994).

Zen-Noh has registered three intervarietal japonica hybrids which yielded 5 to 15 percent more than the in-bred check variety, Todorokiwase. The Ministry has also bred several rice hybrids such as Kanto Cross 1, Ouu Cross 1 and Shu Cross 4781 which have a yield advantage of 16 to 19 percent over the inbred check variety, Akihikari. However, because of their poorer grain quality, not all hybrids are suitable for the Japanese market. Another constraint in the development and use of hybrid rice technology in Japan has been the high cost of seed production.

IRRI

China's successful experience with hybrid rice technology encouraged IRRI in 1979 to explore the prospects and problems of using hybrid rice to increase yields. By 1989, two commercially usable CMS lines, IR58025A and IR62829A with a "WA" cytoplasm, were bred at IRRI and shared with national programmes worldwide (Virmani et al., 1996). IR58025A is stable in sterility in tropical countries, whereas IR62829A has good combining ability but its sterility is not stable enough for hybrid seed production under higher temperatures. In recent years, IRRI's capacity to breed genetically diverse CMS lines has increased and 10 to 20 new CMS lines are now bred annually.

Hybrid seed technology for the tropics has been developed at IRRI in collaboration with national programmes, and the institute's technology packages can now result in a hybrid seed yield of up to 2 tonnes/ha in the tropics.

FAO considers hybrid rice technology a key approach for increasing global rice production to help meet the world's growing food requirements. The Organization is therefore organizing a task force on hybrid rice development in Latin American and Caribbean countries (e.g. Columbia and Brazil) and providing support through Technical Cooperation Programme (TCP) activities in some Southeast Asian countries such as Bangladesh, India, Myanmar, the Philippines and Viet Nam. It is hoped that hybrid rice may play an important role in fighting world hunger in the near future.

CONSTRAINTS AND CHALLENGES

Although tremendous progress has been made in research on the commercial utilization of heterosis in rice over the last 20 years in China, from a strategic point of view, the technology is still in its early stages and the high yield potential of hybrids has not yet been fully tapped. A possible strategy for the development of hybrid rice breeding could follow two phases.

i) Breeding methodology, involving three approaches:

· three-line method or CMS system;

· two-line method or PGMS and TGMS system;

· one-line method or apomixis system.

ii) Increasing the degree of heterosis in rice to increase yield potential using:

· intervarietal hybrids;

· intersubspecific hybrids;

· distant hybrids (interspecific or intergeneric hybrids) .

· If successful, each phase would signify a new breakthrough in rice breeding and would result in considerable yield increases.

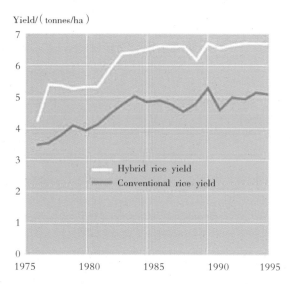

Figure 1   National average yields of hybrid and conventional rice planted in China, 1976—1995

· The existing rice hybrid varieties used in commercial production belong to the category of intervarietal hybrids and use the CMS, or three-line, system. Although research over the years has proved it to be an effective method-and it will continue to play an important role in the development of hybrid varieties this century-the CMS system does meet with certain constraints:

—The yield of existing rice hybrids, including newly developed varieties, has stagnated for some years now (Fig. 1) , meaning a plateau has been reached and, unless new methods and novel materials are found, it will be very difficult to increase yield potential further.

—There are only limited sources of male sterility-inducing cytoplasm (used for developing better CMS lines) . Currently, about 85 percent of the A lines used in commercial production still belong to the "wild aborted" (WA) type. The dominant cyto-sterility situation of the WA type has a latent weakness that, over time, could make the hybrid rice vulnerable to a destructive pest.

—The heterosis level of japonica hybrids is not as good as that of indica hybrids. Furthermore, the sterility of japonica CMS lines (BT type) now being used is not stable enough to produce very pure F_1 seeds. Therefore, the cultivation area of japonica hybrids has been limited to around 0.1 million ha for many years while, what is worse, the area of japonica hybrids is declining.

—There is a lack of very early-maturing combinations with strong heterosis suitable for the first crop in double rice-cropping regions along Yangtse Valley, the main rice production region in China. This is another major reason why the area under hybrid rice cannot be increased further.

PROGRESS IN TWO-LINE HYBRID RICE IN CHINA

The discovery in 1973 of Nongken 58 S, a PGMS/TGMS *japonica* rice line (Shi, 1985), provided the first genetic source for the development of two-line system hybrid rice. The major feature of such PGMS/TGMS lines is that, under longer day length and higher temperatures they show complete pollen sterility, in which case they can be used for hybrid seed production, while under shorter day length and moderate temperatures they show almost normal fertility and thus can multiply themselves by selfing. In 1987, China initiated a collaborative research project involving the exploitation of PGMS/TGMS lines to develop two-line system rice hybrids. This method has the following advantages over the classical CMS or three-line system:

There are no constraints caused by the restoration-maintenance relationship, since the male sterility of PGMS and TGMS is controlled by only one or two pairs of recessive genes. Nor is there any need for special R genes for fertility restoration, so the choice of parents in developing heterotic hybrids is greatly broadened. Experience has shown that more than 95 percent of existing rice cultivars can restore fertility of the PGMS/TGMS lines to a normal degree, while in the CMS system only about 5 percent of the existing lines can be used as restorers.

· The cost of hybrid seed will be cut down because of the simplified production procedure. With the PGMS/TGMS system, no maintainer line is needed in seed multiplication for male sterile lines.

· Negative effects from the sterile cytoplasm will be avoided, and the vulnerability to destructive diseases or insects that is usually induced by a unitary cytoplasmic resource may be eliminated.

· After more than ten years of nationwide collaborative studies, important progress has been made in two-line system hybrid rice, both in theoretical research and practical application, and a yield advantage of 10 percent over three-line system hybrids has been attained. With techniques advancing each day, two-line system hybrid rice is becoming more popular in large-scale application. The area planted to two-line system hybrid rice has reached about 64 000 ha.

· CNHRRDC has successfully bred the first practical and widely used PGMS/TGMS line, Pei'ai 64S, and the first two-line system hybrid combination, Pei'ai 64S/Teqing, certified for commercial use. Pei'ai 64S/Teqing has been expanded to cover more than 2 000 ha (in 1997) and has been ranked first five times in the history of rice production in Hunan as well as the whole country (Table 7).

Table 7　High grain yield records of Pei-ai 64S/Teqing

Location	Area /ha	Time	Cropping	Yield / (tonnes/ha)	Ranked first
Hunan regional trial		1992	Middle	9.47	Hunan yield trials
Yongsheng, Yunnan	0.1	1992	Single	17.11	Actual yield
Qianyang, Hunan	0.15	1994	Middle	12.96	Middle crop in Hunan
Xiangtang, Hunan	0.2	1994	Late	11.63	Late crop in China
Hanshou, Hunan	0.2	1995	Single+ratooning	15.12	Ratoon rice in Hunan

These new records indicate that the two-line system hybrid rice has broken through the old yield plateau established long ago by the three-line system hybrid rice and that it has shown great potential for increasing yields even though it is still in its initial stages.

Up to now, ten hybrid combinations and more than ten commercially used PGMS/TGMS lines have been certified and registered in China. It should be noted, however, that the key selection criterion for PGMS/TGMS lines is that the critical sterility-inducing temperature (CST) must be relatively low and should be based on historical meteorological data from the target areas. For example, in the rice regions in central China, the CST of commercial PGMS/TGMS lines should be limited to about 23.5 °C.

Another important consideration is the drift up of CST when seed multiplication of PGMS/TGMS lines is normally conducted without any artificial selection. A "core seed" concept has been proposed (Yuan, 1994b) and an effective production procedure for core and foundation seeds has been established so as to ensure that the PGMS/TGMS lines are reliable on a commercial scale (Yuan, 1994b; Deng, Fu and Yuan, 1997).

BREEDING FOR SUPER HYBRID RICE

The target of super high-yielding rice

Several research programmes have been dedicated to super high-yielding rice since 1980s but there is apparently no consensus on what the yield level of these varieties should be. Based on the present situation of hybrid rice production and breeding, and especially considering the breakthrough achieved recently in China, CNHRRDC has set up a super hybrid rice breeding programme with the aim of achieving a super hybrid yield of 100 kg/ha per day by the year 2000 (Yuan, 1997).

The morphological features of super high-yielding hybrid rice

Recently, several two-line intersubspecific hybrids with super high-yielding potential have been identified. For instance, the yield trial of Pei'ai 64S/E32 was conducted on a total area of 0.24 ha at three locations in 1997, where its average yield was as high as 13.26 tonnes/ha and its growth duration 130 days. Pei'ai 64S/E32 has reached the standard of super high-yielding rice, although only in trials on small plots, and it is a good example for breeding super high-yielding rice. Recent studies and analyses on Pei'ai 64S/E32 have shown that the most important morphological feature of super high-yielding rice lies in the uppermost three leaves, which should be long, erect, narrow, V-shaped and thick. These morphological features allow the very rich source of assimilates necessary for a super yield.

Another prominent feature is that this hybrid has both large and uniform panicles in population. The number of panicles per m^2 is 260 and the average grain weight per panicle is 5.3 g, while the coefficient of panicle weight variance is only 20 percent. This feature means that the hybrid not only has a very large sink but also a high population photosynthesis efficiency.

The third outstanding characteristic of this hybrid is that it has a taller erect-leaved canopy with drooping panicles. Its canopy height is more than 120 cm while the top of the filled panicles is only 60 to 70 cm above the ground. Such an architecture is more productive in terms of photosynthesis and, at the same time, makes the hybrid highly resistant to lodging, which is also one of the essential characters required for breeding super high-yielding rice (Fig. 2).

Figure 2　Super hybrid rice plant

Strategies for breeding super hybrid rice

ⅰ) Utilization of intersubspecific heterosis. It has been proved that the heterosis of intersubspecific hybrids is much stronger than that of intervarietal hybrids; therefore, utilization of intersubspecific hybrids is the most feasible approach for realizing super yields. At present efforts have been focused on using Pei´ai 64S as the major female parent in the selection of super high-yielding combinations. Because Pei´ai 64S is an intermediate type between *indica* and *japonica*, it has a very wide compatibility. To exploit the heterosis of intersubspecific hybrids and improve the efficiency of super high-yielding hybrid breeding, the emphasis is on the development of various widely compatible lines, especially those that have a broad spectrum of compatibility, including restorer lines and male sterile lines of *indica* type, *japonica* type and the intermediate type with different growth durations. This will ensure abundant parental lines for various super high-yielding hybrids that can be adapted to different ecological environments.

ii) Utilization of favourable genes from wild rice. In 1995, based on molecular analysis and field experiments carried out as part of a cooperative research programme with Cornell University, CNHRRDC identified two favourable QTL genes (*yld1* and *yld2*) from wild rice (*O. rufipogon L.*). Each of the QTL genes contributed to a yield advantage of 18 percent over the high-yielding hybrid V64 (one of the most élite hybrids in China, with a yield potential of 80 kg/ha per day). By means of molecular marker-facilitated backcrossing and selection, the development of near-isogenic lines carrying these two QTL genes is under way (Xiao et al., 1996).

iii) Utilization of IRRI's new plant types. Khush predicted that: "These new plant types are likely to have 20 percent higher yield potential than the existing high-yielding indicas. The new plant types will be employed in developing *indica/japonica* hybrids which may have a yield advantage of 20 to 25 percent over the best inbred lines. A combination of the two approaches may raise the yield potential of tropical rice by 50 percent." It seems that IRRI's new plant type varieties could play an important role in hybrid rice breeding for super yields in the future by using them as parental lines.

iv) Utilization of biotechnology. Genetic engineering techniques, such as anther culture, marker-aided selection and gene transformation, offer reliable opportunities for accelerating breeding progress, increasing selection efficiency and transferring genes across species and generic barriers. These will play an important role in breeding for super hybrid rice. For example, if biotechnology can be used to transfer apomixis to rice from grass species, hybrid rice production will be revolutionized and will reach even higher levels.

CONCLUSIONS

The whole package of hybrid rice technology, mainly developed in China, is now available to most rice-growing countries. Generally speaking, the intervarietal hybrids can outyield improved pure line varieties by a margin of around 20 percent and the intersubspecific hybrids have a further 15 to 20 percent advantage.

The classic three-line method can be applied universally while the two-line method, although more advanced, can only be used in certain areas where the climatic conditions are suitable for both hybrid seed production and PGMS/TGMS line multiplication. China's existing rice hybrids may be adaptable to temperate and subtropical regions for commercial production. In tropical areas it will be necessary to develop locally adaptable parental lines as well as hybrid combinations or to introduce IRRI's lines and hybrids for trial and use.

Those countries whose hybrid rice programmes are still in their initial stages should currently aim at developing intervarietal hybrids by using the three-line method. Some countries such as India and Viet Nam, where the development of hybrid rice technology is under way, could start two-line hybrid and intersubspecific hybrid research programmes in addition to using the three-line method. The one-line method and utilization of distant heterosis are is still at the experimental stage, and only highly advanced biotechnology institutes are likely to undertake research in these areas.

References

Deng, Q.Y., Fu, X.Q. & Yuan, L.P. 1997.On fertility stability of the P/TGMS lines and their identification technology. In *Proceedings of the International Symposium on Two-line System Heterosis Breeding in Crops*, p. 76–85.6–8 September 1997, Changsha, China.

FAO. 1995. *Technology of hybrid rice production*. By L.P. Yuan & X.Q. Fu. Rome. 84 pp.

Fu, X.Q. & Gong, S.W. 1994.The great achievements of hybrid rice in thirty years and its further developing strategies in China. *Hybrid Rice*, 1994 (3–4) : 17–21.

IRRI. 1980.Hybrid rice work revived at IRRI. *IRRI Reporter*.

Kato, H., Maruyama, K. & Uchiyamada, H. 1994.Hybrid rice research in Japan. In S.S. Virmani, ed. *Hybrid rice technology: new developments and future prospects*, p. 149–156.Los Baños, the Philippines, IRRI.

McWilliam, J.R., Ikehashi, H. & Sinha, S.K. 1995. Progress in hybrid rice in India. *Int. Rice Com. Newsl.*, 44: 80–87.

Nguyen, V.L., Minh, H.T. & Suan, N.V. 1994.Hybrid rice research in Vietnam. In S.S. Virmani, ed. *Hybrid rice technology: new developments and future prospects*, p. 187–193.Los Baños, the Philippines, IRRI.

Nguyen, V.L., Nguyen, V.S. & Virmani, S.S. 1995. Current status and future outlook on hybrid rice in Vietnam. In *Vietnam and IRRI-a partnership in rice research*, p. 73–80.

Paroda, R.S. & Siddiq, E.A. 1996.Current status of hybrid rice research and development in India. In M. Rangaswamy, P. Rangaswamy, K. Thiyagarajan & A.S. Ponnuswamy, eds. *Hybrid rice technology*, p.18–30.

Shi, M.S. 1981.The preliminary report on the breeding and application of a late-season *japonica* natural dual-purpose male sterile line in rice. *Hubei Agric. Sci.*, 1981 (7) : 1–3.

Shi, M.S. 1985.Discovery of and preliminary studies on a photosensitive recessively male sterile rice. *Sci. Agric. Sin.*, 18 (2) : 44–48.

Trinh, T.T. 1992.The FAO hybrid rice programme. *Int. Rice Com. Newsl.*, 41: 51–59.

Trinh, T.T. 1993.New developments in hybrid rice. *Int. Rice Com. Newsl.*, 42: 28–34.

Trinh, T.T. 1994.FAO's contribution to hybrid rice development. In S.S. Virmani, ed. *Hybrid rice technology: new developments and future prospects*, p. 267–274.Los Baños, the Philippines, IRRI.

Uehara, Y., Ohta, H., Shimizu, H. & Ostuki, H. 1997.Utilization of environment-sensitive genic male sterility in hybrid rice breeding. In *Proceedings of the International Symposium on Two-line System Heterosis Breeding in Crops*, p. 193–197.6–8 September 1997, Changsha, China.

Virmani, S.S., Casal, C.L., Toledo, R.S., Lopez, M.T., Murakami, H.G. & Manalo, J.O. 1996.Tropical rice hybrids introduced to Asian farmers. *Asian Seed*, 3 (1) : 17–18.

Xiao, J.H., Grandillo, S., Ahn, S.N., McCouch, S.R., Tanksley, S.D., Li, J.M. & Yuan, L.P. 1996.Genes from wild rice improve yield. *Nature*, 384: 223–224.

Yin, H.Q. 1997.*Technical cooperation project on hybrid rice in Viet Nam*. Consultancy Mission Report on FAO Project TCP/VIE/6614 (T) . Rome, FAO.

Yuan, L.P. 1966.A preliminary report on male sterility in rice. *Sci. Bull.*, (4) : 32–34. (in Chinese)

230

Yuan, L.P. 1994a. Development and perspective of hybrid rice research in China. *Hunan Agric. Res. Newsl.*,1 (1) : 7–8.

Yuan, L.P. 1994b. Purification and production of foundation seed of rice PGMS and TGMS lines. *Hybrid Rice*, 1994 (6) : 1–3.

Yuan, L.P. 1996.*Hybrid Rice*, (6) :1–3.

Yuan, L.P. 1997.*Hybrid Rice*, (6) :1–3.

Yuan, L.P. & Virmani, S.S. 1986.Status of hybrid rice research and development. In *Proceedings of the International Symposium on Hybrid Rice*, p. 7–24.6–10 October, Changsha, China.

作者: Yuan Longping

注: 本文为发表于 1998 年 9 月 7—10 日在埃及开罗召开的国际大米委员会上的论文。

杂交水稻选育的回顾、现状与展望

一、简略回顾

水稻是我国主要的粮食作物之一，其播种面积占粮食作物播种面积的 30%，而其产量却占粮食作物总产量的 42.2%，单位面积产量比整个粮食作物平均高出 45.7%。其中一个重要原因是水稻杂种优势广泛应用发挥了举足轻重的作用。我国于 1964 年开始研究杂交水稻，1973 年实现三系配套，1976 年开始大面积推广。实现了我国水稻育种的第二次突破。现在杂交水稻常年种植面积为 1 470 万 hm² 左右，占水稻总播种面积的 50%~55%，比常规稻增产 20%~30%。因此，三系法杂交水稻的培育成功和大面积推广，为大幅度提高我国的粮食产量作出了重大的贡献。但是，随着科学技术的进步、生产的发展、杂交水稻育种水平的不断提高，三系法杂交稻在生产中也暴露出一些缺陷。主要是：①种子生产程序烦琐，既要用雄性不育系（A）与雄性不育保持系（B）杂交以繁殖雄性不育系种子，还要用雄性不育系（A）与雄性不育恢复系（R）杂交制种获得杂交稻种子；②受恢保关系的限制，应用三系法途径不能充分利用自然界与水稻有关的种质资源，在某种程度上增加了进一步利用水稻杂种优势的难度，选育出高产、优质、抗性强的杂交稻组合的概率较低，导致产量徘徊不前；③缺乏高产、优质、熟期适宜的杂交早稻组合。上述原因使得 20 世纪 90 年代初以来，我国杂交水稻的种植面积徘徊不前，针对上述问题，杂交水稻育种在朝着高产、优质、多抗及亚种间杂种优势利用的方向发展。

二、现状与展望

随着光温敏核不育水稻的发现和两系法杂交水稻研究的深入开展，我国在杂交水稻研究方面又取得了重大突破。

1. 两系法杂交水稻研究取得成功

两系法杂种优势利用只需要光、温敏核不育系和恢复系两个育种材料。光、温敏核不育系在低温和／或短日照条件下表现可育、自交结实、繁殖种子；在高温和／或长日照条件下，表现不育，可以用来与恢复系制种，生产杂交种子。由于光、温敏核不育系能一系两用，与三系法相比，两系法种子生产就少了一个环节，简化了种子生产程序，更为有利的是，光温敏核不育系由简单核基因控制。从理论上讲，任何优良的育种材料都可以培育成光、温敏核不育系；而且水稻种质资源中，98% 以上的育种材料都可用作两系法中的恢复系，这就极大地提高了选配杂交稻组合的自由度，也增加了选配优良杂交稻组合的概率，两系法杂交水稻的研究，已取得了令人瞩目的进展。

（1）掌握了光、温敏核不育系育性转换与光、温变化关系的基本规律，从而提炼出了选育实用光温敏核不育系的 4 项指标，使制种风险率降到 1% 以下。

①导致不育的起点温度低。这是最重要、最关键的指标。具体的温度指标因地而异，在长江流域应定为日均温 23 ℃（光温处理时，变幅为 19 ℃～27 ℃），偏北地区须更低，华南地区可稍高。

②光敏温度范围宽（22 ℃～23 ℃至 29 ℃～30 ℃），即光、温敏核不育系在此温度范围内，长日诱导不育，短日诱导可育。

③临界光长短。以 13 小时为宜，这样才有更广泛的应用范围。

④长日对低温、短日对高温的不足补偿作用强。即不育系在长日条件下，遇上低于起点温度的低温仍然表现为不育。或在短日条件下，遇上高于临界高温的温度时仍然表现为可育。

（2）全国已有以培矮 64S 为代表的 10 多个实用光、温敏核不育系通过了省级鉴定。

（3）育成了一批优良两系法杂交稻组合，10 多个组合通过了省级审定。这些组合在大面积示范中表现良好，1998 年示范推广面积达 56.73 万 hm^2，至此全国累计种植面积达 123.43 万 hm^2。1999 年计划种植 70.44 万 hm^2，已开始进入快速发展阶段。与三系法杂交水稻相比，两系法杂交水稻的增产幅度在 10% 左右。其中，以培矮 64S/ 特青表现突出，在我国中南部主要稻区多点、多年试种，处处成功、季季增产，并创造了 5 项高产纪录（表 1），累计试种面积已超过 40 万 hm^2。

表 1　培矮 64S/ 特青试种高产纪录

年份	地点	季别	面积 /m^2	产量 /（kg/ 亩）	纪录类型
1992	湖南省区试	中稻		630	湖南省区试产量纪录
1992	云南永胜	一季稻	1 000	1 141	全国单产纪录
1994	湖南黔阳	中稻	1 534	864	湖南省中稻单产纪录

续表

年份	地点	季别	面积/m²	产量/（kg/亩）	纪录类型
1994	湖南湘潭	晚稻	2 000	775.5	全国晚稻单产纪录
1995	湖南汉寿	一季加再生	2 000	1 008	湖南再生稻单产纪录

（4）在敏感期采用冷水串灌法彻底解决了低温敏核不育系繁殖产量低而不稳的难题。

（5）大面积制种产量已达三系杂交水稻的平均水平。

（6）培矮64S系列组合的不育系繁殖、制种和大田生产面积比例为1∶100∶15 000，高于三系法杂交水稻（1∶50∶6 000）。

（7）建立了以生产核心种子为关键环节的光、温敏核不育系的提纯和原种生产程序，从而能在一定程度上防止逐代升高的不育起点温度飘移现象，保证不育系的不育起点温度始终保持在同一水平上。水稻光、温敏核不育系的提纯和原种生产程序为：单株选择→低温或长日低温处理→再生留种（核心种子）→原原种→原种→制种。

（8）提出了育性转换受1~2对主基因控制，而起点温度的高低为数量性状遗传的假说。该假说不仅圆满地解释了起点温度发生飘移的原因，而且为根除飘移现象指明了选育对策。

以上研究展示了两系法杂交水稻在技术上业已基本成熟，完全可以因地制宜地在大面积上推广应用。

2. 长江流域杂交水稻"优而不早、高而不优"的难题得以解决

长江流域早稻种植面积为530万hm²，其中杂交早稻只占10%左右。与此相反，长江流域晚稻总面积的90%以上为杂交晚稻。究其原因，主要是目前推广的三系法杂交早稻组合产量虽高，但生育期太长，而且品质较差，特别是整精米率低；如大面积推广的威优49的整精米率仅为20%~30%。由于两系法不受恢保关系的限制，配组自由，因而近年来应用两系法技术，在选育长江流域早中熟杂交早稻组合方面取得了长足的进展。新育成的中熟两系杂交早稻组合香125S/D68在湖南省区试中单产498 kg/亩，与三系法迟熟杂交组合威优402持平，日产量居第一位，生育期110天，比威优402短4~5天，品质达部颁二级优质米标准，已于1998年2月破格审定。该组合在湖南、湖北、江西、浙江、安徽、福建等地试种，单产都在500 kg/亩左右。同时，香125S/D69、香125S/139、早S/D68等一批表现更好的苗头组合也正在参加各级试验示范。

3. 超级杂交稻育种的理论体系日趋完善

粮食问题始终是关系到国计民生的头等大事。我国人口众多，人均耕地面积少，解决21世纪的吃饭问题，唯一出路在于依靠科技进步，提高粮食作物的单位面积产量。农业生产和作物育种的实践表明，凡在作物育种上有所突破，就会给作物生产带来一次飞跃。迄今，我国在水稻育种上已有两次突破，并且都处于世界领先水平。第一次是矮秆水稻的培育成功，第二次是杂交水稻的研究成功。两次突破使单产潜力均在原有品种的基础上增加20%左右。现正在启动的超级杂交稻研究，其产量指标是比现有杂交稻增产30%左右。它的实现将是水稻育种上的第三次突破。因此，培育和推广超级杂交稻对于解决我国21世纪的粮食问题具有极其重大的战略意义。

超级稻的选育是当今世界上的热点，也是难点。水稻超高产育种最早由日本人于1980年提出。他们计划用15年时间，在1995年育成比当时对照品种增产50%，糙米产量达到$7.5 \sim 9.5 \ t/hm^2$的超高产水稻品种。但因难度太大，且技术路线不妥，至今没有实现。1990年，国际水稻研究所启动了新株型育种项目，计划到2005年育成生育期120天、产量潜力达到$12 \ t/hm^2$的水稻品种。我国近年来也由农业部牵头，确立了一个超级水稻育种项目，计划在2000年在较大面积上实现单产$600 \sim 700 \ kg/$亩，即$9.0 \sim 10.5 \ t/hm^2$。各家所走的技术路线基本上都是培育新株型或理想株型和利用籼粳亚种间杂交优势，然而，在研究过程中，国际水稻研究所发现新育成的新株型材料虽库容很大，但籽粒充实不良。我国在籼粳亚种间杂种优势利用方面也遇到了同样的难题。袁隆平院士针对这些难题，通过反复实践和思考，认为这与过去育种家片面追求大库、重库轻源有关。有效增源是实现水稻超高产育种的关键。要达此目标，除了稻株形态改良和杂种优势利用，别无其他选择。因此，特提出下面的超级杂交稻的选育目标和超高产稻株的形态模式。

（1）超级杂交稻的选育目标：中晚稻日产量$100 \ kg/hm^2$，早稻日产量$90 \ kg/hm^2$；米质达部颁二级优质米标准；抗2种以上主要病虫害。

（2）超高产稻株的形态模式：株高$100 \ cm$左右；秆高$70 \ cm$左右；叶片长、直、窄、凹、厚。长（上部3片叶长$50 \sim 60 \ cm$），直（剑叶、倒2叶、倒3叶的角度分别为$5°$、$10°$、$20°$左右），窄（宽$2 \ cm$左右），凹（似V字形），厚（上部3片叶每$100 \ cm^2$干重$1 \ g$以上）；株型适度紧凑，分蘖中等，叶下禾，冠层只见叶片，灌浆后不见稻穗；每m^2穗数270个（每亩18万个）左右；每穗重$5 \ g$左右；上部3片叶的叶面积指数为6.5左右；叶粒比 100∶2.3 左右，即生产 2.3 g 稻谷的上部 3 片叶的面积为 100 cm² 左右；收获指数 0.55 以上。

该模式与前人提出的新株型或理想株型的不同之处在于强调有效增源。前人往往强调适度增加株高以提高光合效率，同时增强茎秆硬度以提高抗倒伏能力，但这将使大量的养分分配到茎秆上，从而会降低收获指数，很不经济。本文提出的超高产稻株模式强调叶片长、直、窄、凹、厚、冠层高，而茎秆矮，同时强调穗下垂，这样既充分扩大了有效的光合面积，又能保证茎秆较矮，重心较低，从而可避免倒伏。

选育超级杂交稻可以从以下几方面着手：一是利用籼粳亚种间杂种优势。二是利用野生稻中的有利基因，现已发现野生稻中隐藏着一些高产的优势基因，应用分子标记辅助选择技术，将其转移到杂交稻中可以大幅度提高产量。三是利用国内外的优异种质资源。国际水稻研究所新育成的新株型稻材料很有特色。利用其作亲本培育亚种间强优势杂交组合，可解决增源的问题。因为亚种间杂交稻具有根系发达、光合效率高的优点，完全可以弥补"新株型稻"源不足的缺陷。四是针对超级杂交稻的特点，研究配套的栽培技术模式，使其充分发挥它的增产潜力。

4. 超级杂交稻组合的选育初露端倪

根据上述的超高产稻株形态模式，利用亚种间杂种优势，目前已初步选育出了几个超高产苗头组合。其中最具有代表性的是培矮 64S/E32（该组合系国家杂交水稻工程技术研究中心与江苏农科院合作选配的），1997 年在江苏 3 个点试种 0.24 hm²，平均单产 884 kg/ 亩，生育期 130 天，日产量达到 6.8 kg/ 亩（即 102 kg/hm²）。在小面积上达到了超级杂交稻的选育目标。1998 年又在江苏、湖南 4 个试点种植 2.5 hm²，平均单产都超过了 800 kg/ 亩。培矮 64S/E32 的株叶形态与上面提出的超高产稻株模式相似，每平方米有效穗 260 个左右，单穗重 5.3 g 左右，冠层高 120 cm 左右，但成熟时穗尖离地面高仅 60 cm 左右，抗倒能力特强。1999 年，仅湖南就安排了 30 个百亩片的培矮 64S/E32 超级杂交稻示范。

根据目前的育种进展，超级杂交稻育种计划运转良好。预计到 2001 年可育成 2～3 个准超高产组合（长江流域中、晚稻组合在一个生态区 2 年 2 个百亩片上达到日产 90 kg/hm²，或比当地主栽组合增产 15% 以上；早稻组合在一个生态区 2 年 2 个百亩片上日产达到 80 kg/hm²）和 1～2 个超高产组合（在一个生态区 2 年 2 个百亩片上日产达到

100 kg/hm^2），到 2002 年力争建立超级杂交稻综合育种体系。同时，中试、示范和推广研究都得到加强，使所取得的研究成果与各地高新技术开发区紧密结合，加快成果转化步伐，推进超级杂交稻的产业化。

作者：袁隆平　唐传道

注：本文发表于《中国稻米》1999 年第 4 期。

水稻强化栽培体系

水稻强化栽培体系（System of Rice Intensification，SRI）系 20 世纪 80 年代由 Henri de Laulanie 神父在马达加斯加（Madagascar）提出的一种新的栽培方法。在马达加斯加应用多年获得了很好的增产效果，近年在其他国家的试验结果也显示其具有较大的增产潜力。为使我国农业科技人员了解和应用这一新的方法，现对 SRI 的基本情况作一简要介绍。

1 采用 SRI 方法能获得多高的产量

在马达加斯加积极提倡 SRI 的非政府组织——Tefy Saina 发现，采用 SRI 方法，任何品种的产量至少可增加 1 倍，而且通常比这高得多。在 Ranomafana 国家公园附近，传统栽培法的灌溉水稻产量为 2 t/hm² 左右，而采用 SRI 法，过去 5 年的平均产量是 8 t/hm² 以上。在这段时期，运用 SRI 的农民从 38 户增加到 396 户，面积超过 50 hm²。根据国际自然资源保护委员会田间工作人员报道，在 Zahamena 国家公园北部附近，有些农民获得类似的增产幅度，从每公顷 2 t 增至 8~9 t。上述地区属热带雨林边缘地带。

马达加斯加的其他地方也具有类似的增产纪录。在中央高原（海拔 1 000 m 以上）因为精耕细作和较多的现代化投入，那里的水稻产量比 Ranomafana 和 Zahamena 地区的高。1995—1996 年对 108 户首次采用 SRI 的农民进行调查发现，在首都 Antananarivo 附近，每公顷平均产量从 3.2 t 增至 6.3 t，在 Antsirabe 附近，为 3.9 t 增至 8 t，在北部的 Andapa，一家私营公司（SOAMA）报道，农民采用高产品种和"最佳"施肥的产量为 6.2 t/hm²，27 户农民采用 SRI 的产量达 10.2 t/hm²。另一家公司（FITABE）报道，在西北部 Marovoay 附近，农民应用"现代化"栽培法的产量为 4.8 t/hm²，SRI 法为 7.1 t/hm²。

在 Ranomafana 附近，有些农民能很有效地运用 SRI 技术，他们的产量是每公顷 14~16 t；在 Fianarantsoa 地区一些农民获得的产量更高。1999 年 5 月，N. Uphoff 与马达加斯加农业研究和推广局的一位研究员 B. Andrianaivo（曾在国际水稻研究所培训过的农艺师）访问了在 Soatanana 附近的一位农民，他在 0.125 hm² 的土地上收获了 2 740 kg 稻谷，折合每公顷 21 t，这是该国平均单产的 10 倍。这位农民是第 6 年运用 SRI，已非常熟练地掌握了该项技术。由于施用了大量配制良好的堆肥（40 t/hm²），达到如此的超高产是可能的。从近乎海平面到海拔 1 200 m 以上以及在多种土壤类型中都取得了上述的增产，尽管以排水良好的土壤产量最高，但这并不需要"肥沃"的土壤，在马达加斯加大多数土壤都很贫瘠。

在其他国家有类似产量结果的证据。1999 年，中国南京农业大学农学系用不同的株行距（20 cm×20 cm 到 30 cm×30 cm）进行 SRI 试验，产量为 9.2~10.5 t/hm²，后者是稀植的产量。印度尼西亚农业研究和发展署在 Sukamandi 水稻试验站首次试验的产量为 6.2~6.8 t/hm²（1999 年旱季），第 2 次试验的产量达 9.5 t/hm²（1999—2000 年雨季）。

菲律宾和柬埔寨的非政府组织报道，他们与小农合作，采用 SRI 方法，第 1 年就增产 1 倍以上（分别达 5~6 t/hm²），由于植株的这种生长方式和抗虫抗病，农民非常渴望继续应用 SRI，因此对 SRI 的评价正在高涨。2000 年，国际水稻研究所和世界农业研究和发展署在进行 SRI 试验，斯里兰卡许多农民在其农业部的鼓励下亦在采用 SRI。

2 基本观点

SRI 发现并证明能帮助水稻实现其真正产量潜力的一些重要栽培方法，而这种潜力被以前的栽培措施掩盖了。SRI 所立论的基本观点是：

a）农民首先应了解这些观点，然后选用和评估适合他们自己条件最有利的具体措施。

b）SRI 通过改变对土壤和水分的管理措施，改变稻株的结构——根和分蘖的密度和数量，使它们变得更高产。

c）为使稻株更高产，它们必须每株有更多的分蘖、更多的有效穗、更多的每穗粒数、更大的籽粒。

d）如果水稻是稀植而不是密植，它们就有较大的生长空间，能获得较多的阳光和空气，从而能产生更多的分蘖，这些分蘖大多数会成为有效穗。同时，生长空间大，其根系发达，能更好地从土壤中吸取养分。

少株稀植，每个植株都能苗壮成长，从而就可能生产出更多的稻谷，虽然这有点出人意料。为了使地上部分生长良好，地下部就必须要有强健的根系。SRI 的"诀窍"是每个植株地下部分要有强大的根系和地上部分要有较多的分蘖、叶片和谷粒。

e）密植丛栽，不仅浪费种子，而且个体变小和生产力低。栽较多的植株并不比栽较少但生产力高的植株有利，栽植老秧也消耗其潜力。

采用 SRI 方法，每个单株很容易获得 50 个分蘖，有些农民用此法得到了多达 100 个以上的分蘖。农民在每个有效蘖上能得到 200 粒谷，最好的能多达 400 粒。这不是奇迹。这是来自于对植株、土壤、水分的良好的管理，使其生长和产量潜力得以充分发挥的结果。

3　怎样才能产生更多的分蘖

应用 SRI 成功的关键是幼苗早插，通常秧龄不到 15 d，甚至短到 8~10 d，当秧苗只长出 2 片小叶时就移栽。如果插 3 周、4 周、5 周或 6 周的老秧，就会大大丧失它们大量分蘖的潜力。

秧苗自苗床取出后，如果延迟移栽就会受到损伤，而应在拔秧后半小时内，最好 15 min 内即移栽。如果把秧苗插入本田而不是轻轻摆放，秧苗就要消耗很多的能量来恢复根的生长，干扰其发育。早而仔细地移栽秧苗有助于在大田恢复生长且不会降低其高产的潜力，尤其要重视促进根系的强健生长。

4　怎样促进根系生长强健

4.1　首先是要栽单苗而不是像通常那样，3~4 苗或更多的苗丛栽

当几根苗丛栽在一起时，它们的根必然互相竞争，这正如稻株与杂草长在一起一样，为了争夺养分、水分和阳光而相互竞争。稀植非常重要，至少间隔 25 cm 以上，最好是正方形，以利除草，同时有利于接受更多的阳光和空气。水稻稀植，如果土壤状况良好，根系就有大量伸展的空间，特别是当它们没有相互竞争的时候。单苗稀植，每平方米可以只栽 10 苗或 16 苗而不是 50 苗或 100 苗。获得最高产量的是每平方米仅栽 4 苗即 50 cm × 50 cm，水稻植株长得像灌木丛一样。稀植每公顷可节约多达 100 kg 的种子，同时又能大大提高产量，因为稻株能产生更多的分蘖和籽粒。

4.2　影响稻根粗细和健壮的一个很重要的因素是如何移栽幼嫩秧苗的方法

秧苗直插，根尖会朝上，移栽后的秧像 J 字形。稻根为顶端生长，如果根尖朝上，在恢复生长之前，根必须改变其在土壤中的位置，使之向下。这样，幼弱的根就要消耗很多能量。SIR 不直插秧苗而是轻轻地斜摆，使根平卧在湿润的土壤表层，移栽后的秧像 L 形而不像 J

形。这种 L 形状的秧苗，根尖易往下扎，且耗能少，有利长出更多的新根和促进分蘖。

4.3　与水稻常规栽培措施不同的主要特点是稻田不连续不间断地建立水层

这种革新与移栽幼小苗具有相同的重要性。水稻在生长阶段仅需湿润而且水分不饱和的土壤，稻田应间断地排干到开坼，这一点与多数人对水稻的观点大相径庭，但这是正确的。SRI 的一个重要发现是水稻并非水生植物。虽然水稻能在淹水条件下生存但难以苗壮。水稻在淹水条件下不如根能直接从空气中获得氧气时生长良好。有水层时水稻会对这种环境进行调节，根中形成小气囊（通气组织）使氧气能从地上部分运抵根部。但这不是植物生长的理想条件，它干扰根所吸收的养分运送到分蘖和叶片。

SRI 还发现，水稻抽穗、结实前，在分蘖和长叶期间，土壤仅需保持湿润，但在生殖生长阶段则要建立 1~2 cm 的薄水层。很奇怪，间歇地排干稻田，甚至每周一次，对水稻的生长很有利。这可使更多的氧气进入土壤而达到根部。土壤水分不饱和时，稻根就需深扎以寻求水分。当稻根周围充满大量水分时，根就会变"懒"而不需要大量生长，因而限制了它们从土壤中吸取养分的能力。

当水稻进入抽穗开花期，稻田需保持 1~2 cm 的水层，以支持谷粒形成。收获前 25 d 左右要彻底排干。（对这个问题，本人与 N. Uphoff 教授面谈交换意见时，提出了不同的看法，认为在成熟前 7 d 左右断水为宜，他同意我的观点。）

稻田不长期建立水层，会滋生杂草，因此应尽力消灭杂草使它们不与水稻争夺养料和水分。农民可使用一种很简便的手推除草器（狼牙棒）进行中耕除草。第一次中耕除草应在移栽后 10~12 d 进行，两周后至少要再中耕一次。这样一方面可消灭杂草，另一方面可使更多的空气进入土壤为稻根利用。在始穗前再中耕除草 1~2 次（总共 3~4 次）能给土壤增加更多的氧气，这一点比消除残余的杂草更重要。增加中耕除草次数能大大提高产量。

4.4　由于化肥价格昂贵，农民买不起，而且有时供应不及时，因此 SRI 建议用堆肥或厩肥

由于采用 SRI 法的产量非常之高，就必须增加和补充土壤中的养分。施用堆肥和厩肥来肥沃土壤，可改良土壤结构，因而能促进根的生长。堆肥释放养分比化肥慢得多，因此这种肥源对植物生长更加有利。

制作堆肥和施入稻田很费工夫，但经验表明，这对农民是有利的投入，因为质地好的土壤有利于根系生长和最终表现良好。增施化肥通常能提高产量，但化肥并非有机肥的良好代替品。

以上是改革水稻生产的基本观点，一旦你了解到怎样能促进水稻长出强大的根系以及产生

更多的分蘖，其自然的结果就是你的稻田能产出更多的稻谷。

5　需要——逼迫型对 N 的吸收

现在已知水稻对 N 的吸收效率很低，而且在多量施化学 N 肥时会有极显著的报酬递减效应，施入稻田的 N 只有 30% ~ 40% 为稻株所吸收。Kirk 和 Bouldin 的研究指出，N 的吸收率与根表面周围 N 的浓度无关。稻株对 N 的吸收具有很复杂的机制（意即尚不知道），当稻株内部的 N 充足时，它会下调对 N 的吸收能力，而当它很需要时就会上调这种能力。

由需要来推动 N 的吸收比由供给来推动更加重要，这种模式可以解释施 N 的低效率和报酬递减效应。我们一直过高估计了植物对 N 的需要，很可能是由于我们太致力于对 N（低效）的供应。看来，当需要时水稻能从根际得到足够的 N，植株的生长主要不是受制于 N 肥的施用。

采用 SRI 措施，植株的分蘖和根系呈指数的方式快速增长，这种生长模式可造成对 N 相当大的需求，从而使根系上调吸 N 的能力以满足其需要。

6　SRI 的栽培措施

SRI 在以下几点改变传统的植物—土壤—水分—营养管理方式：

a）不栽 3 ~ 4 周或更长秧龄的壮秧，而是栽 8 ~ 12 d 秧龄，仅长出 2 片小叶，单根且种子还附在其上的极幼嫩的小苗。如果移栽非常仔细，很快就会返青。此法能保持水稻极大的分蘖潜力，每株 30 ~ 80 个分蘖，如果和其他 SRI 措施配套的话，甚至可多达 100 个以上。

b）不是每穴 3 ~ 4 苗或更多的苗丛栽，而是每穴栽 1 苗。这意味着没有多苗根系之间的竞争。种内竞争对植物的生长并不比种间竞争（例如杂草）更有利。这样做的结果是根系十分发达。

c）不密植而是稀植。这会使根系的发育加强，产量提高，因为稀植使分蘖大大增加和籽粒充实度大大提高。但应注意合理稀植而不是越稀越好，因为追求的是单位面积上的产量而不是单株的产量。

SRI 采用正方形而不是条形栽植。最适的距离视土壤和其他条件而定，但最好是从 25 cm × 25 cm 开始，根据试验，或加大或缩小。最高的产量是插 50 cm × 50 cm（每平方米 4 株）。

d）在整个生长期，稻田不建立水层（有水层一直被认为是获得高产所必需的条件），而是在营养生长期保持湿润但不淹水，实行间歇地"轻度"灌溉，土壤水分决不饱和。据 N.

Uphoff 等的观察，如果让稻田不时地干燥几天，达到表面开坼，有助于根际通气，从而使水稻生长良好。

SRI 水分管理制度很难使人接受，因为长期以来一直认为"水稻是一种水生植物"。但是，根据 N. Uphoff 等的观察，水稻虽能在淹水条件下生存但却不能茁壮成长。淹水使土壤不通气，稻根易缺氧。众所周知，在淹水条件下水稻的根在抽穗开花期会早衰。但是，这一直被视为是"自然的"，这种条件是否对水稻有利人们从不提出疑问和评价。

印度的科学家在水分饱和与不饱和的土壤种植同一高产品种（台中本地 1 号），发现在稻穗分化后和谷粒形成及灌浆开始时，在淹水条件下 78% 的根退化了，而在排水良好土壤中，根不退化。很难想象在开始生产谷粒时如此大量的丧失根的能力不会对产量有负作用。

N. Uphoff 等的研究表明，用 SRI 方法在通气条件下种植水稻，根系发育特好。从土壤中拔出稻株所需的力，常规法栽培一丛为 3 株的是 28 kg，SRI 法一株为 53 kg，几乎是前者的 6 倍。

e）SRI 须进行几次中耕除草。在孕穗前至少 2 次，最好是 4 次。对 76 户农民的调查资料分析发现，在中耕 2 次以上的基础上，每增加一次中耕除草每公顷可增产 1~2.5 t。

f）施用堆肥。每公顷产 21 t 稻谷的施用量为 40 t/hm^2，在种前作（旱作）时施下。

7 产 21 t/hm^2 的有关数据

a）面积：0.125 hm^2。

b）品种：台中 16 选系，生育期 120 d。

c）插植规格：50 cm×50 cm，幼嫩单苗浅栽。

d）产量结构：每株 80 个分蘖，70 个有效蘖，每平方米 280 个有效穗，每穗实粒数 260 粒，千粒重 29 g，每穗谷重 7.54 g。

作者：袁隆平

注：本文为作者摘译自美国康奈尔大学 N.Uphoff 教授的有关文章，

发表于《杂交水稻》2001 年第 16 卷第 4 期。

Recent Progress in the Development of Hybrid Rice in China

A.General Situation

· In recent years the total planting area of rice is around 31 million ha and the average yield is 6.2 t/ha in China.

· The acreage under hybrid rice is about 50% of the total rice area and its average yield is 6.9 t/ha while the yield of pure line varieties is 5.4 t/ha.

· Area of hybrid seed production is around 0.14 million ha and the average yield is 2.5 t/ha.

B.Constraints and Challenges

· The expansion of hybrid rice area stands still since 1991.Two major factors limiting further expansion of hybrid rice area.

1.Lack of very early maturity (105-110 days) combinations with good heterosis for the first crop in the middle south part of China. In this region, double cropping rice is practiced and the area under first crop is around 4.5 million ha in which hybrid rice only covers about 10%.

2.The heterosis of *japonica* hybrids is not as good as that of *indica* hybrids. The yield advantage by growing *japonica* hybrids is only around 10%. Besides, the purity is also a problem because the male sterility of BT type A lines is not stable enough to produce very pure hybrid seeds. Therefore farmers are not very much interested in growing *japonica* hybrids. There are six million ha of *japonica* rice in China. Because of the above-mentioned problems, the acreage under *japnica* hybrids is only about 4% of the total *japonica* rice area.

· The yield of the existing three line hybrids including newly developed ones has stagnated for years.

· The grain quality needs to be improved because the living standard of Chinese people has been raised to a rather high level recently.

C.Recent progress

· The development of two line hybrid rice has been achieved.

The planting area of two line hybrids reached 1.6 million ha in 2000 and about 2.5 million ha this year. Generally speaking, its yield is 5%−10% higher than that of the existing three line hybrids.

· Breeding of very early maturity hybrids with high yield potential and good grain quality through two line method is basically succeeded. For example, one hybrid variety−125S/D68, suitable for the first crop in the double cropping region, was released in 1998.Its planting area soon reached 60 000 ha in 2000 and the average yield was around 7.5 t/ha outyielding the hybrid CK and pure line CK by 10% and 20%, respectively.

· Great progress has been made in breeding of super hybrid rice. There are two pioneer super hybrid rice combinations developed recently.

1. Pei'ai 64S/E32

Year	Total area	Average yield	Maximum yield
1998	2.25 ha	12.02 t/ha	
1999	24 ha	11.1 t/ha	17.07 t/ha
2000	3 600 ha	9.8 t/ha	15.1 t/ha

2. Pei'ai 64S/9311

Year	Total area	Average yield	Maximum yield
1998	86 ha	10.79 t/ha	
1999	35 330 ha	10.2 t/ha	14.8 t/ha
2000	233 500 ha	9.6 t/ha	15.1 t/ha

Because of very high yield potential with good grain quality the expansion of these super hybrids is very fast. Their planting area reaches 1.4 million ha this year.

D. The goal of breeding super hybrid rice in the next 5 years (2001−2005)

· Single crop
Large scale: 12 t/ha
Yield potential: about 15 t/ha
· Double crop
1.First (early) season
Large scale: 10.5 t/ha
Yield potential: about 12 t/ha
2.Second (late) season

　　　　Large scale：11 t/ha

　　　　Yield potential：about 13 t/ha

So far, there have been two demonstration locations (7 ha each) where the average is over 12 t/ha. Especially the location in Fujian Province a three line hybrid (97A/Ming86) yielded 12.76 t/ha.

E. Major experience

　　· Raising yield potential

　　1.Extending gentic diversity to enhance heterosis level

　　　Mainly to development of inter-subspecific (*indica/japonica*) hybrids.

　　2.Morphological improvement

　　　The key is to have a huge source on the basis of large sink of *indica/japonica* hybrids, i.e. a taller erect-leaved canopy with drooping panicles. Such architecture is not only more productive in terms of photosynthesis, but also enables the hybrid to be highly resistant to lodging.

　　· Raising seed yield

　　1.Breeding of MS lines with high outcrossing rate

　　* Early and concentrating blooming daily

　　* High rate of exserted stigma

　　* Big stigma size

　　* Exserted panicles

　　2.Improvement of techniques

　　* Heavy dosage of GA3 application

　　* Increasing row ratio

　　* Conducting supplementary pollination on time

　　* More seedlings per hill for MS lines

F. Some policies promoting development of hybrid rice

　　· Based on the great success of large scale demonstration on hybrid rice in multi-locations, by the end of 1975, our central government made an important decision：providing 8 million yuan (equivalent to $ 4 000 000) financial suport for 4 000 ha hybrid seed production in Hainan island to accelerate commercial production of hybrid rice.

　　· Tax-free for seed companies and sometimes subsidized them if their seed yield was poor in the late 1970's.

　　· Establishment of a well organized hybrid seed production system：

　　1. Provincial seed company-resiponsible for purification of parental lines.

2.Prefecture seed company-mutiplication of A lines.

3.County seed company-F$_1$ seed production.

· Establishment of a high level research institute-China National Hybrid Rice R & D Center.

There are eight senior plant breeders, five agronomists, four seed production experts and four scientists in biotechnology.

作者: Yuan Longping

注: 本文为于 2001 年 5 月在孟加拉 "TC of IRRI-ADB" 上发表的论文。

Research and Development of Hybrid Rice in China

Introduction

Researches on developing hybrid rice in China was initiated in 1964. The genetic tools, i.e. A, B and R lines were developed in 1973. Several hybrid combinations with good heterosis and higher yield potential were identified in 1974. The techniques of hybrid seed production were established in 1975. In 1976 hybrid rice was released to farmers for commercial production. Since then the area under hybrid rice had been increasing year after year. Up to 1991, the yearly planting area under hybrid rice reached 17.6 million ha.

It has been proved practically for many years that hybrid rice has 20%-30% yield advantage over conventional pure line varieties. (Table 1)

Table 1　The area and yield of hybrid rice production in China from 1976 to 2000

| Year | Total rice area /mil. ha | Conventional rice | | | Conventional rice | |
		Areas /mil. ha	%	Yield / (kg/ha)	Yield / (kg/ha)	Year of hybrid rice over convent rice/%
1976	36.20	0.138 7	0.38	4 200.0	3 469.5	21.1
1977	35.53	2.066 7	5.82	5 383.5	3 514.5	53.2
1978	34.40	4.266 7	12.40	5 353.5	3 780.0	41.6
1979	33.87	5.000 0	14.76	5 260.5	4 069.5	29.3
1980	33.87	4.813 3	14.21	5 296.5	3 940.5	34.4
1981	29.93	5.133 3	17.15	5 317.5	4 113.0	29.3
1982	33.07	5.600 0	16.93	5 865.0	4 447.5	31.9
1983	33.13	6.733 3	20.32	6 375.0	4 774.5	33.5
1984	33.13	8.866 7	26.76	6 405.0	4 992.0	28.3
1985	32.07	8.400 0	26.19	6 472.5	4 816.5	34.4
1986	32.27	8.933 3	27.68	6 600.0	4 857.0	35.9
1987	32.33	10.933 3	33.82	6 615.0	4 779.0	38.4
1988	32.00	12.666 7	39.58	6 600.0	4 539.0	45.4
1989	32.67	13.000 0	39.79	6 615.0	4 786.5	38.2

Continued

Year	Total rice area /mil. ha	Conventional rice			Conventional rice	
		Areas /mil. ha	%	Yield / (kg/ha)	Yield / (kg/ha)	Year of hybrid rice over convent rice/%
1990	33.07	15.933 3	48.18	6 675.0	5 314.5	25.6
1991	32.60	17.600 0	53.99	6 565.5	4 551.0	44.3
1992	31.73	15.466 7	50.86	6 636.0	4 986.0	33.0
1993	30.00	15.400 0	51.33	6 675.0	4 950.0	34.8
1994	30.17	15.474 2	51.29	6 670.0	5 149.0	30.6
1995	30.74	15.926 3	51.81	6 678.0	5 098.0	31.0
1996	31.41	15.934 2	50.73	6 740.0	5 326.0	26.5
1997	31.76	16.134 1	50.80	6 765.0	5 408.0	25.1
1998	31.21	15.760 8	50.45	6 759.0	5 346.0	26.4
1999	31.28	16.124 8	51.55	6 784.0	5 335.0	27.2
2000	30.19	15.436 1	51.13	6 802.0	5 401.0	25.9

Source: Yuan, 1997; Fu and Gong, 1994.

In recent years hybrid rice covers 50% of the total rice area and its yield has been 6.7 t/ha while that of pure line varieties is 5.1~5.2 t/ha on average. In the meantime, the techniques of hybrid seed production have been well developed to give an average seed yield of 2.3 t/ha nationwide. The field area ratio between A line multiplication, hybrid seed production and F_1 commercial cultivation is about 1 : 50 : 6 000.

Constraints and Challenges

Although researches on the use of rice heterosis have made tremendous achievements during the last 25 years in China, from a strategic point of view, it is still in the juvenile stage. The hybrid rice breeding still possesses a brilliant future since its very high yield potential has not been fully tapped yet. Based on our studies, the development of hybrid rice breeding may involve the following phases:

First, in terms of breeding methodology there could be three approaches:

1. Three-line method or CMS system

2. Two-line method or PGMS and TGMS system

3. One-line method or apomixis system

Second, in terms of increasing the degree of heterosis, it could also be divided into three levels:

1.Inter-varietal hybrids

2.Inter-subspecific hybrids

3.Distant hybrids

Each of the phases marks a new breakthrough in rice breeding and will result in great yield increase if it is attained.

The existing hybrids used in commercial production belong to the category of inter-varietal hybrids by using three-line system. Many years' practices have proved that the three-line system is an effective way to develop hybrid varieties and will continue to play an important role in the next decade. However, there are some constraints in this system.

1.The yield has stagnated for years.

2.Poor source of male-sterility inducing cytoplasm. About 85% of the A lines used in commercial production still belong to "WA" type.

3.Weak heterosis in *japonica* hybrids.

4.Lack of very early maturity hybrids with good heterosis.

Future Strategies of Hybrid Rice Breeding

In order to further increase the yield potential of hybrid rice, we are focusing our effort on the following two research programs.

1.Development of two-line hybrids

2.Development of super hybrid rice

Development of Two-line System Hybrid Rice

This method is based on a new kind of male sterile materials called P/TGMS. The major feature of such P/TGMS lines is that: Under longer day-length and higher temperature (summer season) they show complete pollen sterility, and therefore can be used for hybrid seed production, while under shorter day-length and moderate temperature (autumn) they show almost normal fertility, so can multiply themselves by selfing. Advantages over the three-line method:

1.B line is no longer needed

2.The choice of parents for better combinations is greatly broadened

3.Unitary cytoplasm situation can be avoided

After 9 years of intensive studies, the technology of development of two-line hybrid rice was basically ripened and we began to release two line hybrids to farmers for commercial production in 1995. Since then, the area of two-line hybrids has been increasing year after year. Their yield advantage is 5%–10% over the three-line hybrids. (Table 2)

Table 2　Area and yield of two-line system hybrid rice in China (1995—1998)

Year	1995	1996	1997	1998
Area/1 000 ha	73	200	270	437
Yield/kg/ha	7 215	7 130	7 150	

It has been planned that the area under two line hybrids will reach 1.5 million ha in 2000.

Development of super hybrid rice

Based on the progress of hybrid rice breeding and especially a new breakthrough recently, we have set up a super hybrid rice breeding program. The target is to develop very high yield hybrids that can yield 100 kg/ha per day (The yield potential of the existing best hybrid variety is 75 kg/ha per day), which is expected to be realized by the year of 2000.

Recently, several two-line inter-subspecific hybrids with super high yield potential have been identified. Among them, the most promising one may be Pei′ai 64S/E32.In a yield trial of this hybrid at three locations in 1997, this hybrid yielded averagely as high as 13.26 tons/ha under a total area of 0.24 ha, which outyielded the best CK by 35%, and its growth duration was 130 days. So it has reached the standard of super high yield [100 kg/ (ha · day)] .

This year, we continue to conduct yield trial for this hybrid at 20 locations in two provinces. The area at each location ranges from 0.1 to 2 ha. Now, this hybrid is at ripening stage. Preliminary data show that its yield can be again as high as around 12 tons/ha at most locations. Therefore, we have full confidence that we shall succeed in developing super hybrid rice by the end of this century. It is indeed a great breakthrough in rice breeding.

Taking Pei′ai 64S/E32 as a good example, two approaches are involved in our breeding program to develop super hybrid rice.

Morphological improvement

1.To enrich the source. The upper three leaves should be long, erect, narrow, V-shap and thick.

2.To build a huge sink. The panicle size should be very big and even. The number of panicles per m^2 is about 270 and the average grain weight per panicle is 5 grams while the coefficient of variation of panicle weight is around 20%.

3.To construct a taller erect-leaved canopy with drooping panicles. The height of canopy should be taller than 120 cm while the tip of the filled panicles is about 60 cm above ground at ripe stage. Such architecture is more productive in terms of photosynthesis, and at the same time enables the hybrid highly resistant to lodging.

Utilization of inter-subspecific heterosis

The heterosis of inter-subspecific hybrids is much stronger than that of inter-varietal hybrids. In addition to dominance effects which play a major role in inter-varietal hybrids, the inter-subspecific hybrids also has overdominance effects. These effects generally result in very big panicles and vigorous root system. Therefore, to develop inter-subspecific hybrids is the most feasible approach for super high yield at present. Actually, Pei'ai 64S/E32 belongs to inter-subspecific hybrids.

作者：Yuan Longping

注：本文为袁隆平 2001 年 11 月访问委内瑞拉时发表的文章。

Future Outlook on Hybrid Rice: Research and Development in China

A.Current status

· In recent years, the total planting area of rice is around 30 million ha and the average yield is 6.2 t/ha in China.

· The acreage under hybrid rice is about 50% of the total rice area and the average yield of hybrid rice is 6.9 t/ha while the yield of inbred varieties is 5.4 t/ha.

· The area of two line hybrid rice was 2.5 million ha in 2001. Generally, two line hybrids have 5%-10% yield advantage over three line hybrids.

· The area of seed production is around 115 000 ha and the seed yield is 2.7 t/ha on average.

B.Recent progress

· By way of morphological improvement plus utilization of inter-subspecific (*indica/japonica*) heterosis, several pioneer super hybrid rice varieties had been developed by 2000.There were more than 20 locations with 6.7 ha (100 *mu*) or 67 ha (1000 *mu*) each for super hybrid rice demonstration in 2000, the average yield of each location was over 10.5 t/ha. The average yield of such hybrids was 9.2 t/ha on commercial production (1.2 million ha) in 2001.

· Breeding of early maturity (105-110 days) combinations with good heterosis and better grain quality has been achieved through two line method. These hybrids are suitable for the first crop in Yangtze Valley, the main rice growing region in China, where double cropping of rice is practiced. The area under first crop in this region is around 4.5 million ha in which hybrid rice covers only about 10% for years due to lack of short growth duration hybrids. These new hybrids outyield early maturity inbred varieties by around 15% on large commercial scale.

· Good progress is being made in developing japonica hybrids possessing strong heterosis and good grain quality by way of two line method. The yield of some promising combinations is 12-14 t/ha in experiment plots.

C.Future Outlook

· To develop second generation super hybrid rice which can yield 12 t/ha on large scale, or 15% higher than the pioneer super hybrids. It is expected that the goal can be reached by 2005.

· The area of high yielding early maturity hybrids for the first crop in Yangtze Valley and that of high yielding two line *japonica* hybrids for eastern China can be greatly increased. Because of this, the area under hybrid rice will expand to 60% of the total rice area within 5 years. However, there is still no japonica hybrids adaptable to northern China, which is the main *japonica* rice growing region (about 3 million ha). To develop high yielding *japonica* hybrids for northern China is a big challenge facing rice breeders.

· Molecular biotechnology will play an important role in hybrid rice breeding. At present, it includes:

1.Utilization of favorable genes from wild rice may be a promising way to further enhance rice yield. Based on molecular analysis and field experiments, two yield enhancing QTLs from wild rice have been identified. By means of molecular marker-assisted backcross and field selection, an excellent R line (Q661) carrying one of these QTLs is developed. Its hybrid, J23A/Q661, outyielded CK hybrid by 35% in replicated trial for the second cropping rice in 2001.

2.Transgenic rice plants of some parental lines resistant to stem borer, bacterial leaf blight, blast or herbicide have been obtained. Their effects including combining ability are under evaluation.

3.A draft sequence of genome of an *indica* variety, which is the major parent of super hybrid rice combination, has been completed ("Science" Vol 256). It will build a strong foundation for revealing genetic basis of rice heterosis and bring about a new revolution for breeding of hybrid crops.

作者：Yuan Longping

注：本文发表于 2002 年 5 月 14—17 日在越南河内举行的第四届国际杂交水稻研讨会。

Hybrid Rice Technology for Food Security in the World (A)

As human population continues to expand and arable land is being reduced each year, it will become more and more difficult to feed the world. The current world population is over 6 billion and will reach 8 billion in 2030.Meanwhile, the annual loss of land to other use is 10 to 35 million ha per year, with half of this lost land coming from cropland. In 1960, about 0.5 ha of cropland was available per capita worldwide, now it is reduced to 0.27 ha, especially in China, the available cropland is only 0.08 ha per capita. There are 1 to 2 billion malnourished people in the world, according to the World Bank and United Nations.

Facing such severe situation of population growth pressure plus cropland reduction, it is obvious that the only way to solve food shortage problem is to greatly enhance the yield level of food crops per unit land area through advance of science and technology.

Rice is a main food crop. It feeds more than half of world population. It has been estimated that the world will have to produce 60% more rice by 2030 than what it had produced in 1995.Therefore, to increase production of rice plays a very important role in food security and poverty alleviation. Theoretically, rice still has great yield potential to be tapped and there are many ways to raise rice yield, such as building of irrigation works, improvement of soil conditions, cultural techniques and breeding of high yielding varieties. Among them, it seems at present that the most effective and economic way available is to develop hybrid varieties based on the successful experience in China.

It has been proved practically for many years that hybrid rice has more than 20% yield advantage over conventional pure line varieties. In recent years, hybrid rice covers 50% or 15 million ha of the total rice area. The nationwide average yield of hybrid rice is 6.9 t/ha, about 1.5 t/ha higher than that of pure line varieties (5.4 t/ha). The yearly increased paddy in China due to growing hybrid rice can feed 60 million people each year. Therefore, hybrid rice has been playing a critical role in solving the food problem of China thus making China the largest food self-sufficient country.

China makes increasing progress in development of hybrid rice technology. Following the success of three-line hybrid rice in 1970s, two-line hybrid rice was successfully commercialized in 1995, and now the super hybrid rice is coming out.

The development of two-line hybrid rice has been very fast in these years. The area of two-line hybrid rice was 2.5 million ha, about 17% of total hybrid rice area in 2001. The yield advantage of two-line hybrid rice is 5%－10% higher than that of the existing three-line hybrid rice.

More encouragingly, great progress has been made in developing super hybrid rice varieties since the initiation of the super hybrid rice research program in 1996.Several pioneer super hybrids have a yield advantage of around 20% over current three-line hybrids on commercial scale. The area planted to super hybrid rice was 240 000 ha and the average yield was 9.6 t/ha in 2000.The area under super hybrid rice was increased to 1.2 million ha with an average yield of 9.2 t/ha in 2001.In addition, a two-line super hybrid P64S/E32 and a three-line super hybrid Ⅱ－32A/Min86 created a record yield of 17.1 t/ha in 1999 and 17.95 t/ha in 2001, respectively. In the meantime, the grain quality of the pioneer super hybrid rice varieties is very good. Therefore, the super hybrid rice shows a very good prospect in the future. If super hybrid rice covers an annual area of 13 million ha in China and calculating by a yield increase of 2.25 t/ha, it is expected that the annual increased grains will reach 30 million tons, which means 75 million people more can be fed every year.

Hybrid rice has been proved to be a very effective approach to greatly increase yield not only in China, but also outside China. Vietnam and India have commercialized hybrid rice for years. Now, about 600 000 hectares are covered with rice hybrids in these two countries. On average, the rice hybrids yield 20%－30% higher or 1－2 tons per ha more than the best local check inbred varieties. Besides, many other countries, such as the Philippines, Bangladesh, Myanmar, Indonesia, Pakistan and the USA have also achieved great progress in extending hybrid rice technology. Recently, a number of experimental trials and large scale demonstrations in farmers' field conducted in these countries have shown that hybrid rice can significantly outyield their local CK varieties. These facts clearly indicate that hybrid rice technology is also effective to increase rice yield outside China.

The ever-forwarding technology improvement by scientists, its dissemination by seed industries and extension workers and the policy and financial support by national and local governments contribute greatly to the success in development and use of hybrid rice technology. And firmly believe that hybrid rice, relying on scientific and technological advances and the efforts from all other aspects, will have a very good prospect for commercial production and continue to play a key role in ensuring the future food security worldwide in the new century.

作者: Yuan LongPing

注: 本文发表于 2002 年 5 月 14—17 日在越南河内召开的第四届国际杂交水稻研讨会开幕会议。

The Second Generation of Hybrid Rice in China

Strategy of Hybrid Rice Breeding

Based on our studies, the strategy of the development of hybrid rice breeding may involve following phases.

First, in terms of breeding methodology there could be three approaches:

1. Three-line method or CMS system.
2. Two-line method or PGMS and TGMS system.
3. One-line method or apomixis system.

Second, from the view point of increasing the degree of heterosis, i.e., yield potential, the exploitation of heterosis in rice could also be divided into three levels.

1. Inter-varietal hybrids
2. Inter-subspecific hybrids (indica/japonica hybrids)
3. Distant heterosis (using yield enhancing genes from other species or genera)

Each of the phases marks a new breakthrough in rice breeding and will result in great increase in rice yield if it is attained.

The existing rice hybrid varieties used in commercial production mainly belong to the category of inter-varietal hybrids by using CMS system. Many years' practices and experiences have proved that the CMS system can still play an important role in the future. However, there are some constraints and problems in this system.

1. The yield of the existing three line inter-varietal hybrids including newly developed ones has stagnated for years. Their average yield per ha nationwide had reached 6.6 t in late 1980s but is only about 6.8 t in recent years. This means that they have already reached their yield plateau and it seems very difficult to further increase their yield potential if no new methods and novel materials are invented and adopted.

2. The sources of male sterility inducing cytoplasm which can be used for developing better CMS lines are poor. Currently, about 85% of the CMS lines used in commercial production still belong to "WA" type. The dominant cyto-sterility situation of "WA" type has a latent crisis in the long run which may lead the hybrid rice to a destructive pest.

Recent progress and Achievement

Taking the long-range strategy of rice heterosis breeding into account as mentioned above, many Chinese rice scientists have been making attempts to explore new technological approaches so as to further enhance yield potential of rice. So far, the most successful one is the development of inter-subspecific (*indica/japonica*) hybrids by using two line method plus morphological improvement. This achievement means that hybrid rice breeding has entered a new phase in China.

Experiments, large-scale demonstration and commercial production have proved that the two-line inter-subspecific hybrid rice, or called super hybrid rice by media, can outyield the high yielding three-line inter-varietal hybrid rice by around 20%. For example, there were more than 20 locations with 6.7 ha (100 *mu*) or 67 ha (1 000 *mu*) each in demonstration under a pioneer combination (P64S/9311) in 2000, in which the average yield of each location was above 10.5 t/ha. The average yield of this hybrid in commercial production was 9.6 t/ha (240 000 ha) and 9.2 t/ha (1.2 million ha) in 2000 and 2001, respectively. Another combination, P64S/E32, created a record of 17.1 t/ha in an experimental plot (720 m^2) in 1999.

In addition, a breakthrough has been achieved in breeding of early season hybrid rice. A newly developed short growth duration two-line *indica/japonica* hybrid is being demonstrated. The area under demonstration is 7 ha, it is estimated that its average yield is 9.5 t/ha which may out-yield the CK_1 (three-line inter-varietal hybrid) and CK_2 (inbred variety) by more than 20% and 40%, respectively.

Now, the efforts are focused on developing second-phase super hybrid rice that can yield 12 t/ha on a large scale, or 15% higher than pioneer super hybrids and good progress is being made. There were three new two-line *indica/japonica* hybrid combinations out-yielding the CK (pioneer super hybrid) by 8%−18% in replicated trials in 2001. In addition, two promising combinations yield 13.5 t/ha and 15 t/ha respectively in experimental plots (plot size 100 m^2) last winter season in Hainan island. It is expected the goal of breeding second-phase super hybrid rice can be achieved by 2005.

Technical Approaches

Crop improvement practices have indicated, up to now, there are only two effective ways to increase the yield potential of crops through plant breeding, i.e. morphological improvement and heterosis utilization. However, the potential is very limited by using morphological improvement alone and heterosis breeding will produce undesirable results if it does not combine with morphological improvement. Any other breeding approaches and methods including high technology like genetic engineering must be incorporated into good morphological characters and strong heterosis, otherwise there will be no actual contributions to yield increase. On the other hand, the further development of plant breeding for high target must rely on the progress of biotechnology.

1.Morphological improvement

Good plant type is the foundation for super high yield. Since Dr. Donald proposed the concept of ideotype, many rice breeders have paid great attention to this important topic and proposed several models for super high-yielding rice. Among them the famous one is the "New plant type" proposed by Dr. Khush. Its main features are: ① big panicles, 250 spikelets per panicle; ② less tillers, 3−4 productive tillers per plant; ③ short and sturdy culm. Whether this model can realize super high-yield or not, it should be proved by practices.

Based on our studies, especially inspired by the striking characteristics of a high yielding combination, P64S/E32, which has created 17.1 t/ha grain yield, we have found that the super high yielding rice variety has following morphological features:

A.Tall erect-leaved canopy

The upper three leaf blades should be long, erect, narrow, V-shape and thick. Long and erect leaves not only have larger leaf area but also can accept light on both sides and will not shade each other. Therefore, light is used more efficiently; Narrow leaves occupy relatively small space therefore allow to have a higher effective leaf area index; V-shape makes leaf blade more stiff so that not prone to droopy; Thick leaves have higher photosynthetic function and are not easily senescent. These morphological features mean a huge source of assimilates essential to super high yield.

B.Lower panicle position

The tip of panicle is only 60−70 cm above the ground during ripening stage. Such architecture enables the plant to be highly resistant to lodging, which is also one of the essential characters required for breeding super high-yielding rice variety.

C.Bigger panicle size

The grain weight per panicle is around 5 grams and the number of panicles is about 300 panicles/m^2. Theoretically, the yield potential is 15 t/ha if they are attained.

Grain yield=HI (Biomass). Nowadays the harvest index (HI) is very high (above 0.5). Further lifting of rice yield ceiling should rely on increasing biomass because further improvement of HI is quite limited. From view point of morphology, to raise plant height is the effective and feasible way to increase biomass. However, this approach will cause lodging. To solve this problem, many breeders are trying to make the stem thicker and sturdier, but this approach usually results in HI decrease. Therefore, it is difficult to obtain super high yield by this way. The plant model of taller canopy which consists of leaves can combine the advantages of higher biomass, higher HI and higher resistant to lodging together.

2.Utilization of indica/japonica heterosis

The heterosis level in rice has the following general trend: *indica/japonica>indica/ javanica>japonica/javanica>indica/indica>japonoca/japonica*, according to our studies. *Indica/ japonica* hybrids possess very large sink and rich source, the yield potential of which is 30% higher than inter-varietal *indica* hybrids theoretically. Therefore, efforts have been focused on using *indica/japonica* heterosis to develop super hybrid rice. However, there exists a lot of problems in *indica/japonica* hybrids, especially very low seed set, which must be solved in order to use their heterosis in practice. By means of wide compatibility (*WC*) genes and using intermediate type lines as parents instead of typical *indica* or *japonica* lines, a number of inter-subspecific hybrid varieties with stronger heterosis and normal seed set have been successfully developed as mentioned above.

3.Utilization of favorable genes from wild rice

This is another promising approach to develop super hybrid rice. Based on molecular analysis and field experiments, two yield enhancing QTLs from wild rice (*O. Rufipogon* L.) were identified. Each of the QTL genes contributed to a yield advantage of 18% over the high yielding CK hybrid Weiyou64 (one of the most elite hybrids). By means of molecular marker-assisted backcross and field selection, an excellent R line (Q611) carrying one of these QTLs is developed. Its hybrid, J23A/Q611, outyielded CK hybrid by 35% in replicated trial for the second cropping rice in 2001.Its yield potential on large scale is being evaluated for the time being.

Prospects

The yield standard of second phase super rice (12 t/ha) can be achieved by 2005.By reaching this target, 2.25 t/ha more rice can be produced, which will increase 30 mt of grains yearly and can feed 75 million more people when it is commercialized up to 13 million.ha.

The development of science and technology will never stop. Rice still has great yield potential, it can be further tapped by advanced biotechnology. Excitingly, C_4 genes from maize have been successfully cloned and are being transferred into the parental lines of super rice by cooperative research between HK Chinese University and our center. By using this transgenic lines to develop super hybrid rice, the yield potential of rice could be further increased by a big margin. Relying on this progress, the phase Ⅲ super hybrid rice breeding program is proposed, in which the yield target is 13.5 t/ha on a large scale by 2010.

作者：Yuan Longping

注：本文为 2002 年 7 月 23—26 日在泰国曼谷举行的第 20 届委员会上发表的文章。

A Draft Sequence of the Rice Genome (*Oryza sativa* L. ssp. *indica*)

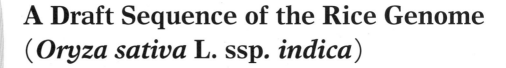

【Abstract】We have produced a draft sequence of the rice genome for the most widely cultivated subspecies in China, *Oryza sativa* L. ssp. *indica*, by whole-genome shotgun sequencing. The genome was 466 megabases in size, with an estimated 46 022 to 55 615 genes. Functional coverage in the assembled sequences was 92.0%. About 42.2% of the genome was in exact 20-nucleotide oligomer repeats, and most of the transposons were in the intergenic regions between genes. Although 80.6% of predicted *Arabidopsis thaliana* genes had a homolog in rice, only 49.4% of predicted rice genes had a homolog in *A.thaliana*. The large proportion of rice genes with no recognizable homologs is due to a gradient in the GC content of rice coding sequences.

Rice is the most important crop for human consumption, providing staple food for more than half the world's population. The euchromatic portion of the rice genome is estimated to be 430 Mb in size (1 – 3), which is the smallest of the cereal crops. It is 3.7 times larger than that of *A.thaliana* (4 – 6), and 6.7 times smaller than that of the human (7, 8). The well-established protocols for high-efficiency genetic transformation, widespread availability of high-density genetic and physical maps (9, 10), and high degrees of synteny among cereal genomes (11 – 15) combine to make rice a unique organism for studying the physiology, developmental biology, genetics, and evolution of plants. The International Rice Genome Sequencing Project (IRGSP) (16) has already delivered a substantial amount of sequence for the *japonica* (*Nipponbare*) subspecies, in bacterial artificial chromosome (BAC) and P1 – derived artificial chromosome (PAC) -sized contigs. Working independently, researches (17, 18) established proprietary working drafts for *japonica*, in April 2000 and February 2001, respectively. The Monsanto sequence has been used to assist in the efforts of the IRGSP.

We are releasing a draft genome sequence for rice from 93 – 11 (19), which is a cultivar of *Oryza sativa* L. ssp. *indica*, the major rice subspecies grown in China and many other Asia-Pacific regions. It is the paternal cultivar of a super – hybrid rice, *Liang – You – Pei – Jiu*

(*LYP9*), which has 20% to 30% more yield per hectare than the other rice crops in cultivation (20). The maternal cultivar of *LYP9* is *Pei - Ai 64s* (*PA64s*), which has a major background of *indica* and a minor background of *japonica* and *javanica*, two other commonly cultivated subspecies. We have also produced a lowcoverage draft sequence for *PA64s*. A preliminary assembly and analysis on a subset of this sequence was published in the *Chinese Science Bulletin* (21). Our discussion will focus largely on the genome landscape of rice, how it differs from that of the other sequenced plant, *A.thaliana*, and how both plant genomes differ from that of the human.

We will show that rice genes exhibit a gradient in GC content, codon usage, and amino acid usage. This compositional gradient reflects a unique phenomenon in the evolutionary history of rice, and perhaps all monocot plants, but not eudicot plants. As a result, about one-half of the predicted rice genes have no obvious homolog in *A.thaliana*, whereas the other half is almost a replica of the *A.thaliana* gene set.

The entire rice genome sequence can be downloaded from our Web site at http: //btn. genomics. org.cn/rice. Following our announcement of the rice genome sequence at the annual Plant, Animal and Microbe Genomes (PAG X) conference, in San Diego, during the ensuing period from 14 January to 2 March 2002, this sequence was downloaded 556 times, and the BLAST search facilities were used 7 008 times by 343 individuals. This sequence has also been deposited at the DNA Data Bank of Japan/European Molecular Biology Laboratory/GenBank under the project accession number AAAA00000000. The version described in this paper is AAAA01000000.

Experimental design. The rice genome project at the Beijing Genomics Institute has been designed in two stages. This is a report on stage I, the primary objective of which was to generate a draft sequence of rice at ; 43 coverage for 93 - 11. A similar amount of data will eventually be generated for *PA64s*, but at present there is only enough data to estimate polymorphism rates between rice cultivars. The sequence reads were acquired on high-throughput capillary machines (MegaBACE 1000, 10 to 11 runs per machine per day). Concurrent with the data acquisition, we developed a software package (22) to identify and mask repetitive sequences and to correctly assemble these sequence reads into contigs and scaffolds, even though cereal genomes contain far more repetitive sequence than many other genomes (23, 24). We generated 87, 842 expressed sequence tags (ESTs), against our ultimate goal of 1, 000, 000 ESTs, to provide confirmatory evidence for gene identification, and for gene expression analysis. Comparing the 93 - 11 contig assemblies with the public data, we generated a set of polymorphic markers for genetic analysis. In stage II of the project, our objective will be to obtain a high-quality sequence, fully integrated with the physical/genetic maps, and with complete gene annotations.

We used a "whole-genome shotgun" approach, as successfully applied to *Drosophila melanogaster* (25) and *Homo sapiens* (8). Our data are complementary to those of the IRGSP, which is sequencing *Nipponbare*, a cultivar of the subspecies *japonica*, with a "clone-by-clone" approach. If we assume a euchromatic rice genome size of 430 Mb, and a Phred Q20 (26, 27) read length of 500 base pairs (bp), then 1 × coverage would be equivalent to 0.86 million sequence reads, or 1 million reads after the typical success rate of 80% to 85% is factored in. Shotgun libraries were constructed with a variety of methods for clone-insert preparation (28 - 30), to minimize the likelihood of systematic biases in genome representation. A total of 55 plasmid libraries were constructed for 93 – 11 and *PA64s*, with a 2kb nominal clone-insert size. Overall, we prepared 2.75 million plasmid DNA samples (31, 32). Sequencing was performed on both ends of the inserts. By the 21 October 2001 freeze, there were 4.62 million successful reads, indicating an 84% success rate. The average Q20 read length was 546 bp.

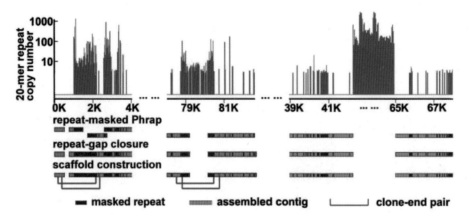

Fig.1 Typical RePS assembly, with 93-11 (*indica*) contigs aligned to pnished BAC sequences from *GLA* (*indica*) (GenBank accession numbers AL442007 and AL512542). Exact 20-mer repeats are indicated by the blue histogram bars, with bar heights proportional to estimated copy number in 93-11 (*indica*). Three stages are shown : repeat-masked Phrap, repeat-gap closure, and scaffold construction. First, we mask exact the 20-mer repeats and use Phrap to assemble the data on the basis of the unique sequence. Second, we use the clone-end pairing information to close smaller repeat masked gaps (RMGs) ignored by Phrap because of the masking. However, larger RMGs and gaps due to sampling statistics, Lander-Waterman gaps (LWGs), cannot be so closed. Third, we use the clone—end pairing information to construct scaffolds—sets of nonoverlapping contigs linked together in the correct order and orientation. A LWG at 0.5 kb is scaffolded over. RMGs at 1.5 and 2.5 kb are closed, and another at 80 kb is scaffolded over. The RMG between 42 and 65 kb is too large to scaffold across given a clone-insert size of 2 kb.

Assembling the draft. Genomic studies of grasses, especially the cereal crops, have indicated that the intergenic regions between genes are inhabited by clusters of nested retrotransposons (23, 24, 33), which compose almost half of the rice genome, and substantially larger fractions of other crop plants like *Zea mays* (maize) and *Triticum aestivum* (wheat). Our sequence assembler software was designed to handle highly repetitive genomes without having to first characterize the repeats in any traditional biological sense. The focus was on contiguity at the scaffold level, instead of complete assembly across all the repeats. However, error probabilities would be computed for every base that was successfully assembled.

A typical assembly, based on our software RePS (Repeat-masked Phrap with Scaffolding) (22), is shown in Fig.1. We began by computing the number of times that any 20 – bp sequence (20 – nucleotide oligomer, 20 – mer) appeared in the data set. Those 20 – mers that appeared more often than a fixed threshold were flagged as mathematically defined repeats (MDRs). RePS made no effort to identify biologically defined repeats (BDRs), because if a 20 – mer was repeated in the MDR sense, it would complicate the sequence assembly，regardless of its biological context（e.g.，microsatellites，transposable elements，multigene families，recently duplicated chromosomal segments，or pseudogenes）.Instead，it masked the MDRs，so that they were invisible to the sequence assembler Phrap（34）. This reduced the computational load by many orders of magnitude，while minimizing the likelihood of making a false join. However，it also introduced another class of gaps，repeat masked gaps（RMGs），distinct from the Lander-Waterman gaps（LWGs）that are usually encountered in sequencing. In a RMG，the gap sequence is actually in the data set，but it was not usable because it was made invisible to Phrap by the masking. In a LWG，the gap sequence is missing，as a result of sampling statistics（35）. Some of the RMGs could be closed with the clone-end pairing information，assuming that both clone ends were not fully masked. After repeat-gap closure，and regardless of the nature of the remaining gaps，RePS was used to analyze the clone-end pairing information to construct scaffolds—nonoverlapping contigs linked together in the correct order and orientation.LWGs were usually small，easy to close by polymerase chain reaction. Gaps larger than a few kb were usually RMGs due to the nested retrotransposons in the intergenic regions between genes. Whether these gaps should be closed or not remains to be resolved.

Table 1　Sequence assembly statistics for 93-11（*indica*）.

Basic shotgun data	
Total genome size/Mb	466
Number of reads	3 565 386
Q20 read lengths/bp	546
Shotgun coverage	4.2
Exact 20-nt oligomer repeats	
Length of fraction masked	42.2%
No. of fully masked reads	18.7%
Sequence assembly	
Total contig size/Mb	361
N50 contig size/kb	6.69
Total scaffold size/Mb	362
N50 scaffold size/kb	11.76
Unassembled data	
Fully masked reads/Mb	78
All other reads/Mb	26

Note：The Q20 read lengths refer to the usable part of the sequence with error probabilities less than 10^{-2}. Masking 20-mer repeats eliminated 42.2% of the sequence by length. Some reads were partially masked, but 18.7% of reads were fully masked. The N50 contig or scaffold sizes define that size above which 50% of the assembly was found. To estimate the assembled-equivalent size of the unused reads, we divided total Q20 lengths by the 4.2× depth of reads in the assembled contigs. This resulted in an assembled-equivalent size of 104 Mb, of which 78 Mb was fully masked reads. The total genome size was thus estimated to be 466 Mb.

Shotgun data for 93-11 (Table 1) and *PA64s* were assembled separately, to allow for large differences in their genome sequences.In 93-11, there were 3.57 million sequence reads after removal of the ones containing mitochondrial, chloroplastic, and bacterial sequence. Our RePS assembly yielded 127,550 contigs with an N50 size (i.e., the size above which 50% of the total assembly is found) of 6.69 kb. The total contig length was 361 Mb. These contigs were linked into 103,044 scaffolds with an N50 size of 11.76 kb, or a 1.8-fold increase over the initial contigs. The total scaffold length was 362 Mb. In contrast, for the *PA64s* data set, we had only 1.05 million sequence reads. With such low coverage, the N50 contig and scaffold sizes were much smaller, at 1.88 kb and 1.97 kb, respectively. These statistics differ slightly from those reported in the *Chinese Science Bulletin* (21), because of improvements in the RePS software. Remaining gaps between scaffolds are probably larger than the clone-insert size of 2 kb; otherwise, we would have been able to bridge them. We cannot provide a gap size distribution, but in the rice BACs that have been sequenced, repeat cluster sizes up to 25 kb have been observed.

The total contig and scaffold lengths fall far short of the previously estimated euchromatic genome size of 430 Mb. Where is the missing DNA? In the initial phase of the RePS assembly, 42.2% of the sequence was identified as a MDR and masked. A total of 18.7% of all the reads were fully masked and not immediately usable. Even though some were later incorporated into the assembly, with the clone-end pairing information, a large number of fully masked reads, and some partially masked reads, remained unused.To estimate the effective-assembled size of the unused reads, we defined an empirical coverage based on the depth of reads in the assembled contigs, $4.2\times$. The effective-assembled size for the unused fully masked and partially masked reads was thus estimated as 78 Mb and 26 Mb, respectively, resulting in a total genome size of 466 Mb.That this is larger than the previous estimates is reasonable, given that whole-genome shotgun data inevitably contain some amount of heterochromatin DNA.

Quality assessments. We assumed that any large cluster of MDRs was an intergenic region and that we could safely avoid having to assemble across such a region. If so, then most of the "functional sequence" that encodes genes, and their immediate regulatory elements, should lie in our 361 Mb of assembled contigs. To confirm that this was indeed the case, we gathered all the publicly available sequence-tagged sites (STSs) and fulllength cDNA sequences, as well as our own ESTs, and searched for them in our assembled contigs, using BLAST (36). Fortunately, a dense physical map of STS markers had already been established (37) for *japonica*. A total of 2 845 markers were analyzed, and on the basis of sequence identity, 91.5%of their total length could be found in our contigs.Similarly, 24,776 UniGene clusters were assembled from 87,842 ESTs for 93-11, and 93.8% of their total length could be found in our contigs. Finally, 907 nonredundant cDNA sequences were extracted from Gen-Bank release 125 (15 August 2001), and 90.8% of their total length could be found in our contigs. Averaged across these three data sets, the functional coverage was 92.0%.

The quality metrics that matter for gene identification are (i) contiguity on the length scale of a gene, (ii) single-base error probability, and (iii) contig assembly accuracy on the length scale of a gene. As will be detailed in a later section, the mean gene size for rice is about 4.5 kb. Considering that our N50 scaffold size is only 11.76 kb, larger scaffolds would reduce the number of genes that are split across scaffolds, and this is a key objective in stage II of the project. The number most often cited is the single-base error probability, which the International Human Genome Sequencing Consortium (7) determined should be 10^{-4} or better, based on a human polymorphisms rate of 10^{-3}. Actually, as is detailed in a later section, rice polymorphism rates are closer to 10^{-2}, so an error rate of 10^{-4} is better than needed. On the basis of Phrap estimates (26, 27, 34), 94.2%, 90.8%, and 83.5% of the 93-11 sequence had an error rate of better than 10^{-2}, 10^{-3}, and 10^{-4}, respectively. However, most of the problematic bases were at the ends of the contigs. When we restricted this calculation to contigs greater than 3 kb and ignored bases within 500 bp of the ends, 97.3%, 96.1%, and 92.5% of the 93-11 sequence had an error rate of better than 10^{-2}, 10^{-3}, and 10^{-4}, respectively. It is important to bear these error rates in mind when comparing two sequences to estimate polymorphism rates.

Assembly accuracy is an often overlooked but nevertheless important quality metric. When the sequence reads are joined together in the wrong order or orientation, some of the exons will be arranged in the wrong order or orientation. This will confuse any gene-annotation program. For example, a 2-kb segment that is flanked by a pair of inverted repeats might be assembled in the wrong orientation. Comparison of independently assembled BACs would not necessarily detect the mistake, because the problem is due to sequence content, not data quality, and the same mistake could be made in both BACs. Comparison with existing physical or genetic maps validate assembly accuracy on the Mb length scale, but that is much larger than the size of most genes. Clone-end pairing information does validate a contig assembly on the kb length scale of the genes. However, when the clone ends are also used to assemble the sequence, they do not qualify as an independent confirmation. To address this problem, we aligned cDNA sequences (i.e., experimentally derived transcripts) with the genome sequence.

We removed obvious redundancies by eliminating any cDNA that was more than 90% contained inside another. Transposon sequences identified by RepeatMasker (38), generally in the 3'-untranslated region, were trimmed off to minimize the number of ambiguous hits. Alignments were allowed to span multiple contigs. Within any one contig, a putative misassembly was flagged whenever an exon was missing from the middle of the chain, in the wrong order, or in the wrong orientation. Missing splice sites resulting from minor sequencing errors, and partial alignments resulting from missing sequences at the end of a contig, were not counted. All putative misassemblies were validated by visual inspection, to ensure that no better alignments could be found. If in the end, the best alignment remained problematic, we concluded that there must have been a misassembly. One might think that lower quality cDNA sequences would contribute to the problematic alignments, and that this procedure would only set an upper bound on the number of misassemblies. However, we doubt that this is a serious problem. Substitutional errors might be common in cDNA sequences, but they would not trigger our detection algorithm. Only exon-sized rearrangements, especially those that change the order and orientation, would do so, but such rearrangements are rare in cDNA sequences.

We benchmarked our misassembly detection procedure on two of the most recently completed model organism genomes: *A.thaliana*, which is of finished quality (4), and *Drosophila melanogaster*, from the Celera 13× whole-genome-shotgun sequence (25). For *A.thaliana*, we detected problems in 0.2% of 4804 genes, and for *D. melanogaster*, we detected problems in 1.1% of 1889 genes.

For 93-11 contigs, we detected problems in 1.1%of 907 genes, which was comparable to the *D.melanogaster* data.

Compositional gradients. The rice genome has compositional properties that differentiate it from the other sequenced plant genome, *A.thaliana*, and introduce unique difficulties for genome analysis. Here, we show data on exon, intron, and gene sequences derived from alignment of cDNAs with genomic sequence. Indeed, for Figs. 3 through 7, all of the gene models were derived from cDNA alignments, not gene-prediction programs. GenBank release 125 (15 August 2001) was used for the *A.thaliana* figures, and for the rice cDNAs. The rice genome sequence was our 93-11 assembly.The human cDNA sequence was downloaded on 2 March 2001 from NCBI-RefSeq ftp: //ncbi.nlm.nih.gov/ refseq/H_sapiens, and the human genome sequence was downloaded on 27 February 2001 from ftp: //ncbi.nlm.nih.gov/genomes/H_sapiens, immediately after the initial annotation papers.

*Genomic, exon, and intron GC contents.*The average genomic GC content for prokaryotes and eukaryotes varies widely. It ranges from less than 22% in the human malaria parasite, *Plasmodium falciparum*, to more than 68% in the large amplicon of *Halobactrium* sp. NRC1 (39). Local heterogeneity in GC content can be enormous, ranging from 26 to 65% in the human genome alone. In contrast, AG content (purine) is homogeneous (40-43), fluctuating by just a few percent about a mean of 50%. Compositional heterogeneity has been debated for more than 30 years (44-47). Discussions have focused on the characterization of the human genome as a mosaic of GC-rich and AT-rich "isochores," which are observed in warm-blooded vertebrates, but not in coldblooded vertebrates. More recently, an elevated GC content in the *Gramineae* (grass) genomes was reported, extending perhaps to all monocot genomes, but not to eudicot genomes (48). It is not known whether or how this phenomenon is related to isochores.

Major differences between sequence content in *A.thaliana*, rice, and human are observable even at the simplest level, from distributions of genomic GC content. Traditionally, GC content was computed on a large window size, typically in the 100s of kilobases, to mimic the original Cs_2SO_4 density gradient experiments (49, 50). We have found that smaller windows are more informative, because when these windows are larger than a typical gene size, they obscure differences between intergenic DNA and genes. We used a 500-bp window size, to obtain a smaller size than that of most plant genes (Fig.2). As previously reported (51), the *A.thaliana* distribution displayed a "shoulder" on the AT-rich side, which could be attributed to the sizable fraction of the genome that was in intergenic DNA. The primary peak at 0.382 was nearly identical to the 0.388 GC content of the average *A.thaliana* gene. In contrast, no shoulder was observed in rice. However, a "tail" was apparent on the GC-rich side. The human distribution also displayed no shoulder, but a minor tail might have been present. To analyze these features, we plotted GC content distributions for exons and introns (Fig.3). Rice exons exhibited a GC-rich tail, but rice introns did not, indicating that the GC-rich tail in the rice genomic distribution was primarily due to the exons.

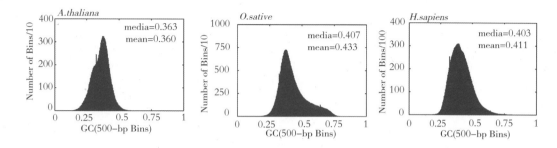

Fig.2　Distributions for genomic GC content in *A.thaliana*, *O.sativa*, and *H.sapiens*, computed over a bin size of 500 bp. Note that for bins/10=100, the number of bins with that GC content is 1 000.

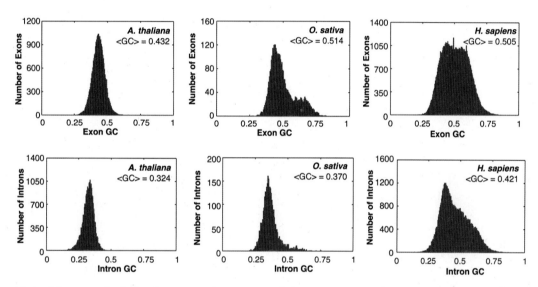

Fig.3　GC content distribution for exons and introns in *A. thaliana*, *O. saiva*, and *H. sapiens*. All exon and intron sequences were derived from cDNA-to-genomic alignments. Mean GC content is computed on a length-weighted basis as $<GC> = \sum_i L_i \cdot GC_i / \sum_i L_i$ where GC_i and L_i are the GC content and length for the *i*th segment (exon or intron).

Variation in GC content within genes. The key question is whether the increase in exon GC content was due to many genes with a few GC-rich exons or to a subset of GC-rich genes. Equivalently, was most of the variation in exon GC content within genes or between genes? After the GC contents of individual exons and introns were plotted as a function of genomic length (i.e., the sum of exon and intron lengths), it was apparent that most of the variation was within genes (Fig.4). Contrary to the expectation that, in the human genome, large genes are on average more AT-rich than small genes, we found that at least one exon of exceptionally high GC content could be found in almost every rice gene, including the largest ones. Moreover, when the GC content of the protein-coding regions was plotted as a function of position along the direction of transcription, starting from the 5′ end, we observed a negative gradient in the GC content of rice genes (Fig.5). Typically, the 5′ end was up to 25% richer in GC content than the 3′ end. These gradients would extend to about 1 kb from the 5′ end, before finally petering out. The magnitudes of these gradients varied. A few genes had zero gradients, but almost no genes had positive gradients. In contrast, for *A.thaliana*, no comparable gradients were observed. Examining hundreds of best available homologs (i.e., possible orthologs), we found that the GC content of rice genes was equal to or exceeded that of their *A.thaliana* counterparts, at all positions along the coding region.

A more detailed analysis of this compositional gradient will be presented elsewhere (52). The important point is that, not only is there a gradient in GC content, but there are also gradients in the patterns of codon and amino acid usage. The former is a novel challenge for the ab initio gene-prediction programs that rely on codon-usage statistics, and the latter makes it more difficult to do protein homology searches across the monocot-eudicot divide.

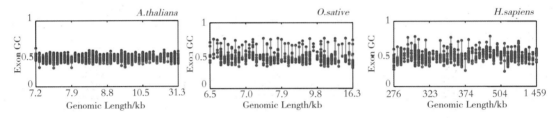

Fig.4 GC content for individual exons as a function of their gene size, in *A.thaliana*, *O.sativa*, and *H.sapiens*. All exon and intron sequences were derived from cDNA-to-genomic alignments. Each data point is a single exon.Exons for the same gene are plotted at the same abscissa and connected by a vertical line. The genes are sorted by size, where gene size is defined as the sum of exon and intron lengths. To make the Tgure legible, we use constant spacing between genes, thus resulting in nonuniform abscissa labels. We show only the 41 largest genes for which the entire cDNA could be aligned to genomic sequence. Given the draft nature of the rice genome, some of the largest rice genes had to be omitted.

Genic and intergenic DNA. We examined exon and intron distributions for every plant, vertebrate, and invertebrate organism with more than a hundred or so genes in GenBank, either by cDNA alignments or by parsing annotations. Numerical summaries (Web Supplement 1) are available on *Science* Online at www.sciencemag.org/cgi/content/full/296/5565/79/DC1. Exon sizes are

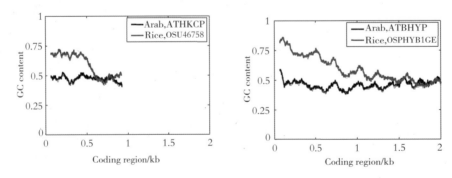

Fig.5 GC content for homologous genes in *A.thaliana* and *O.sativa* as a function of gene position from the 5' to 3' end, computed on a sliding 129-bp window (equal to the median exon size in rice). Only the coding region is shown. GenBank locus identiTers are specified in the legend. The smaller gene is "potassium channel beta subunit," and the larger gene is "phytochrome B."

narrowly constrained, but intron sizes can be highly variable within and between organisms. Intron-size distributions tend to be bimodal, weakly (most organisms) or strongly (human). There is always a sharp "spike" at some organism-specific minimum size, which is about 90 bp for plants and vertebrates (Fig.5). There is also a broad "hump" due to the larger introns. The magnitude of this hump is highly variable between organisms and can be difficult to ascertain precisely, because of the systematic biases against the complete sequencing of large genes.

Although the existence of this acquisition bias is known, the magnitude of its effect on our perception of intron and gene sizes is not well appreciated. For example, in the initial annotation of the human genome, the reported mean gene size of 27 kb turned out to be an underestimate by a factor of 3 (53). To correct for this bias, one need only realize that the bias against complete sequencing of large genes is equivalent to the bias against production of large genomic contigs. Thus, the correction can be made by restricting the computation of the mean gene size to cDNA alignments in contigs above a minimum size, and extrapolating to the limit of infinite contigs. For the human genome, the extrapolated mean gene size was 72 kb. In *A.thaliana*, there was no appreciable bias, because most of the contigs were already much larger than the genes, which had a mean size of 2.4 kb. This was larger than the published gene size of 2.0 kb, but only because we included UTRs, whereas the published numbers did not. In rice, there was a small acquisition bias, given the draft nature of our sequence. Nevertheless, the extrapolated mean gene size was only 4.5 kb, much smaller than in the human, consistent with the relatively small hump in the rice intron-size distribution.

A preliminary gene count can be estimated from the mean gene size, for comparison against the number of genes identified by the gene-prediction programs. Our estimated 4.5 kb mean gene size for rice is similar to the maximal gene density of one per 4 to 5 kb, based on analyses of syntenic loci across many plant species (54). Assuming that the rice intergenic fraction is equal to the 42.2% of the sequence that was in MDRs, and taking 466 Mb as the genome size, the estimated number of rice genes is 59, 855. One could also include CpG islands in the gene count, although not every CpG island is associated with a gene, so that this number can at best be considered an upper bound (55). Including both assembled contigs and unused reads, 138, 485 CpG islands were identified by the standard algorithm (56). Either way, rice almost certainly has more genes than *A.thaliana* (4), which has only 25, 498. It might even have more genes than the human (7, 8), which has 30, 000 to 40, 000, although the actual gene count remains controversial. The idea that plants might have more genes than humans is not new, as it was predicted before our analysis (57, 58).

Where did the transposons end up? The significance of these size distributions is that a prominent "hump" in the intron-size distribution, as observed for the human, is evidence of extensive transposon activity in the evolution of intron size. RepeatMasker (38) identified at least one transposon, and often many more than one, in almost every human intron larger than 1 kb. It rarely found a transposon in smaller introns less than 1 kb, not only in the human, but in every organism analyzed. The negative result might have been due to the incomplete status of the transposon database on which RepeatMasker relies. To support this claim, we introduce an argument that does not rely on knowledge of the sequences of all the extant transposons.

The main assumption was that any transposon should have inserted into the genome many times before becoming inactive. Despite subsequent degradation of these transposon sequences, portions should remain in many different places throughout the genome.RePS computed the copy number for every 20-mer sequence in the genome, indicating how many times each occurred in the genome. We could therefore determine the copy number required to account for all of a particular sequence data set. These data sets would include all exons, introns, and known transposons (Fig.7) . For the exons and introns, we used cDNA-to-genomic alignments.For transposons, we used RepBase 6.6 (59), a database of consensus sequences for every known family or subfamily of transposons. In plants, exons and introns were fully accounted for by 20-mer with copy numbers of less than 10. Transposons required much higher copy numbers of 10 to 10^2 in *A.thaliana* and 10^2 to 10^3 in rice. One could legitimately ask if the absence of large MDR clusters in our rice assemblies was a confounding factor in the intron analysis. We therefore performed an analysis on introns from finished BAC sequences and found no detectable differences. Strikingly, in the human, extremely large copy numbers of 10^4 to 10^5 were required to fully account for the introns, as was observed with the transposons. Human exons, however, were found at the same low copy numbers as in plants.

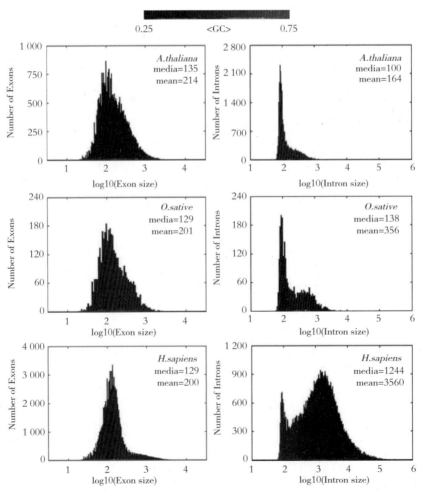

Fig.6　Exon-and intron-size distributions for *A.thaliana* , *O.sativa* , and *H. sapiens* , with color indicating averaged GC content for exons or introns at that size range. All exon and intron sequences were derived from cDNA-to-genomic alignments.

The copy number analysis shows that few plant transposons are in the introns, and by definition, plant transposons must be located in the intergenic regions between genes. Conversely, analyses of gene size show that most human transposons are in the introns (53). We believe that this dichotomy in where the transposons ended up reflects a fundamental difference in plant and vertebrate genomes. The dichotomy is not due to any lack of transposons in plants, because plant genomes contain many transposons. At least 24.9% of the rice genome was identifiably of transposon origins, based on a weighted average of assembled contigs and unused reads, but the correct percentage is likely to be much higher, because the transposon databases on which RepeatMasker relied were incomplete. *A.thaliana*, being a more compact genome, had a reported transposon fraction of 10%, although we suspect that this too is an underestimate.

Repetitive sequences. We deal with three classes of repeats: simple repeats [e.g., $(CAG)n$], complex repeats (i.e., transposable elements or TEs), and mathematically defined repeats (MDRs). Here, we focus on the first two classes, which we called biologically defined repeats (BDRs). As with intron and gene sizes, acquisition biases must be factored in, so that we do not introduce additional discrepancies among the published studies. For example, a survey of 73,000 sequence-tagged connectors, totaling 48 Mb of sequence from *japonica* (60), found that 63% of identified TEs were retrotransposons (e.g., *copia* and *gypsy*). However, a survey of 910 kb of rice genomic sequence (61) found that 18.6% of identified TEs were retrotransposons. Most of the remainder were MITEs, or miniature inverted-repeat TEs, which accounted for 71.6% of identified TEs. Similarly large discrepancies were encountered in analyses of microsatellite distributions using mixed data from BAC ends, ESTs, and finished BAC/PACs (62).

Table 2　Simple repeats.

	93-11 assembled contigs				93-11 fully masked reads				Full-length cDNAs			
	n=6-11		n>11		n=6-11		n>11		n=6-11		n>11	
	% GC	% of data set	% GC	% of data set	% GC	% of data set	% GC	% of data set	% GC	% of data set	% GC	% of data set
Mononucleotides	7.63	1.784 7	27.65	0.068 0	20.34	0.695 3	21.02	0.015 4	24.08	0.630 3	1.31	0.712 5
Dinucleotides	35.77	0.090 4	13.08	0.084 7	41.86	0.055 3	4.38	0.029 4	46.85	0.057 3	31.11	0.039 4
Trinucleotides	71.79	0.045 4	10.08	0.010 6	67.20	0.009 8	20.81	0.001 2	83.05	0.133 5	66.67	0.004 3
Tetranucleotides	28.77	0.007 2	24.90	0.003 2	37.35	0.002 0	31.90	0.001 0	50.00	0.001 8	0.00	0.000 0
All periods		1.927 7		0.166 5		0.762 4		0.046 9		0.822 9		0.756 1

Note: Shown are tandem repeats with periods 1 to 4 (mono-, di-, tri-, and tetranucleotide) and the totality of repeats with all periods. The index n is the number of periodic units. For example, AGTTAGTT is a tetranucleotide otide of n=2. We compute mean GC contents of the observed repeats in each category. Repeat content is then given as a percentage by length, normalized with respect to the data set (assembled contigs, fully masked reads, or cDNAs).

For the 93-11 sequence, it is particularly important that we analyze those sequence reads that were not assembled into contigs. Tables 2 and 3 thus summarize repeat contents in the two largest components: 361 Mb of assembled contigs and 78 Mb of unused fully masked reads. Weighted averages for the entire rice genome were also computed.For comparison, we show repeat content in 907 nonredundant full-length cDNAs from GenBank release 125 (15 August 2001). Absolute numbers are not listed because, with so much of the genome in unassembled reads, and with so many of the transposons nested inside some other transposon, accurate counts were not feasible. Results are listed as a fraction of total sequence length.

Simple sequence repeats (SSRs). SSRs are particularly useful for developing genetic markers. They are believed to vary through DNA replication slippage (63-65), and are related to genetic instability (66). In Table 2, we describe SSR content for two sectors, $n=6$ to 11 units and $n>11$ units, to emphasize that the number of SSRs dropped substantially after 11 units. The SSR content for 93-11was 1.7% of the genome, lower than in the human, where it was 3% (7). The overwhelming majority of rice SSRs were mononucleotides, primarily $(A)_n$ or $(T)_n$, and with $n=6$ to 11. In contrast, for the human, the greatest contributions came from dinucleotides.Notably, trinucleotides with $n=6$ to 11were a barometer of gene content. The basic effect was captured by the ratio of trinucleotide otide to dinucleotide content, which was 2.33, 0.50, and 0.18 in cDNAs, assembled contigs, and fully masked reads, respectively. As required for a barometer, these numbers are well correlated with presumed gene content. In addition, the GC content of these trinucleotides was high, consistent with the high GC content of many rice exons.

Complex sequence repeats (TEs). Transposons identified by RepeatMasker (38) were assigned into three classes. Class I repeats are retrotransposons, primarily *Ty1/Copia*-like and *Ty3/Gypsy*-like. Class II repeats are DNA transposons, including *Ac/Ds*, *En/Spm*, *Mariner*-like, and Mutator elements. Class III repeats are a previously unknown type of short DNA transposons called MITEs (67, 68). The two common examples are *Stowaway* and *Tourist*. Recently, an active family of *tourist-like* MITEs was identified in maize (69). Programs like RepeatMasker identify sequences that share at least 50%identity with a known TE. Because TEs are under no selective constraints after they insert in a genome, they tend to diverge from their ancestral sequence, and become unrecognizable over a time scale of a hundred million years (70). Identifiable repeat content is thus a function of TE age and completeness of the TE databases. The numbers listed in Table 3must therefore be considered underestimates.

Fully masked reads were composed of 59% identifiable TEs. Assembled contigs were only 16%. Of these TEs, the amount in class I and class III repeats was 97 and 1%, respectively, for fully masked reads, but 42 and 40% for assembled contigs. This extremely biased distribution is notable, because class I repeats reportedly inhabit the intergenic regions (23, 24), and class III repeats are found near, although not necessarily in, the genes (71). Thus, we had 92.0% functional coverage despite having only 361 Mb in assembled contigs, in a genome of total size 466 Mb. The reason class I repeats failed to assemble is apparent when one examines their mean size. Class III repeats were usually smaller than 671 bp, but class I repeats were as large as 7 kb. Our ability to close repeatmasked-gaps, or RMGs, was limited by the clone-insert sizes. For this assembly, the clone-insert sizes were only 2 kb, although we plan to use larger sizes for the next stage of the rice genome project.

　　Finally，the TEs in rice cDNAs constituted only 1% of the sequence，which is much lower than the 4% that was reported for human genes (72)．Gene-associated TEs，in human and other vertebrates，have been proposed to play crucial roles in creating new genes (73) and in changing the regulatory circuitry to promote evolution in the host genome (74)．

　　Rice gene annotations. Gradients in GC content and codon usage for rice genes create special problems in the gene-annotation process (52)．Because rice genes have different compositional properties at their 5′ and 3′ ends，it is difficult to train a program to perform well under all circumstances. Some ab initio gene-prediction programs can use different codon-usage statistics for different genes，on the basis of regional GC content，but none use different codon-usage statistics at different positions along the same gene. Unless the gradient is explicitly modeled，or perhaps，codon-usage statistics are abandoned altogether，performance will be subject to the vagaries of the training process. With this in mind，we set out to survey all of the programs trained for rice：FGeneSH (75)，GeneMark (76)，GenScan (77)，GlimmerM (78)，and RiceHMM (79). Strictly speaking，GenScan was trained for maize，another monocot with GC content gradients.

Table 3　Complex repeats.

		Number	Total /bp	Mean /bp	93-11 assembled contigs		93-11 fully masked reads		Full-length cDNAs	
					% of data set	% of repeats	% of data set	% of repeats	% of data set	% of repeats
Class I	LINEs	5	18 997	3 799	1.190 5	7.43	0.131 8	0.22	0.025 7	2.51
	SINEs	7	1 254	179	0.088 8	0.55	0.004 7	0.01	0.026 8	2.61
	gypsy–like	19	105 614	5 559	3.728 5	23.28	41.689 4	70.35	0.123 8	12.07
	copia-like	5	35 151	7 030	1.717 5	10.72	15.850 6	26.75	0.086 9	8.47
	Subtotal				6.725 4	41.99	57.676 6	97.33	0.263 1	25.65
Class II	*Ac/Ds* TEs	3	1 567	522	0.109 9	0.69	0.014 5	0.02	0.000 0	0.00
	En/Spm TEs	3	5 558	1 853	0.259 0	1.62	0.277 0	0.47	0.000 0	0.00
	MULEs	22	25 800	1 173	2.450 0	15.30	0.637 8	1.08	0.180 7	17.62
	Subtotal				2.819 0	17.60	0.929 3	1.57	0.180 7	17.62
Class III	Stowaway–like	70	16 112	230	2.237 0	13.97	0.124 7	0.21	0.191 0	18.62
	tourist-like	77	19 933	259	3.740 5	23.35	0.322 8	0.54	0.345 1	33.65
	Unknown MITEs	2	1 341	671	0.495 0	3.09	0.208 0	0.35	0.045 8	4.46

Continued

	Number	Total /bp	Mean /bp	93–11 assembled contigs		93–11 fully masked reads		Full-length cDNAs	
				% of data set	% of repeats	% of data set	% of repeats	% of data set	% of repeats
Subtotal				6.472 5	40.41	0.655 6	1.11	0.581 8	56.73
Grand Total	213	231 327	1 086	16.016 9	100.00	59.261 5	100.00	1.025 5	100.00

Note: Transposons identified by RepeatMasker are assigned to three classes. Each class has a number of families (e.g., *tourist*-like MITEs), and each family has a number of different subfamilies.The number of subfamilies is listed, as well as their total and mean size. Repeat content for each family is given as a percentage by length, normalized with respect to the data set (assembled contigs, fully masked reads, or cDNAs) or with respect to all identified transposons.

Assessment of gene-prediction programs. All the gene-prediction programs were pretrained by the authors and tested against our cDNA-to-genomic alignments. These comparisons may favor the program that was trained on the largest and most recent data set, but that information was not available to us. Performance was measured at the base pair and the exon levels, and then plotted as a function of position from 5′ to 3′ end (Fig.8) .Sensitivity is the probability that the actual coding region is correctly predicted (1 minus false-negative rate) . Specificity is the probability that the predicted coding region corresponds to the actual coding region (1 minus false-positive rate) . Some programs, including GenScan, had sensitivities that were extremely dependent on position, although this was not the case when we applied these performance metrics to human genes. This suggests that the compositional gradients were indeed a source of error. That GenScan would be affected is significant because, in the most recent comparative analysis of human genes (80), two of the most successful programs were FGeneS (a variant of FGeneSH) and GenScan. For rice, however, FGeneSH is the most successful program. It is not obvious why, although the documentation states that FGeneSH places more weight on signal terms (e.g., splice sites, start and stop codons) than on content terms (i.e., codon usage) .

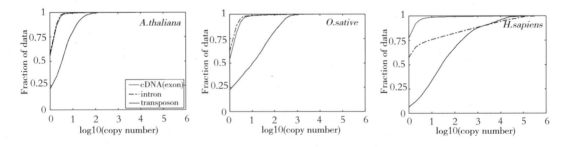

Fig.7 Cumulative copy numbers for exons , introns , and known transposons in *A.thaliana* , *O.sativa* , and *H.sapiens*. We determined the copy number of 20-mers in each genome , and then mapped these 20-mers back to exons , introns , and known transposons for each genome.All exon and intron sequences were derived from cDNA-to-genomic alignments. The analyzed transposons were the consensus sequences for the known families or subfamilies of transposons. We show here the fraction of each data set that is in 20-mers up to the indicated copy numbers.

Submitting our 93-11 assembly to the FGeneSH Web site returned 75, 659 predictions.However, only 53, 398 were complete, in the sense that initial and terminal exons were both present; 7489 had only an initial exon, 11, 367 had only a terminal exon, and 3405 had neither. When we include predictions without both an initial and terminal exon as only half a gene, we obtain an upper bound of 64, 529 genes. Without correcting for sensitivity or specificity, the estimated gene count is 53, 398 to 64, 529. This is similar to the 59, 855 genes that we predicted from considerations of gene size and repeat content. How good are these predictions? We have reservations about the absolute value of the performance metrics, because FGeneSH was probably trained on a gene set with considerable overlap to our reference cDNAs. These metrics may not tell us how well FGeneSH performs for rice genes with substantially different compositional properties. However, their relative values should be interpretable. Namely, base-level specificities were better than base-level sensitivities, indicating that false-negatives are more likely to be a problem than false-positives. The program is more likely to miss an exon fragment than to label something part of an exon by mistake. Sensitivities and specificities were much worse at the exon level, implying that that the exon-intron boundaries are not precisely defined, even when the presence of a gene is correctly detected.

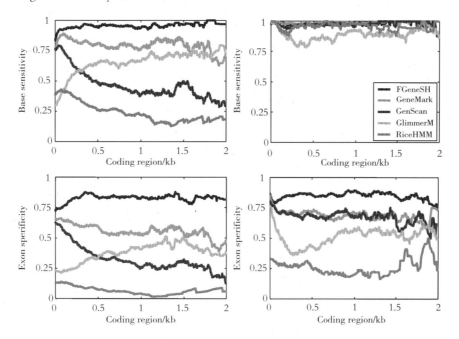

Fig.8　Performance metrics for ab initio gene-prediction programs , as a function of gene positionfrom 5′ to 3′ end , when compared against cDNA-to-genomic alignments at the same loci.Sensitivity is the probability that the coding region is correctly predicted (1 minus false-negativerate). SpeciTcity is the probability that the predicted coding region is real (1 minus false-positiverate). At the exon level , both splice sites must be correctly predicted for an exon to be counted ascorrect.

Two pieces of evidence qualify our level of confidence in the gene predictions. First, if the sensitivity is really as good as suggested, then we ought to be able to find most of the ESTs in the

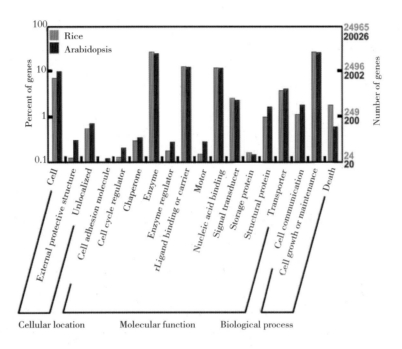

Fig.9 Functional classiTcation of rice genes , according to Gene Ontology Consortium , andassigned by homology to categorized *A.thaliana* genes. In this ontology , "biological process ," "cellular location ," and "molecular function" are treated as independent attributes. Only 36.3%ofthe 25 , 426 predicted genes for *A. thaliana* are classified. For rice , only 20.4% of the 53 , 398 complete predictions , with both initial and terminal exons , could be classified.

predicted gene set. We thus performed a comparison against the 24 , 776 UniGene clusters assembled from our 87 , 842 ESTs. The result was that only 77.3% of these clusters could be found in the FGeneSH predictions. Second , the mean size of the predicted coding regions in rice was only 328 residues , or 73.5% of the predicted coding regions in *A.thaliana* , which averaged 446 residues. This was the case even though we restricted the mean to complete genes with initial and terminal exons. Although it is possible that rice genes are intrinsically smaller than *A.thaliana* genes , we believe that this discrepancy reflects a deeper problem that is related to the compositional gradients , as will be explained below.

*Functional classification of rice genes.*Although 25 , 426 genes have been identified in *A.thaliana* , fewer than 10% have been documented experimentally (81) . Consequently , functional classification of plant genes must rely heavily on homology , coupled with a few nonhomology-based methods , such as phylogenetic profiling , correlated gene expression , and conserved gene orders. Only 27.3 and 36.3% of *A.thaliana* genes have been classified by InterPro (82) and Gene Ontology Consortium (83) , respectively.To establish functional classifications for rice genes , we performed protein-to-protein sequence comparisons against *A.thaliana* annotations , and adopted classifications from the best match to *A.thaliana*. We considered only those 53 , 398 predictions from FGeneSH with initial and terminal exons. When multiple hits were found , we selected the one with the longest extent of homology.We required that at least 25% of the protein length be matched. This is a low-threshold setting , but as we will explain below, it was necessary. In total, 15.9% and 20.4% of rice gene predictions were classified by InterPro and Gene Ontology Consortium, respectively.As a percentage of classified genes, the predicted gene sets for rice and *A.thaliana* are similarly distributed among different functional categories (Fig.9). We depict Gene Ontology Consortium because more genes were classified. Tables of predicted rice genes and their functional classifications (Web supplement 2), as well as InterPro figures (Web supplement 3), are available on *Science* Online at www.sciencemag.org/cgi/content/full/296/5565/79/DC1.

We advise extreme caution in interpreting minor differences in functional classification between the predicted gene sets for rice and *A.thaliana*. With such a large fraction of the genes unclassified, intrinsic uncertainties in any classification scheme are amplified into artifactual differences. For example, the largest difference for InterPro was in signal transduction genes, but no notable difference was observed for Gene Ontology Consortium. Furthermore, focusing on small differences that had a high likelihood of being artifactual would distract from the major difference between rice and *A.thaliana*, which as we will show next, lies almost entirely in those genes with no functional classification.

A.thaliana comparisons. In general, there are two ways to compare gene sets: through colinearity and homology. Colinearity of plant genomes has been studied extensively (84, 85). For analyses done within a plant family, high degrees of colinearity have been consistently observed. Across the monocot-eudicot divide, with rice and *A.thaliana* as representative species, observed degrees of colinearity have been considerably lower (86-89). For example, an analysis of a 340-kb segment on rice chromosome 2 identified 56 putative genes (88). Homologs for 22 (39%) of them were identified in *A.thaliana*, but were distributed among 5 chromosomal segments, with several small-scale inversions. Another study of 126 rice BACs, totaling 20 Mb of sequence and with 3011putative genes, identified homologs in *A.thaliana* for 1747 (58%) of these genes (89).Typically, each 150-kb BAC mapped to three or more chromosomes. Notwithstanding the absence of colinearity, the finding that only half of the rice genes had a homolog in *A.thaliana* was unexpected. Although these analyses were based on predicted genes, which have not yet been confirmed, we do not believe that this was why so few rice genes had a homolog in *A.thaliana*, because a similar analysis was done with 27 294 unique ESTs from *Z. mays* (maize), and only 62 % of the open reading frames had a homolog in *A.thaliana* (90).

We focus exclusively on homology, rather than orthology, because extensive gene duplications in *A.thaliana* (4, 91) and rice make strict one-to-one pairing relations, the classic definition for orthology (92), difficult to determine. A mere 35% of *A.thaliana* genes are unique and 37.4% belong to gene families with more than five members. Segmental duplications larger than 100 kb in size constitute 58% of the genome, and 17% of the genes are arranged in tandem arrays. In comparisons of rice with *A.thaliana*, and vice versa, we sought to compute the degree of homology in each direction, and the extent to which gene duplications in *A.thaliana* are replicated in rice when decomposed by functional classification. Even this modest objective was not easy to accomplish, because of unexpected complications introduced by the compositional gradients in rice.

Homology between monocots-eudicots. The complete set of 25, 426 annotated *A.thaliana* genes was downloaded from the Arabidopsis Information Resource Web site (93) on 29 November 2001. As a control, 1441 proteins were downloaded from SwissProt (94) on the same day. The rice genes were restricted to the 53, 398 predictions from FGeneSH with initial and terminal exons. We compared protein sequence to all six reading frames of the genome sequence by means of TblastN (36). Therefore, if the homology search failed, it would not be due to a gene being missing from the annotation of the target genome. The expectation value cutoff was set to 10^{-7}. This was not a sensitive parameter, as most hits were either very good or very bad. What mattered was the "coverage rule." We projected every hit back to the protein query, and unless a minimum fraction of the protein was covered, none of the hits were accepted. The hits had to occur in the same order in both the query and the target, and they all had to be in the same orientation. When a homolog spanned more than one scaffold, the coverage rule was imposed on each scaffold. From this rule, we estimated the number of homologs per gene, the extent of the homology, and the percentage amino acid identity (95).

The asymmetry in the monocot-eudicot analysis was striking (Fig.10). About 80.6% of *A.thaliana* genes had a homolog in rice.The mean extent of homology was 80.1% of the protein length, and there was 60.0% amino acid identity. If instead of the full set of annotated genes, we had used SwissProt genes, 94.9% of the genes would have had a homolog, across 86.7% of the protein length and at 72.9% amino acid identity. Presumably, there were more homologs in the SwissProt data because they were more biased toward highly conserved proteins. In contrast, only 49.4% of predicted rice genes had a homolog in *A.thaliana*. The mean extent of homology was 77.8% of the protein length, and there was 57.8% amino acid identity.For brevity, predicted rice genes with a homolog in *A.thaliana* are called WH genes, and those with no homologs are called NH genes. We identified two distinct problems in this analysis, both attributable to the compositional gradients in rice. One was the poor quality of the FGeneSH predictions for NH genes, and the other was related to the probability of identifying a TblastN hit even with a perfect gene annotation. We did use ESTs to confirm that NH genes were not false predictions, but first, we will discuss what we believe to be the true problems.

We had previously observed that rice gene predictions were only 73.5% the size of *A.thaliana* gene predictions. This discrepancy is not due to the WH half of the rice genes. It is due to the NH half, which was on average 49.4% smaller than the WH half (Fig.11). To analyze the problem, we randomly sampled 3000 WH genes and 3000 NH genes, and applied the analyses of Fig.3 to Fig.7. In general, WH genes resembled the "gold standard" based on alignment of cDNA to genomic sequence. NH genes exhibited a number of striking differences. First, the decreased coding region size was clearly due to a decrease in the number of exons, not to a decrease in the size of the exons. The GC-rich tail in NH gene exon distribution was twice as large as normal (Fig.11), suggesting that NH genes had more pronounced GC content gradients than either WH genes or those cDNAs retrieved from GenBank. It is plausible that FGeneSH performance would have faltered on NH genes, because NH genes did not resemble those genes on which FGeneSH was presumably trained. NH genes also had twice as many introns as normal in the 200-to 2000-bp range (Fig.11). This would be consistent with some of these missing exons being combined with their flanking introns.The preponderance of anomalous subminimal introns would be consistent with exon fragments being mistakenly called introns. However, NH genes could not be transposon sequences, because a 20-mer analysis confirmed that their constituent sequences were found in the genome at low copy numbers, much like WH and cDNA-derived genes.

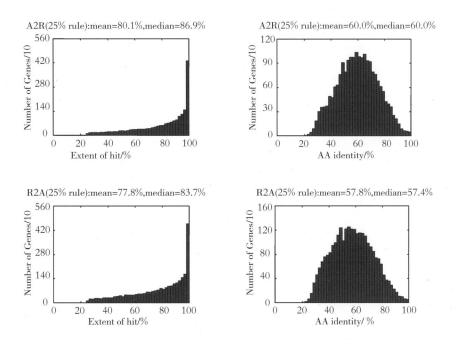

Fig.10　Distributions in extent of homology and maximum amino acid identity , for *Arabidopsis*-to-rice and rice-to-*Arabidopsis* comparisons. These values are based on a comparison of predictedprotein sequence against all six reading frames of the target genome sequence.

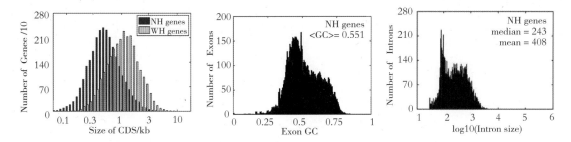

Fig. 11　Size distribution of predicted rice genes with a homolog (WH), and with no homolog (NH), in *A. thaliana*, plus exon GC content and intron size for a random sampling of 3000 NH genes. Gene size refers to the size of the predicted coding region.

Although we do not entirely ascribe the small size of NH genes to a failure by FGeneSH to detect exons, it is likely that more exons were missed than for WH genes.Thus, it would be more difficult to identify a homolog for these genes in *A.thaliana*. However, even for experimentally derived gene sequences, like cDNAs, the probability of identifying a TblastN hit, as a function of the position, dropped precipitously near the 5′ end of the genes (52). Far from the end, the probability was about 90%, but within the first few hundred bases near the 5′ end, the probability dropped to less than 50%. This was another consequence of the compositional gradients in rice. The magnitude of the effect was unexpected. We had thought that selective constraints on coding sequences would have limited the number of amino acid changes, despite pressure from rising GC content. However, this was not the case. Homology searches were more likely to fail with the smaller NH genes because the problematic region was a larger fraction of their total length, and our "coverage rule" required that the TblastN hits cover a minimum fraction of the coding region. Even in the Arabidopsisto-rice analysis, where the gene predictions were more reliable, 83.2%, 80.6%, 69.5%, and 48.5% of *A.thaliana* genes had a homolog in rice, for coverage rules of 0%, 25%, 50%, and 75%. We had to use a relatively low coverage rule of 25%. Given the typical protein and protein domain sizes of 446 and 100 residues (96-98), respectively, this was equivalent to one protein domain.

Alternatively, what if the problem were due to scaffold size? Half of the NH genes were identified in a scaffold that was smaller than 7.1 kb. However, as a function of scaffold size, predicted coding regions for NH genes were almost always the same size. NH genes found in scaffolds greater than 7.1 kb were only 7% larger than those found in scaffolds less than 7.1 kb. Scaffold size could not have been responsible for the small size of the NH genes. Perhaps NH genes are not real genes at all. Are they even expressed? Looking back at our EST confirmation analysis, we found that 42.9% of WH genes were confirmed by a UniGene cluster, compared with 15.4% of NH genes. Assuming that all WH genes are real, this would imply that $(15.4/42.9) \times 100\% = 35.9\%$ of NH genes are real. However, if we adjust for their being 49.4% smaller than normal, attributing this size deficit to missed exons, then 72.7% of NH genes are real. Certainly, not every NH gene is real, but many are. To be conservative, we can adjust our gene count estimates by a factor of $(0.494 + 0.727 \times 0.506)$, resulting in a revised gene count of 46,022 to 55,615.

Considering the relatively recent divergence between monocots and eudicots, 145 to 206 million years ago, it is surprising to find so many genes in rice with no homolog in *A.thaliana*. Even more intriguing, this absence of homology for NH genes extended to other sequenced organisms, including *D.melanogaster*, *Caenorhabditis elegans*, *Saccharomyces cerevisiae*, and *Schizosaccharomyces pombe*. Although WH genes had a 30.5% probability of being homologous to at least one gene in one of these organisms, NH genes had a 2.4% probability. Hence, the major difference between rice and *A.thaliana* gene sets lies in that half of the predicted rice gene set with essentially no homologs in any organism, and whose functions are largely unclassifiable.

Duplication between monocots-eudicots. Having established the major difference between the gene sets for rice and *A.thaliana*, we now consider the similarity. We had reported that 80.6% of the predicted *A.thaliana* genes, and 94.9% of the SwissProt genes, had a homolog in rice. The actual number is likely to be even higher, because the gradients kept us from identifying potential homologs for smaller genes. We know that, within *A.thaliana*, the genes are highly duplicated. Are these genes duplicated in the same manner when mapped to rice? As a proxy for the number of gene homologs within and between genomes, we used the "hits per gene," as defined in the notes (95). Considering that, in the *Arabidopsis*-to-rice comparison, we used a low coverage rule of 25% to compensate for the gradients, it was inevitable that we would experience more difficulty than usual in distinguishing between duplicated domains and duplicated genes. Thus, the number of hits per gene is an overestimate of the number of gene homologs.

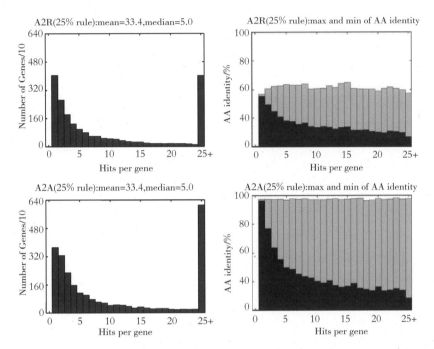

Fig. 12　Distributions in number of hits per gene and maximum-versus-minimum amino acid identity, for *Arabidopsis*-to-rice and *Arabidopsis*-to-*Arabidopsis* comparisons. "Hits per gene" is a proxy for the number of gene homologs, between and within genomes.

　　Comparing *Arabidopsis-to-Arabidopsis*（A2A），the mean and median hits per gene were 38.2 and 6.0，similar to the mean and median of 33.4 and 5.0 that we observed comparing *Arabidopsis*-to-rice （A2R）（Fig.12）. That the A2R numbers would be slightly smaller makes sense，given the 145 to 206 million years of divergence. We further note that the means were large only because of a few outliers，some with up to 1000 hits. The identity of these outliers included protein kinase，cytochrome P450，putative disease resistance，and many "unknown" genes. It is difficult to draw any conclusions about the last category，but the others are highly duplicated gene families，which confirms that these outliers were not computational artifacts. The maximum amino acid identity was independent of the number of hits，but the minimum amino acid identity decreased with the number of hits，which would be consistent with an increasing occurrence of hits to ever larger families of related but divergent genes. Although the number of hits was dependent on the functional classification，it was similarly distributed among the different functional categories for A2R and A2A （Fig.13）. Therefore，not only was it possible to identify a homolog in rice for almost every *A.thaliana* gene，but the patterns of gene duplication in one were largely replicated in the other.

　　The most parsimonious explanation is that the rice gene set is essentially a "superset" of the *A.thaliana* gene set. However，we are unable to say how many of these additional genes that are unique to rice are functionally novel, or merely unrecognizable, because of gradients in rice amino acid usage. It does seem unlikely that so many novel genes would arise within only 145 to 206 million years, and therefore, we suspect that a massive duplication event (or a series of duplication events) occurred, after which many of the rice genes were rendered unrecognizable by compositional gradients. Some may have been inactivated, and now exist only as pseudogenes.However, until we can compensate for the confounding effects of compositional gradients, we cannot explore the extent to which rice (99) and many other plants, including *A.thaliana*, are hybrid (100) or allopolyploid (101, 102) in origin.

Rice polymorphisms. Differences between subspecies or cultivars of rice must be described at two levels, gross and nucleotide.At the gross level, we found kilobase-sized regions of high similarity interspersed with kilobase-sized regions of no similarity. One such example is shown in Fig.14, which was based on a comparison of two overlapping BACs from *indica and japonica*. Every unalignable region coincided with a cluster of MDRs, traceable to length differences of 0.7 to 25 kb between the two source sequences, distributed in almost equal proportions between insertions and deletions. To the extent that BDRs could be identified, in roughly half of the unalignable regions, they belong to the class of nested retrotransposons that inhabit the intergenic regions between genes. This is another confirmation of the observation that genome sizes change rapidly in grasses (103) .On the basis of the available 259 kb of overlapping finished BAC sequences, from *Nipponbare (japonica) and GLA (indica)*, all on rice chromosome 4, we would estimate that 16% of the *indica and japonica* genome is unalignable by this definition.

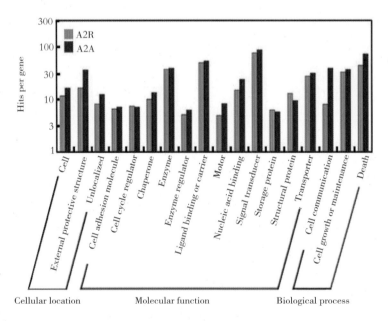

Fig.13 Distributions in the number of hits per gene , sorted according to Gene Ontology Consortium , for *Arabidopsis*-to-rice and *Arabidopsis*-to-*Arabidopsis* comparisons. This Tgure showsonly the 36.3% of predicted *A. thaliana* genes that are classified.

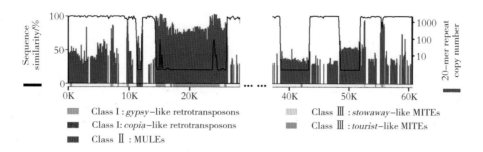

Class I : *gypsy*–like retrotransposons　　　Class Ⅲ : *stowaway*–like MITEs
Class I : *copia*–like retrotransposons　　　Class Ⅲ : *tourist*–like MITEs
Class Ⅱ : MULEs

Fig.14　Comparison of *GLA*（*indica*）and *Nipponbare*（*japonica*）BAC sequences（GenBank accession number AL442110 and AL606449 , respectively）. Exact 20-mer repeats are indicated by bluehistogram bars , with bar heights proportional to copy number in 93-11（*indica*）. Sequence similarityis almost 100% , or unalignable and set to 20% by BLAST. Every unalignable region coincides witha cluster of MDRs , but Repeat Masker fails to identify a BDR in half of these regions.

At the nucleotide level , excluding the unalignable regions , we define polymorphism rates for repeated and unique sequence , partitioned in single-base substitutions（single nucleotide polymorphisms , SNPs）and insertion-deletion polymorphisms（InDels）. By repeated sequence , we mean MDRs. Three different comparisons are shown in Table 4.Two are based on the alignment of 93-11 contigs to finished BAC sequences from *Nipponbare*（*japonica*）and *GLA*（*indica*）, totaling 11.8 Mb and 0.9 Mb , respectively. The other is a comparison of 93-11 and *PA64s* contigs.One might question the accuracy of a polymorphism rate based on rough draft sequence , particularly the low-coverage *PA64s* sequence. However , as we noted in our "quality assessments" section , most of the errors are in the small contigs and at the ends of the contigs. Thus , we restricted this analysis to contigs larger than 3 kb , with 500 bp trimmed off both ends. Overall , there was twice as much variation in the repeated regions as in the unique regions. Substitution rates were two to three times as large as InDel rates.Remarkably , there was very little difference among the three pairwise comparisons. For 93-11 to *PA64s* , averaged over repeated and unique regions , the SNP and InDel rates were 0.43 and 0.23% , respectively. Combining the SNP and InDel rates , we obtained an overall rate of 0.67%. Although the numbers are not exactly comparable , the measured polymorphism rate in maize was 0.96%（104）.

Table 4　Polymorphism rates relative to 93-11（*indica*）.

	Nipponbare（*japonica*）	*PA64s*	*GLA*（*indica*）
SNPs in repeated sequence/%	0.88	0.68	0.65
InDels in repeated sequence/%	0.33	0.45	0.27
SNPs in unique sequence/%	0.50	0.35	0.50
InDels in unique sequence/%	0.14	0.16	0.15
Repeated sequence fraction/%	24.1	25.5	22.8
Unique sequence fraction/%	74.8	74.3	74.1

Continued

	Nipponbare (*japonica*)	PA64s	GLA (*indica*)
Parts unalignable by BLAST/%	1.1	0.3	3.1

Note: Comparisons were made to Tnished BAC sequences from *GLA* (*indica*) *and Nipponbare* (*japonica*), as well as to *PA64s contigs*. Rates werecomputed for repeated and unique regions, in single-base substitutions (SNPs) and insertion-deletions (InDels). The numbers given for "unalignable" are a gross underestimate because RePS assemblies omitmany of the fully masked reads that correspond to the unalignable regions of Fig.14.

SNPs are useful in genetic mapping (105), and are either directly applicable to phenotypes or indirectly applicable through linkage and association studies. Polymorphisms in the unique regions are particularly useful because, unlike those in the repeated regions, they are more reliably genotyped with existing high-throughput technologies, which always involve some sort of hybridization step.We expect that genome-wide SNP mapping in plants (106) will become more popular as new technologies become available, especially as some are customized for plants (107).

Concluding remarks. In the initial annotation of the human genome (7, 8), alternative splicing was proposed as a method by which protein diversity could be generated from the surprisingly small number of genes that were identified. The idea that there is extensive alternative splicing in human genes has been supported by analyses of EST data (108-112). Alternative splicing is often associated with the exon recognition model (113) of pre-mRNA splicing. Exon recognition is facilitated by exonic splicing enhancers-short, degenerate sequences located in the exons that are recognized by a multitude of RNA binding factors (114, 115). Because it is the exons that are recognized by the splicing machinery, the intron sequence content is less critical, and transposon insertions into the intron are more readily tolerated. Thus, the preponderance of large transposons-filled introns in the human genome is consistent with extensive alternative splicing.

The presence of relatively few transposons inside plant introns suggests that exon recognition is not a common process for plant genes. Indeed, exonic splicing enhancers have yet to be identified in plants (116). The corollary is that there should be relatively little alternative splicing in plant genes. Analysis of the EST data confirms that *A.thaliana* has substantially less alternative splicing than vertebrates or invertebrates (117). However, protein diversity must be generated for the organism to evolve. Our analysis has demonstrated extensive gene duplications in rice and *A.thaliana*, which are highly correlated with each other when decomposed by functional classification. The conclusion is that protein diversity in plants is generated primarily through gene duplications, whereas in vertebrates, it is generated through gene duplications and alternative splicing. This would explain why rice has so many genes.However, as a method of generating protein diversity, gene duplications come at the cost of an increase in transcriptional noise (118).Perhaps, at some level of complexity, alternative splicing becomes preferred.

Looking to the future, we intend to improve our draft sequence by adding more reads from large-insert clones, filling any gaps that are likely to contain genes, and integrating the resultant sequence with existing physical and genetic maps. The large-insert clones are necessary to correctly assemble across the large repeat clusters that are sprinkled throughout the rice genome. Until then，the BAC-end sequences (119) may not be useful because they are too large to bridge adjacent contigs，and instead skip intervening contigs，resulting in a morass of interleaving scaffolds. One should also be wary of large-scale differences between *indica* and *japonica*. In any event，the final assembly will be made freely available to the research community. We will then apply the experiences gained from the rice genome project to other agriculturally important crops，including Z.*mays* (maize) and *T. aestivum* (wheat) .

References

1. T. Sasaki, B. Burr, *Curr. Opin. Plant Biol.* 3, 138 (2000) .

2. N. A. Eckardt, *Plant Cell* 12, 2011 (2000) .

3. K. Arumuganathan, E. D. Earle, *Plant Mol. Biol. Rep.* 9, 208 (1991) .

4. The *Arabidopsis* Genome Initiative, *Nature* 408, 796 (2000) .

5. B. Martienssen, W. R. McCombie, *Cell* 10, 571 (2001) .

6. M. Bevan *et al., Curr. Opin. Plant Biol.* 4, 105 (2001) .

7. International Human Genome Sequencing Consortium, *Nature* 409, 860 (2001) .

8. J. C. Venter *et al., Science* 291, 1304 (2001) .

9. Q. Tao *et al., Cell Res.* 4, 127 (1994) .

10. Y. Umehara, A. Miyazaki, H. Tanoue, *Mol. Breed.* 1, 79 (1995) .

11. M. Bevan, G. Murphy, *Trends Genet.* 15, 211 (1999) .

12. G. L. Wang *et al., Plant J.* 7, 525 (1995) .

13. M. D. Gale, K. M. Devos, *Proc. Natl. Acad. Sci. U.S.A.*95, 1971 (1998) .

14. J. Messing, V. Llaca, *Proc. Natl. Acad. Sci. U.S.A.* 95, 2017 (1998) .

15. S. Goff , Curr. *Opin. Plant Biol.* 2, 86 (1999) .

16. Major Web sites for rice genome data：http：//rgp. dna.affrc.go.jp；http：//www.genome.clemson.edu；2http：//ars-genome.cornell.edu；http：//www.tigr.org/tdb/e2k1/osa1/BACmapping/description.shtml.

17. R. J. Davenport, *Science* 291, 807 (2001) .

18. D. Dickson, D. Cyranoski, *Nature* 409, 551 (2001) .

19. Z. Y. Dai, B. H. Zhao, X. J. Liu (in Chinese), *Jiangsu Agric.* Sci. 4, 13 (1997) .

20. L. P. Yuan (in Chinese), *Hybrid Rice* 1, 1 (1997) .

21. J. Yu *et al., Chin. Sci. Bull.* 46, 1937 (2001) .

22. J. Wang *et al., Genome Res.*, in press. The software can be obtained by e-mailing the authors at reps@ genomics.org.cn.

23. J. L. Bennetzen, *Plant Cell* 12, 1021 (2000).

24. A. Kumar, J. L. Bennetzen, *Annu. Rev. Genet.* 33, 479 (1999).

25. M. D. Adams *et al.*, *Science* 287, 2185 (2000).

26. B. Ewing, L. Hillier, M. C. Wendl, P. Green, Genome Res. 8, 175 (1998).

27. B. Ewing, P. Green, *Genome Res.* 8, 186 (1998).

28. For the plasmid shotgun libraries, a DNA isolation protocol was modiped from Sambrook and Russell (29). Fresh leaves at the seeding stage were ground in liquid nitrogen before complete lysis (30). Puriped high-molecular weight genomic DNA was sonicated and sized on agarose gels, selecting for fragments of size 1.5to 3.0kb. QIAEX Gel Extraction Kit (QIAGEN) was used to purify DNA from the gel slices.Genomic fragments were ligated to SmaI-linearized pUC18plasmids and transformed into DH10B-competent cells by electroporation.

29. J. Sambrook, J. D. Russell, *Molecular Cloning*, (Cold Spring Harbor Laboratory Press, Cold Spring Harbor, NY, ed. 3, 2001).

30. S. Hatano, J. Yamaguchi, A. Hirai, *Plant Sci.* 83, 55 (1992).

31. Single colonies were grown in 96-deep-well plates, and plasmid DNA was prepared by alkaline lysis (32). Quality of DNA and insert sizes were examined by agarose gel electrophoresis. Puriped plasmid DNA (200ng; Amersham Pharmacia Biotech, Beijing) was used for the sequencing reactions. DNA sequencing was done with MegaBACE 1000capillary sequencers (Amersham Pharmacia Biotech, Beijing). Machine parameters were adjusted for high output (10 to 11runs a day on average).

32. H. C. Birnboim, *Methods Enzymol.* 100, 243 (1983).

33. N. Jiang, S. R. Wessler, *Plant Cell* 13, 2553 (2001).

34. P. Green, http: //www.phrap.org.

35. E. S. Lander, M. S. Waterman, *Genomics* 2, 231 (1988).

36. S.F.Altschul, W. Gish, *Methods Enzymol.* 266, 460 (1996).

37. Sources for STS, STR, restriction fragment length polymorphism sequences: http: //ars-genome.cornell. edu/; http: //www.ncbi.nlm.nih.gov/; http: //rgp.dna. affrc.go.jp/publicdata/geneticmap2000/index.html.

38. A.F.Smit, P.Green, http: //ftp.genome.washington. edu/RM/RepeatMasker.html.

39. S.P.Kennedy, W.V. Ng, S.L.Salzberg, L.Hood, S.DasSarma, *Genome Res.* 11, 1641 (2001).

40. E.Chargaff, *Experientia* 6, 201 (1950).

41. R.Rolfe, M.Meselson, *Proc.Natl.Acad.Sci.U.S.A.* 45, 1039 (1959).

42. N.Sueoka, J.Marmur, P.Doty, *Nature* 183, 1427 (1959).

43. D.R.Forsdyke, J. R. Mortimer, *Gene* 261, 127 (2000).

44. G.Bernardi, *Gene* 259, 31 (2000).

45. A.Eyre-Walker, L.D.Hurst, *Nature Rev. Genet.* 2, 540 (2001).

46. C.Gautier, *Curr. Opin. Genet. Dev.* 10, 656 (2000).

47. S.Karlin, A.M.Campbell, J.Mrazek, *Annu. Rev. Genet.* 32, 185 (1998)

48. N.Carels, G.Bernardi, Genetics 154, 1819 (2000).

49. J.Filipski, J.P.Thiery, G.Bernardi, *J.Mol. Biol.* 80, 177 (1973).

50. G.Bernardi *et al.*, *Science* 228, 953 (1985).

51. G.K.S Wong, D.A.Passey, Y.Z.Huang, Z.Yang, J.Yu, *Genome Res.* 10, 1672 (2000).

52. G.K.S.Wong *et al.*, *Genome Res.*, in press.

53. G.K.S.Wong, D.A.Passey, J.Yu, *Genome Res.* 11, 1672 (2001).

54. C.Feuillet, B.Keller, *Proc. Natl. Acad. Sci. U.S.A.* 96, 8265 (1999).

55. I.Ashikawa, Plant J. 26, 617 (2001).

56. F.Larsen, G.Gundersen, L.Lopez, H.Prydz, *Genomics* 13, 1095 (1992).

57. J. Messing, *Trends Genet.* 6, 196 (2001).

58. R. Sánchez-Fernández, P. A. Rea, T. G. E. Davies, J. O. D. Coleman, *Trends Plant Sci.* 6, 348 (2001).

59. J. Jurka, *Trends Genet.* 16, 418 (2000); http://www.girinst.org/index.html.

60. L. Mao et al., *Genome Res.* 10, 982 (2000).

61. K. Turcotte, S. Srinivasan, T. Bureau, *Plant J.* 25, 169 (2001).

62. S. Temnykh et al,. *Genome Res.* 11, 1441 (2001).

63. R. I. Richards, G. R. Sutherland, *Cell* 70, 709 (1992).

64. ____, *Nature Genet.* 6, 114 (1994).

65. C. Schlötterer, D. Tautz, Nucleic Acids Res. 20, 211 (1992).

66. D. Tautz, M. Trick, G. Dover, *Nature* 322, 652 (1986).

67. S. A. Surzycki, W. R. Belknap, *Proc. Natl. Acad.*

Sci. U.S.A. 97, 245 (2000).

68. R. Tarchini, P. Biddle, R. Wineland, S. Tingey, A.Rafalski, *Plant Cell* 12, 381 (2000).

69. X.Zhang et al., *Proc. Natl. Acad. Sci.* U.S.A. 98, 12572 (2001).

70. A. F. Smit, *Curr. Opin. Genet. Dev.* 6, 743 (1996).

71. T.E.Bureau, P. C. Ronald, S. R. Wessler, *Proc. Natl.Acad. Sci. U.S.A.* 93, 8524 (1996).

72. A. Nekrutenko, W. H. Li, *Trends Genet.* 17, 619 (2001).

73. M.Long, *Curr. Opin. Genet. Dev.* 11, 673 (2001).

74. F.Jacob, Science 196, 1141 (1977).

75. A.A.Salamov, V.V.Solovyev, *Genome Res.* 10, 516 (2000); http://www.softberry.com/berry.phtml?topic=gpnd.

76. A.V.Lukashin, M.Borodovsky, *Nucleic Acids Res.* 26, 1107 (1998). http://opal.biology.gatech.edu/GeneMark/eukhmm.cgi? org=O.sativa.

77. C.Burge, S.Karlin, *J.Mol. Biol.* 268, 78 (1997); http://genes.mit.edu/GENSCAN.html.

78. A.L. Delcher, D. Harmon, S. Kasif, O.White, S.L.Salzberg, *Nucleic Acids Res.* 27, 4636 (1999); http://www.tigr.org/softlab/glimmerm.

79. K. Sakata *et al.*, *Abstracts of 4th Annual Conference on Computational Genomics (2000)*, p. 31; http://rgp.dna.aff rc.go.jp/RiceHMM.

80. S. Rogic, A. K. Mackworth, F. B. Ouellette, *Genome Res.* 11, 817 (2001).

81. P. Breyne, M. Zabeau, *Curr. Opin. Plant Biol.* 4, 42 (2001).

82. R. Apweiler et al., *Nucleic Acids Res.* 29, 44 (2001).

288

83. The Gene Ontology Consortium, *Nature Genet.* 25, 25 (2000).

84. M. D. Gale, K. M. Devos, *Science* 282, 656 (1998).

85. R. Schmidt, *Curr. Opin. Plant Biol.* 3, 97 (2000).

86. K. M. Devos, J. Beales, Y. Nagamura, T. Sasaki, *Genome* Res. 9, 825 (1999).

87. A. van Dodeweerd *et al.*, *Genome* 42, 887 (1999).

88. K. Mayer *et al.*, *Genome Res.* 11, 1167 (2001).

89. H. Liu, R. Sachidanandam, L. Stein, *Genome Res.* 11, 2020 (2001).

90. V. Brendel, S. Kurtz, V. Walbot, *Genome Biol.* 3, reviews 1005.1 (2002).

91. M. Devan *et al.*, *Curr. Opin. Plant Biol.* 4, 105 (2001).

92. W. Fitch, *Syst. Zool.* 19, 99 (1970).

93. *Arabidopsis* genome annotations: http://www.arabidopsis.org.

94. E. Gasteiger, E. Jung, A. Bairoch, *Curr. Iss. Mol. Biol.* 3, 47 (2001); http://www.expasy.com.

95. For each protein query, we created an array with one element for each amino acid position. Blast_hits ( ) recorded the number of times that each position was covered by a TblastN hit. Each hit had associated with it a score for the percentage of identically matched amino acids. AA_identity ( ) recorded the maximum and minimum score at each postion, across all TblastN hits. "Extent of hit," quoted as a percentage of the protein length, is the number of nonzero elements in Blast_hits ( ). "AA identity" and "hits per gene" are the median values of AA_identity and Blast_hits, computed over positions with one or more hits. We used the median, instead of the mean, to minimize the likelihood of counting a highly duplicated domain when the entire protein is not duplicated.

96. S. A. Islam, J. Luo, M. J. Sternberg, *Protein Eng.* 8, 513 (1995).

97. R. Sowdhamini, S. D. Rupno, T. L. Blundell, *Fold Des.* 1, 209 (1996).

98. S. J. Wheelan, A. Marchler-Bauer, S. H. Bryant. *Bioinformatics* 16, 613 (2000).

99. S. Ge, T. Sang, B. R. Lu, D. Y. Hong, *Proc. Natl. Acad.Sci. U.S.A.* 96, 14400 (1999).

100. L. H. Rieseberg, *Annu. Rev. Ecol. Syst.* 28, 359 (1997).

101. J. Masterson, *Science* 264, 421 (1994).

102. L. Comai, *Plant Mol. Biol.* 43, 387 (2000).

103. C. M. Vicient, M. J. Jaaskelainen, R. Kalendar, A. H.Schulman, *Plant Physiol.* 125, 1283 (2001).

104. M. I. Tenaillon *et al.*, *Proc. Natl. Acad. Sci. U.S.A.* 98, 9161 (2001).

105. L. Kruglyak, *Nature Genet.* 17, 21 (1997).

106. R. J. Cho *et al.*, *Nature Genet.* 23, 203 (1999).

107. E. Drenkard *et al.*, *Plant Physiol.* 124, 1483 (2000).

108. J. Hanke *et al.*, *Trends Genet.* 15, 389 (1999).

109. A. A. Mironov, J. W. Fickett, M. S. Gelfand, *Genome Res.* 9, 1288 (1999).

110. L. Croft *et al.*, *Nature Genet.* 24, 340 (2000).

111. D. Brett *et al.*, *FEBS Lett.* 474, 83 (2000).

112. B. Modrek, A. Resch, C. Grasso, C. Lee, *Nucleic Acids Res.* 29, 2850 (2001).

113. S. M. Berget, *J. Biol.* Chem. 270, 2411 (1995).

114. B. J. Blencowe, *Trends Biochem.* Sci. 25, 106 (2000).

115. M. L. Hastings, A. R. Krainer, Curr. Opin. *Cell Biol.* 13, 302 (2001).

116. Z. J. Lorkovic, D. A. Wieczorek Kirk, M. H. Lambermon, W. Filipowicz, *Trends Plant Sci.* 5, 160 (2000).

117. D. Brett *et al.*, *Nature Genet.* 30, 29 (2002).

118. A. P. Bird, *Trends Genet.* 11, 94 (1995).

119. Source for BAC-end sequences: http://www.genome.clemson.edu/projects/rice/fpc.

120. We are indebted to faculty and staff at the Beijing Genomics Institute, whose names were not listed, but who also contributed to the team effort (www.genomics.org.cn). We are indebted to our scientiTc advisors, M. V. Olson, L. Bolund, R. Waterston, E.Lander, and M-C King, for their long-term support.We are grateful to R. Wu and C. Herlache for editorial assistance on the manuscript. We thank Amersham Pharmacia Biotech (China) Ltd., SUN Microsystems (China) Inc., and Dawning Computer Corp. for their support and service. This project was jointly sponsored by the Chinese Academy of Science, the Commission for Economy Planning, the Ministry of Science and Technology, the Zhejiang Provincial Government, the Hangzhou Municipal Government, the Beijing Municipal Government, and the National Natural Science Foundation of China. The analysis was supported in part by the National Institute of Environmental Health Sciences (grant 1 RO1 ES09909).

作者: Jun Yu[#]　Songnian Hu[#]　Jun Wang[#]　Gane Ka-Shu Wong[#]　Songgang Li　Bin Liu　Yajun Deng　Li Dai　Yan Zhou　Xiuqing Zhang　Mengliang Cao　Jing Liu　Jiandong Sun　Jiabin Tang　Yanjiong Chen　Xiaobing Huang　Wei Lin　Chen Ye　Wei Tong　Lijuan Cong　Jianing Geng　Yujun Han　Lin Li　Wei Li　Guangqiang Hu　Xiangang Huang　Wenjie Li　Jian Li　Zhanwei Liu　Long Li　Jianping Liu　Qiuhui Qi　Jinsong Liu　Li Li　Tao Li　Xuegang Wang　Hong Lu　Tingting Wu　Miao Zhu　Peixiang Ni　Hua Han　Wei Dong　Xiaoyu Ren　Xiaoli Feng　Peng Cui　Xianran Li　Hao Wang　Xin Xu　Wenxue Zhai　Zhao Xu　Jinsong Zhang　Sijie He　Jianguo Zhang　Jichen Xu　Kunlin Zhang　Xianwu Zheng　Jianhai Dong　Wanyong Zeng　Lin Tao　Jia Ye　Jun Tan　Xide Ren　Xuewei Chen　Jun He　Daofeng Liu　Wei Tian　Chaoguang Tian　Hongai Xia　Qiyu Bao　Gang Li　Hui Gao　Ting Cao　Juan Wang　Wenming Zhao　Ping Li　Wei Chen　Xudong Wang　Yong Zhang　Jianfei Hu　Jing Wang　Song Liu　Jian Yang　Guangyu Zhang　Yuqing Xiong　Zhijie Li　Long Mao　Chengshu Zhou　Zhen Zhu　Runsheng Chen　Bailin Hao　Weimou Zheng　Shouyi Chen　Wei Guo　Guojie Li　Siqi Liu　Ming Ta　Jian Wang　Lihuang Zhu[*]　Longping Yuan[*]　Huanming Yang[*].

注: 本文发表于 *Science*, 2002 年第 296 卷。

Development of Hybrid Rice in China

A.Introduction

The success achieved in development of hybrid rice is a great breakthrough in rice breeding which provides an effective way to markedly enhance rice yield on a large scale. China is the first country in the world to exploit rice heterosis commercially. Research on hybrid rice was intiated in 1964. The genetic tools, viz, cytoplasmic male sterile, maintainer and restorer lines (or A, B and Rlines) essential for producing F_1 hybrid seed were developed by 1973.Several hybrid combinations with good heterosis and higher yield potential were identified in 1974.In 1976 hybrid rice was released for commercial production, since then, the area planted to hybrid rice was increased year after year. It has been proved by practices for many years that hybrid rice has 20% yield advantage over modern semi-dwarf inbred varieties. In recent years the acreage under hybrid rice is around 15 million ha, about 50% of the total rice area in China. The average yield of hybrid rice is 7 t/ha while that of the inbred rice is 5.6 t/ha.

In the meantime the techniques of hybrid seed production are well developed to give an average seed yield of 2.7 t/ha nation-wide (0.14 million ha) . The field area ratio between A line multiplication, seed production and F_1 cultivation is 1 : 50 : 6 000.

B. Recent progress

1.Development of two line system hybrids succeeded

The existing hybrid rice used in commercial production mainly belongs to the category of inter-varietal hybrids by using CMS or three line system. Many years' practices and experiences have proved that three line system is an effective way to develop hybrid varieties and will continue to play important roles. However there are some constraints in this system, especially the yield of the three line inter-varietal hybrids including newly developed ones has stagnated for years. This means that they have already reached their yield plateau and it seems very difficult to further increase their yield potential if no new methods and novel materials are invented and adopted.

Taking the long-range strategy of rice heterosis breeding into account, many Chinese rice scientists have been making attempts to explore new technological approaches to further enhance rice yield. One of them achieved is the development of two line system hybrids.

This method is based on a new kind of rice mutant called photo-thermo-sensitive genic male sterile, P (T) GMS, rice. Its male sterility is controlled by one or two pairs of nuclear recessive genes, and has no relation to cytoplasm. Exploitation of the P (T) GMS lines to develop rice hybrids has the following advantages over the classical three-line method:

(1) The mantainer line is avoided. Under long daylength and higher temperature conditions (summer) the P (T) GMS lines show complete pollen sterility, thus they can be used for hybrid seed production in summer season. Under short daylength and moderate temperature conditions (autumn) they become almost normal fertility, thus can multiply themselves by selfing.

(2) The choice of parental lines to develop heterotic hybrids is greatly broadened. Studies showed that over 95% varieties tested can restore P (T) GMS lines' fertility, while for CMS lines it is less than 5%. In addition, P (T) GMS genes can be easily transferred into almost any rice lines with desirable charateristics.

The development of two line hybrid rice was succeeded in 1995.The planting area of two line hybrids was 2 million ha last year. Generally speaking, two line hybrids outyield three line hybrids by 5% – 10%.

2.Development of super hybrid rice

A super rice breeding program was set up by China Ministry of Agriculture in 1996 with the following yield targets on large scale:

Phase I (1996—2000): 10.5 t/ha.

Phase II (2001—2005): 12 t/ha.

By means of morphological improvement plus utilization of inter-subspecific (indica/japonica) heterosis, several pioneer two line super hybrid rice varieties had been developed by 2000, which attained the yield target of the Phase I. There were more than 20 demonstration locatons with an area of 6.7 ha or 67 ha each where the average yield was over 10.5 t/ha in 2000.The average yield per ha was 9.6 t in commercial production (235 000 ha) in 2000 and 9.2 t (1.2 million ha) in 2001. A combination, P64S/E32, created a record yield of 17.1 t/ha in an experiment plot (720 m^2) in 1999.

Now efforts are focused on breeding of Phase II super hybrid rice and good progress is being made. There were three two line indica/japonica hybrid combinations outyield the CK (pioneer super hybrid rice variety) by 6%-18% in replicated trials at our center in 2001.A three line super hybrid, II −32A/ Ming 86, yielded 12.7 t/ha at a location with 7 ha in Fujian Province in 2001, and it also created a new record yield (17.9 t/ha) in an experiment plot (800 m^2) in 2002.Another promising two line indica/ japonica hybrid, P88S/0293, yielded 12.3 t/ha at a location with 8 ha in Hunan Province in 2002.

C.Technical approaches for breeding super hybrid rice

1.Morphological improvement

Good plant type is the foundation for super high yield. Since Dr. Donald proposed the concept of ideotype, many rice breeders have paid great attention to this important topic and proposed several models for super high-yielding rice. Among them the famous one is the "New plant type" proposed by Dr. Khush. Its main features are: ① big panicles, 250 spikelets per panicle; less tillers, ② 3 – 4 productive tillers per plant; ③ short and sturdy culm.

Based on our studies, especially inspired by the striking characteristics of a high yielding combination, P64S/E32, which has created 17.1 t/ha grain yield, we have found that the super high yielding rice variety has following morphological features:

(1) Tall erect-leaved canopy.

The upper three leaf blades should be long, erect, narrow, V-shape and thick. Long and erect leaves not only have larger leaf area but also can accept light on both sides and will not shade each other. Therefore, light is used more efficiently; Narrow leaves occupy relatively small space therefore allow to have a higher effective leaf area index; V-shape makes leaf blade more stiff so that not prone to droopy; Thick leaves have higher photosynthetic function and are not easily senescent. These morphological features mean a huge source of assimilates essential to super high yield.

(2) Lower panicle position.

The tip of panicle is only 60 – 70 cm above the ground during ripening stage. Such architecture enables the plant to be highly resistant to lodging, which is also one of the essential characters required for breeding super high-yielding rice variety.

(3) Bigger panicle size.

The grain weight per panicle is around 5 grams and the number of panicles is about 300 panicles/m^2. Theoretically, the yield potential is 15 t/ha if they are realized.

Grain yield = HI×Biomass. Nowadays the harvest index (HI) is high (above 0.5). Further lifting of rice yield ceiling should rely on increasing biomass because further improvement of HI is quite limited. From view point of morphology, to raise plant height is the effective and feasible way to increase biomass. However, this approach will cause lodging. To solve this problem, many breeders are trying to make the stem thicker and sturdier, but this approach usually results in HI decrease. Therefore, it is difficult to obtain super high yield by this way. The plant model of taller canopy which consists of leaves can combine the advantages of higher biomass, higher HI and better highly resistance to lodging together.

2.Raising heterosis level

The heterosis level in rice has the following general trend: *indica/japonica>indica/javanica>japonica/javanica>indica/indica>japonoca/japonica*, according to our studies. *Indica/japonica* hybrids possess very large sink and rich source, the yield potential of which is 30% higher than inter-varietal indica hybrids theoretically. Therefore, efforts have been focused on using *indica/japonica* heterosis to develop super hybrid rice. However, there exists a lot of problems in *indica/japonica* hybrids, especially very low seed set, which must be solved in order to use their heterosis in practice. By means of wide compatibility (WC) genes and using intermediate type lines as parents instead of typical *indica* or *japonica* lines, a number of inter-subspecific hybrid varieties with stronger heterosis and normal seed set have been successfully developed as mentioned above.

3.Utilization of favorable genes from wild rice

This is another promising approach to develop super hybrid rice. Based on molecular analysis and field experiments, two yield enhancing QTLs from wild rice (O. *Rufipogon* L.) were *identified*. Each of the QTL genes contributed to a yield advantage of 18% over the high yielding CK hybrid Weiyou64 (one of the most elite hybrids). By means of molecular marker-assisted backcross and field selection, an excellent R line (Q611) carrying one of these QTLs is developed. Its hybrid, J23A/Q611, outyielded CK hybrid by 35% in replicated trial for the second cropping rice in 2001.Its yield potential on lare scale is being evaluated for the time being. Prelimilary data show its estimated yield is 13 t/ha in experimental plot and 11 t/ha on farmers' field planted as second cropping.

4.Using genomic DNA from barnyard grass (*Echinochioa crusgalli*) to create new source of rice

Total DNA of barnyard grass was introduced into a restoring line (R207) by Spike-stalk Injection Method, variants occurred in D1.From these variants, new elite stable R lines have been developed. The most outstanding one is RB207-1, its agronomic characters including number of spikelets per panicle and grain weight are much better than those of the original R207.Particularly, its hybrid, GD S/RB207-1, has good plant type and very strong heterosis with big panicles and heavy grains. Its yield potential is over 15 t/ha in experimental plot.

D. Prospects

The yield standard of second phase super rice (12 t/ha) can be achieved by 2005.By reaching this target, 2.25 t/ha more rice can be produced, which will increase 30 million tons of grains yearly and can feed 75 million more people when it is commercialized up to 13 million.ha.

The development of science and technology will never stop. Rice still has great yield potential, it can be further tapped by advanced biotechnology. Excitingly, C_4 genes from maize have been successfully cloned and are being transferred into rice plant by HK Chinese University and our Center. By using this transgenic plant as donor to introduce C_4 genes into super hybrid rice, the yield potential of rice could be further increased by a big margin. Relying on this progress, the Phase Ⅲ super hybrid rice breeding program is proposed, in which the yield target is 13.5 t/ha on a large scale by 2008.

作者：Yuan Longping　Wu Xiaojin

注：本文发表于 2003 年 3 月在乌拉圭召开的第三届温带水稻会议与水稻信息。

Recent Progress in Breeding Super Hybrid Rice in China

China's current population is near 1.3 billion, with less than 0.1 ha of arable land for each person. This population is expected to reach 1.6 billion and crop 1 and will decrease to about 0.07 ha per capita. Because of this population growth pressure and reduction in arable land, to feed all the Chinese people in the new century, a super rice breeding program was set up by China's Ministry of Agriculture in 1996, with the yield targets for hybrid rice listed in Table 1.

Table 1　Yield standard of super rice in China

Phase	Hybrid rice[a]			
	Fitst Cropping	Second cropping	Single season	percent increase
1996 level	7.50	7.50	8.25	0
Phase I (1996−2000)	9.75	9.75	10.25	More than 20%
Phase II (2001−2005)	11.25	11.25	12.00	More than 40%

Note: a. In t ha^{-1} at 2 locations with 6.7 ha each in 2 consecutive years.

With morphological improvement plus the use of intersubspecific (*indica/ japonica*) heterosis, several pioneer two-line super hybrid rice varieties had been developed by 2000, which attained the phase I yield standard of single-season rice. There were more than 20 demonstration locations with 6.7 ha or 67 ha each, where their average yield was more than 10.5 t · ha^{-1} in 2000. The average yield was 9.6 t · ha^{-1} in commercial production (235 000 ha) in 2000 and 9.2 t · ha^{-1} (1.2 million ha) in 2001. (The average rice yield has heen 6.3 t · ha^{-1} nationwide recently.) One combination, P64S/E32, had a record yield of 17.1 t · ha^{-1} in an experimental plot (720 m^2) in l999.

Efforts now focus on breeding phase II super hyhrid rice and good progress is being made. Three two-line *indica/japonica* hybrid combinations outyielded the check (CK, the pioneer super hybrid rice variety) by 6%−18% in replicated trials at our center in 2001. A three-line *indica/japonica* hyhrid, II − 32A/Ming 86, yielded 12.76 t · ha^{-1} at a demonstration location with 7 ha in Fujian Province in 2001, and it also produced a new record yield (17.95 t · ha^{-1} in an experimental plot (800 m^2) in Yunnan Province in 2001.

Based on progress in 2001, some promising combinations were prepared for demonstration at multiple locations with 7 – 8 ha each in 2002.Among them, the best one is P88S/0293, which yielded 12.3 t · ha^{-1} on average in Longshan County, Hunan. This combination yielded 12.4 t · ha^{-1} in Hainan Province again in 2003 and had a record yield in the province. In addition, three combinations are performing very well in our experimental plots (plot size 700 – 800 m^2) and their estimated yield is around 13 t · ha^{-1}.

A breakthrough has been achieved in breeding the first cropping of super hybrid rice. A newly developed short–growth–duration two–line *indica/japonica* hybrid (HY-S/F49) was demonstrated near Changsha in 2002.The area under demonstration was 7 ha and its average yield was 9.1 t · ha^{-1}, which outyielded CK$_1$ (three-line intervarietal hybrid) and CK$_2$ (inbred variety) by 20% and 40%, respectively. There are two demonstration locations with 7 – 8 ha each in 2003 and their estimated yield is around 10 t · ha^{-1}.

Technical approaches

Crop improvement practices have indicated, up to now, that there are only two effective ways to increase the yield potential of crops through plant breeding, that is, morphological improvement and the use of heterosis. However, the potential is very limited when using morphological improvement alone and heterosis breeding will produce undesirable results if it is not combined with morphological improvement. Any other breeding approaches and methods, including high technology such as genetic engineering, must be incorporated into good morphological characters and strong heterosis; otherwise, there will be no actual contributions to a yield increase. On the other hand, the further development of plant breeding for a high yield target must rely on progress in biotechnology.

Morphological improvement

A good plant type is the foundation for super high yield. Since Dr. Donald proposed the concept of ideotype in 1968, many rice breeders have proposed models for super high-yielding rice. Among these is the "new plant type" proposed by Dr. Khush at IRRI. Its main features are (1) large panicles, with 250 spikelets per panicle; (2) fewer tillers, 3 – 4 productive tillers per plant; (3) a short and sturdy culm. Whether this model can realize super high yield or not has yet to he proved.

Based on our studies, especially inspired by the striking characteristics of the high-yielding combination P64S/E32, which has had a record yield of 17.1 t · ha^{-1}, we have found that the super high-yielding rice variety has the following morphological features:

1.Tall erect-leaf canopy

The upper three leaf blades should be long, erect, narrow, V-shaped, and thick. Long and erect leaves have a larger leaf area, can accept light on both sides, and will not shade each other. Therefore, light is used more efficiently. Narrow leaves occupy a relatively small space and thus allow a higher effective leaf area index. A V-shape makes the leaf blade stiffer so that it is not prone to be droopy. Thick leaves have a higher photosynthetic function and are not easily senescent. These morphological features signify a large source of the assimilates that are essential to super high yield.

2.Lower panicle position

The tip of the panicle is only 60 – 71 cm above the ground during the ripening stage. Because the plant's center of gravity is quite low, this architecture enables the plant to be highly resistant to lodging. Lodging resistance is also one of the essential characters required for breeding a super high-yielding rice variety.

3.Bigger panicle size

Grain weight per panicle is around 5 g and the number of panicles is about 300 m^2. Theoretically, yield potential is 15 t $\cdot$ ha^{-1}.

Grain yield=biomass×harvest index. Nowadays, the harvest index (HI) is very high (above 0.5). A further raising of the rice yield ceiling should rely on increasing biomass because further improvement of the HI is limited. From the view-point of morphology, to increase plant height is an effective and feasible way to increase biomass. However, this approach will cause lodging. To solve this problem, many breeders are trying to make the stem thicker and sturdier, but this approach usually results in a decrease in HI. Therefore, it is difficule to obtain a super high yield in this way. The plant model of a taller canopy can combine the advantages of a higher biomass, higher HI, and higher resistance to lodging.

Raising the level of heterosis

Heterosis in rice has the following general trend—*indica/japonica>indica/javanica>japonica/javanica>indica/indica>japonica/japonica*—according to our studies. *Indica/japonica* hybrids possess a very large sink and rich source, the yield potential of which is 30% higher than that of intervarietal *indica* hybrids being used commercially. Therefore, efforts have focused on using *indica/japonica* heterosis to develop super hybrid rice. However, many problems exist in *indica/japonica* hybrids, especially their very low seed set, which must be solved to use their heterosis. With wide compatibility (WC) genes and using intermediate-type lines as parents instead of typical *indica* or *japonica* lines, several intersubspecific hybrid varieties with stronger heterosis and normal seed set have been successfully developed.

Biotechnology

This is another important approach for developing super hybrid rice. So far, two very promising results have been obtained in this research area.

1.The use of favorable genes from wild rice

Based on molecular analysis and field experiments, two yield-enhancing QTLs from wild rice (*Oryza rufipogon* L.) were identified. Each of the QTLs contributed to a yield advantage of 18% over the high-yielding check hybrid Weiyou64 (one of the most elite hybrids). By means of molecular marker-assisted backcrosses and field selection, an excellent R line (Q611) carrying one of these QTLs was developed. Its hybrid, J23A/Q611, outyielded the check hybrid by 35% in a replicated trial for the second rice crop in 2001.Its yield potential on a large scale is now being evaluated. Preliminary data shows that its estimated yield is $13 \text{ t} \cdot \text{ha}^{-1}$ in experimental plots and $11 \text{ t} \cdot \text{ha}^{-1}$ in farmers' fields planted as second rice crop.

2.Using genomic DNA from barnyardgrass (*Echinochloa crus – galli*) to create a new source of rice

The total DNA of barnyardgrass was introduced into a restoring line (R207) by the spike-stalk injection method and variants occurred in the D_1.From these variants, new elite stable R lines have been developed. The most outstanding one is RB207-1, and its agronomic characters such as number of spikelets per panicle and grain weight are much better than those of the original R207.Particularly, its hybrid. GD S/RB207-1, has a good plant type and very strong heterosis. Its estimated yield was more than $15 \text{ t} \cdot \text{ha}^{-1}$ in our experimental plot in 2002.

Prospects

The yield standard of phase II super rice ($12 \text{ t} \cdot \text{ha}^{-1}$) can he achieved by 2005.By reaching this target, $2.25 \text{ t} \cdot \text{ha}^{-1}$ more rice can be produced, which will increase grain by 30 metric tons annually and feed 75 million more people when it is commercialized up to 13 million ha.

The development of science and technology will never stop. Rice still has a large yield potential, which can be exploited by advanced biotechnology. It was exciting to learn that C_4 genes from maize have been successfully cloned and transferred into the rice planl by the HK Chinese University. Using this transgenic plant as a donor to introduce C_4 genes into super hybrid rice parents is under way. If this approach is successful, the yield potential of rice could be further increased by a large margin. Relying on this progress, the phase Ⅲ super hybrid rice breeding program is proposed, in which the yield target is $13.5 \text{ t} \cdot \text{ha}^{-1}$ on a large scale by 2010.

作者: Yuan Longping

注: 本文发表于 *Science Progress in China* 2003 年。

Hybrid Rice Technology for Food Security in the World (B)

The current world population is over 6 billion and will reach 8 billion in 2030.Meanwhile, the annual loss of land to other use is 10 to 35 million ha, with half of this lost land coming from cropland. Facing such severe situation of population growth pressure plus cropland reduction, it is obvious that the only way to solve food shortage problem is to greatly enhance the yield level of food crops per unit land area through advance of science and technology.

Rice is a main food crop. It feeds more than half of world population. It has been estimated that the world will have to produce 60% more rice by 2030 than what it produced in 1995.Therefore, to increase production of rice plays a very important role in food security and poverty alleviation. Theoretically, rice still has great yield potential to be tapped and there are many ways to raise rice yield, such as building of irrigation works, improvement of soil conditions, cultural techniques and breeding of high yielding varieties. Among them, it seems at present that the most effective and economic way available is to develop hybrid varieties based on the successful experience in China.

It has been proved practically for many years that hybrid rice has more than 20% yield advantage over improved inbred varieties. In recent years, hybrid rice covers 50% or 15 million ha of the total rice area in China. The nationwide average yield of hybrid rice is 7 t/ha, about 1.4 t/ha higher than that of inbred varieties (5.6 t/ha). The yearly increased paddy in China due to growing hybrid rice can feed 60 million people each year. Therefore, hybrid rice has been playing a critical role in solving the food problem of China thus making China the largest food self-sufficient country.

China makes inceasing progress in development of hybrid rice technology. Following the success of three-line hybrid rice in 1970s, two-line hybrid rice was successfully commercialized in 1995.The extension of two-line hybrid rice has been very fast in these years. The area of two-line hybrid rice was 2.6 million ha, about 18% of total hybrid rice area in 2002.The yield advantage of two-line hybrid rice is 5% - 10% higher than that of the existing three-line hybrid rice.

More encouragingly, good results have been achieved in developing super hybrid rice varieties since the initiation of the super rice research program in 1996.Several pioneer super hybrids have a yield advantage of around 20% over current three-line hybrids on commercial scale. The area planted to super hybrid rice was 240,000 ha and the average yield was 9.6 t/ha in 2000.The area under super hybrid rice was increased to 1.4 million ha with an average yield of 9.1 t/ha in 2002.In addition, a two-line super hybrid P64S/E32 and a three-line super hybrid Ⅱ－32A/Ming86 created a record yield of 17.1 t/ha in 1999 and 17.95 t/ha in 2001, respectively. In the meantime, the grain quality of the super hybrid rice varieties is very good. Now efforts are focused on developing second generation super hybrid rice. Its yield target is 12 t/ha on large scale demonstration, and good progress has been made. Last year there were five locations with 7 ha each in Hunan Province, where the average yield was over 12 t/ha. 30 locations with 6.7 ha each are arranged in the southern promises this year. Among them three locations so far harvested, their average yield is above 12 t/ha. Therefore, the super hybrid rice shows a very bright future. If super hybrid rice covers an annual area of 13 million ha in China and calculating by a yield increase of 2.25 t/ha, it is expected that the annual increased grains will reach 30 million tons, which means 75 million people more can be fed every year.

Hybrid rice has been proved to be a very effective approach to greatly increase yield not only in China, but also outside China. Vietnam and India have commercialized hybrid rice for years. Last year about 600,000 hectares were covered with rice hybrids in Vietnam. On average, the yield of rice hybrids is 6.3 t/ha while that of the inbred varieties is 4.5 t/ha. Because of planting hybrid rice on large-scale commercial production, Vietnam becomes the second largest rice export country in Asia. Besides, many other countries, such as the Philippines, Bangladesh, Indonesia, Pakistan, Ecuador, Guinea and the USA, have also achieved great progress in extending hybrid rice technology. Recently, a number of experimental trials and large-scale demonstrations in farmers' field conducted in these countries have shown that hybrid rice can significantly outyield their local CK varieties. For example, in Philippines, under technical assist by FAO, IRRI and our Center, hybrid rice has been commercialized two years ago. Especially a super hybrid rice variety called SL－8 has been developed by my assistant in Philippines, it was planted to about 3 000 ha in 2003 and the average yield was 8.5 t/ha, more than doubled the country's average yield. Based on this achievement, the Philippines government has made an ambitious plan, in which the goal is to plant 3 million ha of hybrid rice by 2007.The above facts clearly indicate that hybrid rice technology developed by China is also effective to greatly increase rice yield worldwide.

The ever-forwarding technology improvement by Scientists, its dissemination by seed industries and extension workers and the policy and financial support by national and local governments contribute greatly to the success in development and use of hybrid rice technology. And I firmly believe that hybrid rice, relying on scientific and technological advances and the efforts from all other aspects, including scientists, seed industries, governments and particularly from FAO and IRRI, will have a very good prospect for commercial production and continue to play a key role in ensuring the future food security worldwide in the new century.

Finally, as the Director General of the China National Hybrid Rice R&D Center, I am very glad to announce here that we will try our best to help other countries to further speed up the development of hybrid rice in the whole world.

作者：Yuan Longping

注：本文发表于 2004 年 2 月 13 日在罗马举行的国际大米研讨会。

应用 SSR 分子标记鉴定超级
杂交水稻组合及其纯度

【摘 要】应用 SSR 分子标记技术对超级杂交稻 5 个组合（HYS-1/R105、培矮 64S/E32、两优培九、88S/0293、J23A/Q611）及其 9 个亲本进行了鉴定。用 144 对 SSR 引物进行筛选，有 47 对能够在实验材料中显示较好的多态性，其中，RM337 与 RM154 呈现丰富多态性，可鉴别供试组合并分别与其亲本区分开。对于水稻的每一条染色体，各筛选出两条产生多态性的引物，共 24 对，并提供一组作为鉴定参考的图谱；通过杂种表现为父母本互补带型的特点，找到在杂交稻组合及其亲本间具有多态性的引物，筛选出 5 对引物分别作为鉴定上述 5 个超级杂交稻组合的特异引物，进而针对杂交稻不同的纯度问题设计鉴定方法。

【关键词】分子标记；微卫星标记；超级杂交稻；品种鉴别；纯度鉴定

【Abstract】Five super hybrid rice combinations and their parental lines were tested by means of SSR analysis. A total of 144 SSR primer pairs distributing on 12 rice chromo somes were used，47 o f them showed polymorphism. Among all these primers，RM337 and RM154 produced polymo rphic patterns in four or more of the tested experimental materials respectively，and they could distinguish almost all the rice genotypes tested. Twenty-four primer pairs，two on each rice chromosome，were selected to make a reference SSR marker-based fingerprinting for the rice lines. For most of primer pairs，F_1 hybrids mainly showed complementary pattern of both parents，which could be very useful to distinguish the F_1 from its parental lines. In addition，five primer pairs we re selected as special primer pairs for five hybrid rice combinations respectively. By combining the rapid，simple method on DNA ex traction，it is suggested that SSR technique has wide prospective in va riety authentication and purity identification.

【Key words】molecular markers; simple sequence repeats; super hybrid rice; variety identification; purity test

分子标记基于对 DNA 分子内部组织结构的研究而产生的多态性，是一类可通过一定的分子生物学技术来识别的遗传标记。与形态标记、细胞学标记和生化标记等三类遗传标记相比，DNA 标记具有明显的优点：数量极多，遍及整个基因组；多态性水平高，可揭示自然存在的许多等位性变异；许多分子标记表现为共显性，可以鉴别出纯合基因型和杂合基因型；表现为"中性"，无表现型效应，无上位性，不受环境条件和发育时期影响；DNA 分子标记以其高度品种特异性和环境稳定性，并具有检测快速、结果准确等优点，极其适合于品种鉴定和新品种登记、品种纯度和真实性检验等工作[1-4]。

目前，分子标记的方法主要有 RFLP、RAPD、SSR 和 AFLP 等[5-9]。由于 SSR 标记在单个座位上检测到的多态性远高于其他任何一种分子标记，且广泛、随机、均匀地分布于整个基因组，具共显性遗传特点，因此成为品种鉴定理想的分子标记[10-15]。在近年的研究中，于永红等[16]用 SSR 技术建立了杂交水稻不育系宁 2A 以及宁 2B 的 DNA 指纹图谱，并用 2 个引物有效地把宁 2A、宁 2B 与参试的其他杂交水稻亲本区别开来；詹庆才等[17]利用 SSR 标记对 6 个杂交稻组合及其亲本用 178 对引物进行筛选，有 52 对能分别在一个或多个组合中显示稳定的多态性，杂种表现为父母本的带型，并用多态性好的两对引物分别对参试的 V46 和金优 207 进行了纯度鉴定；彭锁堂等[18]选用分布于水稻 12 条染色体上的 26 对 SSR 引物对 9 个杂交稻组合及其亲本进行了 SSR 标记分析，能够有效地区分所有的恢复系和大部分不育系，并用杂交稻组合汕优 63 和两优培九进行了单粒种子 SSR 鉴定，所测纯度与田间纯度非常接近，认为 SSR 标记是适合于实施品种鉴定的技术。

本研究中，我们应用 SSR 标记技术对目前我国杂交水稻领域取得重大和最新进展的超级杂交稻先锋和苗头组合及其亲本进行了鉴定，以期提供一套可用于鉴定这些杂交稻组合及其亲本的参考图谱，为供试杂交稻及其种子纯度的简便、高效鉴定提供基础。

1　材料与方法

1.1　水稻材料

供试的 5 个超级杂交稻组合为：HYS-1/R105、培矮 64S/E32、两优培九（培矮 64S/9311）、88S/0293、J23A/Q611；9 个亲本为：HYS-1、R105、培矮 64S、E32、9311、88S、0293、J23A、Q611。

在 5 个参试的超级杂交稻组合中，两优培九和培矮 64S/E32 为超级杂交稻先锋组合，已于 2000 年达到农业部制定的中国超级稻第一期大面积每亩产量 700 kg 的指标，其中，培矮 64S/E32 曾于 1999 年在云南省永胜县创造了每亩产量 1 139 kg 的超高产纪录。近几年来，年推广面积达 133 万 hm² 以上。新的超级杂交稻组合 88S/0293 在 2002 年至 2004 年间，已先后在 8 个百亩（6.67 hm²）示范片上实现每亩产量超过 800 kg 的目标，是前景看好的第二期超级杂交稻达标组合。HYS-1/R105 和 J23A/Q611 均连续数年小面积达到每亩产量 650～680 kg，为具有超高产潜力的超级杂交稻早、晚稻组合。

1.2 水稻总 DNA 的提取

比较了 FastDNA Kit 法、CTAB 法和酸碱快速提取 DNA 法 3 种不同的提取水稻总 DNA 的方法。FastDNA Kit 是 BIO101,Inc. 公司生产的快速提取实验生物 DNA 的专用试剂盒，在实验过程中按照试剂盒提供试剂和规定的步骤进行操作，需用到 Cylindrical beads、Fast Prep 仪和可控温振荡器等较特殊物品和仪器。CTAB 法按下述步骤操作，取种植于温室生长至 5～6 叶的水稻材料幼叶各 0.1 g，加液氮研磨，用含有 2-巯基乙醇的缓冲液于 60 ℃下间隙混匀 30～45 min，再加体积比为 24 : 1 的氯仿和异戊醇混合液混匀 5 min，4 ℃、13 000 r/min 下离心 5 min，吸取上清液，加入等体积的异丙醇混匀置于 -20 ℃下 30～60 min，4 ℃、13 000 r/min 下离心 10～15 min，用 70% 乙醇清洗沉淀物，离心、真空抽干沉淀物，于含 10 ng RNA 酶的 TE 中溶解 1 h 后，储存于 -20 ℃备用。酸碱快速提取法根据詹庆才介绍的方法略作改进[17]。取 37 ℃温箱中催芽 12 h 的芽或催芽 2～3 d 的芽，加入 40 μL ddH$_2$O，放入沸水中 5 min，再加入 40 μL（0.25 mol/L）NaOH，再放在沸水中 30 s，最后加入 80 μL HCl（0.5 mol/L）和 40 μL 缓冲液 Tris（pH 7.0，0.5 mol/L），继续置于沸水中 2 min，DNA 样品每 3～5 μL 可用作一次 PCR 的模板。

1.3 PCR 扩增

引物购自 Proligo LLC.，共 144 对，所用 PCR 扩增体系总体积为 25 μL，其中含 20 ng/μL 的模板 DNA1 μL（与简易提取法 DNA 模板量有异），40 μmol/L 的正反向引物各 0.125 μL，10×PCR 缓冲液 2.5 μL，10 mmol/LdNTPs 0.5 μL，20 mmol/L MgCl$_2$ 2.0 μL，Taq 聚合酶 0.1 μL（5U/μL），超纯水 17.65 μL。PCR 反应在 PTC 200 PCR 热循环仪上进行，继 94 ℃下预变性 2～5 min 后，94 ℃下变性 1 min、55 ℃下退火 1 min、72 ℃下延伸 2 min，共 35 个循环，最后 72 ℃下延伸 10 min。取扩增产物 7～7.5 μL，加 2 μL 溴酚蓝（6×），在 2% 琼脂凝胶中以 1×TBE 为电泳缓冲液、100～120 V/cm 电压下电泳 100～120 min。电泳之后的凝胶用溴化乙锭染色后，经 Gel Doc 2000 凝胶成像系统成像，紫外灯下观察并照相。根据扩增产物的电泳结果，将各引物的扩增产物情况进行记录。

2　结果与分析

2.1　不同提取 DNA 方法的效果

在所使用的 3 种 DNA 提取方法中，相对而言以使用 FastDNA Kit 方法提取的 DNA 质量最好，在使用这种方法提取时，以经 37 ℃暗箱培养 4~6 d 的水稻黄化苗提取的 DNA 质量最好，提取操作过程时间为 3~4 h，经 PCR 扩增能得到光亮清晰的条带，有利于获得理想的鉴定结果；其次是 CTAB 方法提取的 DNA，也能获得较为满意的带型，但提取过程中需要使用液氮进行研磨，而且操作过程较麻烦，耗时费力；酸碱快速提取 DNA 的方法是一种十分快捷方便、易于操作的方法，整个提取过程较为粗放，在 30 min 左右即可完成，及时进行 PCR 扩增，也能得到良好的效果。

2.2　SSR 分析

通过分布于水稻第 1~12 染色体的 144 对 SSR 引物针对供试杂交稻组合及其亲本进行的筛选，结果有 109 对引物能够扩增出条带，占检测引物的 75.7%；在 109 对引物中，有 47 对引物能够在实验材料中显示稳定的多态性，占检测引物的 32.6%。其余不能扩增出条带的引物，在改变 PCR 反应条件的情况下，也能将供试试验材料全部或部分扩增出条带。

2.2.1　最少且能区别供试杂交稻组合及其亲本的引物多态性分析

在能产生多态性的引物中，根据对比分析，发现两对引物 RM337 和 RM154 扩增出多态性很丰富的条带，而且各有其特异性（图 1）。从图 1 中看到，RM337 在不同的组合中扩增出分子量存在差异的 5 种不同带型，分别在 170 bp、190 bp、250 bp、440 bp、530 bp 的位置。它对于 HYS-1、培矮 64S、88S 产生 440 bp 的一条带；对 R105、9311、0293、Q611 产生 170 bp 的 1 条带；对 E32、J23A 产生 190 bp 的 1 条带；而对 HYS-1/R105、两优培九、88S/0293 则产生 170 bp、440 bp、530 bp 的 3 条带；对培矮 64S/E32 产生 190 bp、440 bp、530 bp 的 3 条带；对 J23A/Q611 却又产生另外类型的 3 条带（170 bp、190 bp、250 bp）。因此，RM337 的丰富多态性，可以区分开 HYS-1、培矮 64S、88S，R105、9311、0293、Q611，HYS-1/R105、两优培九、88S/0293、培矮 64S/E32 以及 J23A/Q611 四个组别的杂交稻组合和亲本。观察 RM154 扩增带型的胶像，也呈现出很好的多态性，非特异性条带除外，可以扩增出 160 bp、180 bp、190 bp、200 bp 共 4 种条带，如 HYS-1、培矮 64S、88S 的条带分别处于 180 bp、190 bp、160 bp 的位置，可以很容易地区分 HYS-1、培矮 64S 与 88S；

同样地，可以区分 R105、9311 与 Q611，HYS-1/R105 与两优培九。

综上所述，仅用两对引物，就可以将 5 个杂交稻组合分别与其亲本区分开。

2.2.2　筛选分布于水稻 12 条染色体上多态性的引物，建立标准图谱

研究中发现，大多数引物在两个或两个以上的杂交稻组合及其亲本中产生多态性。从理论上说，运用越多的多态性引物，鉴定结果越准确。为此，本研究筛选了分布于水稻第 1~12 染色体上的多态性丰富的引物 24 对，该 24 对多态性引物共扩增出 78 条条带，平均每对引物扩增 3.25 条。同时，24 对多态性引物在供试的杂交稻组合及其亲本扩增条带可建立鉴定参考图谱（图 2）。参照他人研究结果，利用各种特异标记的图谱，可以和系谱、产量性状联系起来用于知识产权的登记。

2.3　SSR 实施种子纯度鉴定的可操作性分析

进行种子的纯度鉴定，对于两系杂交稻而言，一方面是考虑鉴定由于低温引起的不育系自交结实导致杂交种子的不纯；另一方面则是要鉴定制种过程中可能引起的串粉或收获过程中机械混杂而引起的不纯。这两种类型的纯度鉴定，可以通过以下的两个步骤来实施。

2.3.1　筛选在杂交稻组合及其亲本间具有多态性的引物

实验显示，许多引物能在杂交稻组合及其亲本之间显示多态性，且杂交种的多态性为父母本的互补带型。利用这一特点，则可从它们的图谱上方便地区分出杂种及其亲本。

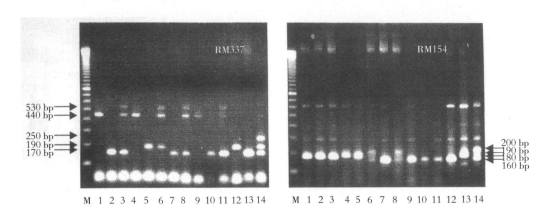

图 1　引物 RM337 和 RM154 扩增多态性结果

注：每个电泳图从左至右 PCR 反应的模板依次为：1-HYS-1；2-R105；3-HYS-1/R105；4- 培矮 64S；5-E32；6- 培矮 64S/E32；7-9311；8- 两优培九；9-88S；10-0293；11-88S/0293；12-J23A；13-Q611；14-J23A/Q611。图 2 同。

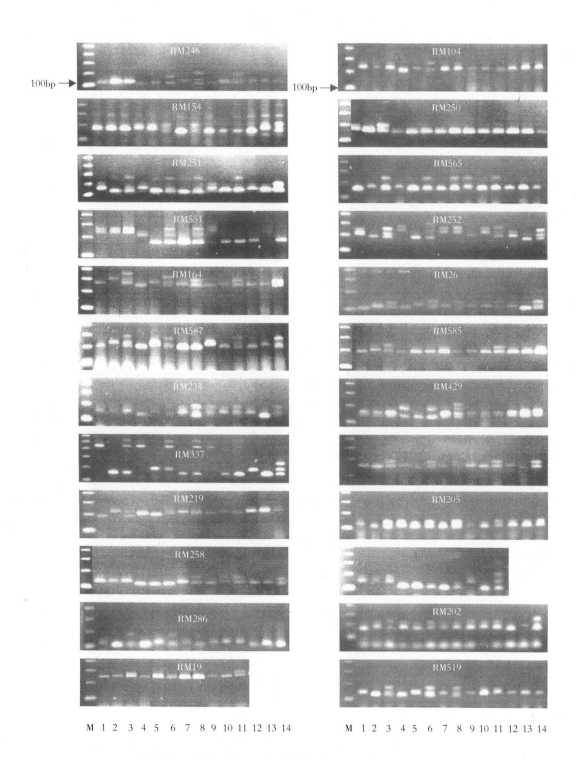

100bp →　　　　　　　100bp →

M 1 2 3 4 5 6 7 8 9 10 11 12 13 14　　　M 1 2 3 4 5 6 7 8 9 10 11 12 13 14

图 2　分布于水稻 12 条染色体上的 24 对引物扩增图谱

本研究筛选到的针对各个供试杂交稻及其亲本所对应的部分具有多态性的引物见表1。根据这些引物扩增鉴定的结果，可以通过分析引物与杂交稻组合及其亲本间的多态性，有助于鉴定混有不育系或恢复系的杂交稻种。

表1　对各供试杂交组合及其亲本间貝多态性的引物

杂交组合	引物
HYS-1/R105	RM545、RM 251、RM519、RM218、RM250、RM565、RM551、RM206、RM337、RM219、RM202、RM21、RM286、RM587、RM505
培矮64S /E32	RM34、RM 164、RM 585、RM251、RM429、RM519、RM154、RM520、RM565、RM234、RM337、RM152、RM264、RM286、RM21、RM304、RM587、RM219
两优培九	RM34、RM251、RM585、RM429、RM258、RM519、RM154、RM472、RM520、RM565、RM234、RM337、RM152、RM219、RM206
88S/0293	RM154、RM251、RM34、RM520、RM505、RM234、RM337、RM519、RM250
J23A/Q611	RM154、RM587、RM585、RM295、RM337、RM202、RM481、RM106

2.3.2　筛选特定杂交组合的特异多态性引物

杂交稻组合的特异引物，就是只在特定的杂交稻中扩增出特有的多态性的引物，可使该杂交稻与其他组合和亲本区分开。这种引物在进行某个特定杂交稻的纯度鉴定时，可以把混入的其他杂交稻等区别出来，达到纯度鉴定的目的。如用酸碱快速提取DNA的方法提取有混杂的HYS-1/R105种子100粒，用RM250扩增进行纯度鉴定，部分图谱（图3）可以看到纯的杂种HYS-1分别产生160 bp、200 bp两条带，带型为160 bp的第5号种子为混杂的恢复系R105种子，带型为200 bp的第10号种子则为混杂的不育系或其他杂交稻种子。

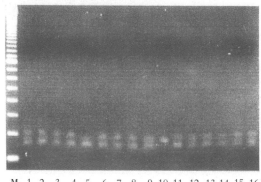

图3　利用酸碱快速提取DNA法提取HYS-1/R105单粒种子DNA及其特异性引物RM250扩增结果

经比较分析，各供试组合的特异引物如下：HYS-1/R105 为 RM250；培矮 64S/E32
为 RM337；两优培九为 RM429；88S/0293 为 RM228；J23A/Q611 为 RM337。

3　讨论

本研究对分布于水稻 12 条染色体上的 144 对 SSR 引物进行了筛选，得到 RM337 与
RM154 的扩增产物呈现丰富多态性，可将供试的材料一一加以鉴别，并提供了一组适于水稻
的每一条染色体鉴定的参考图谱。通过杂种表现为父母本互补带型的特点，找到了在杂交稻组
合及其亲本间具有多态性的引物，以及筛选针对特定组合产生多态性的特异引物，进而针对杂
交稻不同的纯度问题设计了鉴定方法。通过改进获得一种酸碱中和提取 DNA 的新方法，该方
法简单、快捷，可有效实施鉴定杂交稻组合和亲本及其纯度的工作。

我国已加入 WTO，经济的发展不仅需要健全良好的市场环境秩序，而且要实现与国际的
顺利接轨。为了适应经济发展新形势，杂交水稻研究与开发中的知识产权保护工作尤显重要，
特别是新的具有广阔应用前景的超级杂交稻成果的转化和开发，尤需知识产权的登记和保护。
为此，加强与改进品种鉴定工作是形势所需。把先进的 DNA 分子标记技术应用于杂交水稻特
别是超级杂交稻品种及其纯度鉴定，其鉴定结果具有法律效应，并为国际植物种子新品种产权
保护组织所认可，这不仅可以防止假冒种子充斥市场，而且将为中国杂交水稻在走向世界的进
程中，维护其在国际市场上的权威性提供重要的保障。本研究所探索的结果有助于新育成的超
级杂交稻组合及其纯度的鉴定，可为知识产权的保护提供具有司法效力的技术支持。

本研究选用 SSR 标记分析技术作为超级杂交水稻组合及其亲本的鉴定手段。由于 SSR 标
记多态性十分丰富，且目前像水稻等许多物种已有现成的、商品化的 SSR 引物，因此，对于
一般的实验室而言，只需利用现成的 SSR 引物进行 PCR 扩增，技术操作难度低、成本不高，
是十分理想的品种鉴定的分子标记方法。

在研究所使用的 3 种 DNA 提取方法中，酸碱快速提取 DNA 的方法对于简化工作，特
别是在实际应用中进行规模化的种子检测时，可以大大节省时间和精力，取得快速、经济的
效果，具有很好的应用前景。但由于提取过程没有除杂而使获得 DNA 的稳定性相对而言低
一些，选取的材料可以是种子，但最好是催芽 12 h 后的芽或已长 3 d 左右的芽，这种方法需
注意试剂和操作得当，特别是随提随用。对于 PCR 扩增产物的电泳分离和染色，本研究采用
2% 的琼脂凝胶 100～120 V/cm 电压电泳，随后进行 2.5%EB 染色，可以显示出良好的扩
增产物带型，而不需进行较繁杂的聚丙烯酰胺凝胶电泳及银染显色过程，实用性较强。

在实验中，杂种的带型除共显亲本带型以外，往往还可见杂种多于亲本所具有的一条带，

这可能是杂种的杂合链所产生的条带。另外，由于所选引物的限制，对于 0293 和 9311 这对材料而言，只有 RM587 检测到细微差别。0293 与 9311 之所以难以区分，可能与两者亲缘关系相近有关。如果拓宽待检水稻材料，尚需对 SSR 标记进一步筛选。

———————————— References ————————————

参考文献

［1］王艳，张爱民.DNA 分子诊断技术在品种纯度检验中的应用.见：中国农学会.种子工程与农业发展.北京：中国农业出版社，1997：740-745.

［2］赵中秋，郑海雷，张春光.分子标记的发展及其在植物研究中的应用.生命科学研究，2000，4（2）：68-72.

［3］方宣钧，刘思衡，江树业.品种纯度和真伪的DNA 分子标记及其应用.农业生物技术学报，2000，（2）：106-110.

［4］梁明山，曾宇，周翔，等.遗传标记及其在作物品种鉴定中的应用.植物学通报，2001，18（3）：257-265.

［5］Williams J G, Kubelik A R, Livak K J, et al. DNA polym orphisms amplified by arbitrary primers are useful as genetic markers. *Nucl Acid s Res*, 1990, 18（22）：6531-6535.

［6］杨剑波，李莉，汪秀峰，等.利用 RAPD 技术检测杂交水稻种子纯度（Ⅰ）—汕优 63 与其三系DNA 扩增产物的区别.安徽农业科学，1996，24（3）：193-195.

［7］陈洪，钱前，朱立煌，等.杂交水稻汕优 63 杂种纯度的 RAPD 鉴定.科学通报，1996，41（9）：833-836.

［8］杨蜀岚，伏健民，杨仁崔，等.培矮 64S 为不育系的两系杂交稻杂种纯度的 RAPD 鉴定.福建农业大学学报，1998，（1）：6-9.

［9］陈一华，贾建航，李传友，等.通过 AFLP-DNA 指纹的计算机分析进行水稻种子鉴定.见：21世纪水稻遗传育种展望——水稻遗传育种国际学术研讨会论文集.北京：中国农业科技出版社，1999.

［10］Pan aud O, Chen X, McCouch S R. Development of microsatellite markers and ch aracteri zation of simple sequence leng th polymorphism（SSLP）in rice（*Oryza sativa L.* ）.*Mol Gen Genet*, 1996, 252（5）：597-607.

［11］McC ou ch S R, Chen X, Panaud O, *et al*.Microsatellite marker development, mapping and applications in rice genetics and breeding. *Plant Mol Biol*, 1997, 35（1-2）：89-99.

［12］Wu K S, Tank sley S D. Abu ndance, poly morphism and genetic mapping of microsatellites in rice. *Mol Gen Genet*, 1993, 241：225-235.

［13］Akagi H, Yok oz eki Y, Inagaki A, *et al*. A codominant DNA marker closely linked to the rice nuclear resto rer gene, *RF-1*, identified with in ter-SSR fingerp rinting. *Genome*, 1996, 39（6）: 1205-1209.

［14］Yan g GP, Maroof M A, Xu C G, *et al.* Comparative analys is of microsatellite DNApoly-morphism in landraces and cultivars of rice. *Mol Gen Genet*, 1994, 245（2）: 187-194.

［15］15Li L, Yang J B, Wang X F. Identification of Xieyou combinations of hybrid rice and their three parents by using RAPD and microsatellite mark ers. *Chinese JRice Sci*, 2000, 14（4）: 203-207.

［16］16 于永红，李云海，马荣荣，等.用微卫星DNA标记建立宁2A的指纹图谱.中国水稻科学，2001，15（3）: 216-217.

［17］17 詹庆才.利用微卫星DNA标记进行杂交水稻种子纯度鉴定的研究.杂交水稻，2002，17（5）: 46-50.

［18］18 彭锁堂，庄杰云，颜启传，等.我国主要杂交水稻组合及其亲本SSR标记和纯度鉴定.中国水稻科学，2003，17（1）: 1-5.

作者: 辛业芸　张展　熊易平　袁隆平

注: 本文发表于《中国水稻科学》2005年第19卷第2期。

Developments in Super Hybrid Rice : Achievements and Prospects

Background

World population will reach 8 billion in 2030. With rice being a main food crop and feeding more than half of the world's population, it has been estimated that there should be 60 percent more rice produced by 2030 than the production rate in 1995. Theoretically, rice still has great yield potential and there are many ways to raise yield. Among them is to develop hybrid varieties based on the successful experience in China.

Hybrid rice has more than 20 percent yield advantage over improved inbred varieties and in recent years, hybrid rice has covered 50 percent of 15 million ha of the total rice area in China. The nationwide average yield of hybrid rice is 7 t/ha, about 1.4 t/ha higher than that of inbred varieties which is 5.6 t/ha. Hybrid rice has been playing a critical role in solving the food problem of China and with the yearly increased paddy, the country can now feed 60 million people each year.

There has been a clear demonstration of progress in the development of hybrid rice technology in China. Following the success of three-line hybrid rice in 1970s, two-line hybrid rice was successfully commercialized in 1995. The extension of two-line hybrid rice has been very fast in these years. The area of two-line hybrid rice was 2.6 million ha. About 18 per cent of total hybrid rice area in 2002. The yield advantage of two-line hybrid rice is 5 to 10 percent higher than that of the existing three-line hybrid rice.

Hybrid Rice in Asia

Hybrid rice has been proven to be a very effective approach to greatly increase yield not only in China, but also in other parts of Asia and in other regions. According to IRRI data, hybrid rice production now covers an area of 1.46 million ha outside China.

Viet Nam and India have commercialised hybrid rice for years. In 2003, about 600 000 ha were covered with rice hybrids in Viet Nam. On average, the yield of rice hybrids is 6.3 t/ha while that of the inbred varieties is 4.5 t/ha. Rice production in Vietnam from 1975–1988 was not enough to supply the demand of the population and the country had to import about 200 000 - 400 000 tons of rice between 1984-1988. When the government adopted the "Renovation" policy in 1988, production increased substantialially. There was also efforts in promoting hybrid rice on large-scale commercial production, making Viet Nam to become the second largest rice exporting country in Asia.

Many other countries-such as the Philippines, Bangladesh, Indonesia, Pakistan, Ecuador, Guineas and the United States of America (USA)—have also achieved great progress in extending hybrid rice technology. Recently, a number of experimental trials and large-scale demonstrations in farmers' field conducted in these countries have shown that hybrid rice can significantly outyield their local CK varieties.

Table 1   Area cultivated with hybrid rice in Asia (ha)

Country	1997	2001	2003	Hybrid rice area percent of total rice area in 2003
China	17 708 000.00	15 821 000.00	15 210 000.00	52
VietNam	187 000.00	480 000.00	600 000.00	8
India	120 000.00	200 000.00	<200 000	<1
Philippines	5 000.00	90 000.00	107 000.00	3
Bangladesh	—	20 000.00	49 655.00	<1

Source: FAO / IRRI / Country Data.

In the Philippines, under the technical assistance of the United Nations Food and Agriculture Organisation, the International Rice Research Institute (IRRI) and the China National Hybrid Rice R & D Centre, hybrid rice has been commercialised and the government has made an ambitious plan to plant 3 million ha of hybrid rice by 2007.

According to the Philippine Sino Center for Agricultural Technology, the Philippines will start using four new hybrid seeds from China by 2007. The new hybrid varieties, which are undergoing field trials in various parts of the country, yield about 20 percent more rice compared to existing hybrid varieties. The varieties also showed stronger resistance to bacterial leaf blight. Philippine overall rice production reached a record high 14.4 millions tons in 2004 mainly because of the use of high-yielding hybrid seeds and more favourable weather conditions.

Development of super hybrid rice

In order to meet the food requirement for all Chinese people in the 21st century, a super rice breeding program was set up by China Ministry of Agriculture in 1996. The yield targets for hybrid rice are listed below.

Table 2   Yield standards of super rice in China

Phase	Hybrid rice[a]			percent increase
	First Cropping	Second Cropping	Single cropping	
1996 level	7.50	7.50	8.25	0
Phase I （1996−2000）	9.75	9.75	10.50	More than 20%
Phase II （2001−2005）	11.25	11.25	12.00	More than 40%

Note: a In t/ha at two locations with 6.7 ha each in two consecutive years.

With morphological improvement plus the use of intersubspecific (*indica/japonica*) heterosis, several pioneer two-line super hybrid rice varieties had been developed and met the Phase I yield standard of single cropping rice by 2000. There were more than 20 demonstration locations with 6.7 ha or 67 ha each, where their average yield was more than 10.5 t/ha in 2000. A combination, P64S/9311, was released for commercial production in 2001. Since then the area under this hybrid has expanded very fast and reached 2 million ha this year. Its average yield is 8.4 t/ha while that of the nationwide rice was 6.3 t/ha. Another combination, P64S/E32, had created a record yield of 17.1 t/ha in an experimental plot (720 m^2) in 1999.

The efforts now are focused on breeding Phase II super hybrid rice and good progress has been made. Some promising combinations with 13 t/ha of yield potential have been developed recently. Among them, the best one is P88S/0293, which yielded over 12 t/ha at four locations in 2003 and at 12 locations in 2004.

In addition, a breakthrough has been achieved in breeding the second cropping of super hybrid rice. A newly developed short-growth-duration hybrid (J23A/Q611) was put on demonstration in 2004. The area under demonstration was 8.5 ha and its average yield was 10.3 t/ha, which outyielded CK1 (three-line intervarietal hybrid) and CK2 (inbred variety) by 20 percent and 40 percent, respectively. There are three demonstration locations with 7 ha or 70 ha each this year and their estimated yields are around 10.5 t/ha.

Technical approaches

Crop improvement practices are currently showing that there are only two effective ways to increase the yield potential of crops through plant breeding—the morphological improvement and the use of heterosis. However, the potential is very limited when using morphological improvement alone and heterosis breeding will produce undesirable results if it is not combined with morphological improvement. Any other breeding approaches and methods, including high technology such as genetic engineering, must be incorporated into good morphological characters and strong heterosis; otherwise, there will be no actual contributions to a yield increase. On the other hand, further development of plant breeding for a super yield target must rely on progress in biotechnology.

Morphological improvement

A good plant type is the foundation for super high yield. Since Dr. Colin Donald proposed the concept of ideotype in 1968, many rice breeders have also suggested various models for super high-yielding rice. Among the well-known models is the "new plant type" proposed by Dr. Gurdev Khush at IRRI.

Its main features are: (1) large panicles, with 250 spikelets per panicle; (2) fewer tillers, 3 – 4 productive tillers per plant; and (3) a short and sturdy culm. It has yet to be proven if these models can realise super high yields.

Based on our studies, especially inspired by the striking characteristics of the high-yielding combination P64S/E32, which has made a record yield of 17.1 t/ha, we have found that the super high-yielding rice variety has the following morphological features:

1.Tall erect-leaf canopy

The upper three leaf blades should be long, erect, narrow, V-shaped, and thick. Long and erect leaves have a larger leaf area, can accept light on both sides, and will not shade each other. Therefore, light is used more efficiently. Narrow leaves occupy a relatively small space and thus allow a higher effective leaf area index. A V-shape makes the leaf blade stiffer so that it is not prone to be droopy. Thick leaves have a higher photosynthetic function and are not easily senescent. These morphological features signify a large source of the assimilates that are essential to super high yield.

2.Lower panicle position

The tip of the panicle is only 60 – 70 cm above the ground during the ripening stage. Because the plant's center of gravity is quite low, this architecture enables the plant to be highly resistant to lodging. Lodging resistance is also one of the essential characters required for breeding a super high-yielding rice variety.

3.Bigger panicle size

Grain weight per panicle is around 5 g and the number of panicles is about 300 m^2. Theoretically, yield potential is 15 t/ha.

Grain yield=biomass×harvest index. Nowadays, the harvest index (HI) is very high (above 0.5). A further raising of the rice yield ceiling should rely on increasing biomass because further improvement of the HI is limited. From the viewpoint of morphology, to increase plant height is an effective and feasible way to increase biomass. However, this approach will cause lodging. To solve this problem, many breeders are trying to make the stem thicker and sturdier, but this approach usually results in a decrease in HI. Therefore, it is difficult to obtain a super high yield in this way. The plant model of a taller canopy can combine the advantages of a higher biomass, higher HI, and higher resistance to lodging.

Raising the level of heterosis

According to our studies, heterosis in rice has the following general trend: *indica/japonica>indica/javanica>japonica/javanica>indica/indica>japonica/japonica*.

Indica/japonica hybrids possess a very large sink and rich source, the yield potential of which is 30 percent higher than that of intervarietal *indica* hybrids being used commercially. Therefore, efforts have been focused on using *indica/japonica* heterosis to develop super hybrid rice. However, many problems exist in *indica/japonica* hybrids, especially with their very low seed set, which must be solved to use their heterosis. Making use of wide compatibility (WC) genes and selecting intermediate-type lines as parents instead of typical *indica* or *japonica* lines, several intersubspecific hybrid varieties with stronger heterosis and normal seed set have been successfully developed.

By means of biotechnology

This is another important approach for developing super hybrid rice. So far, three encouraging results have been obtained in this research area.

1.The use of favorable genes from wild rice

Based on molecular analysis and field experiments, two yield-enhancing QTLs from wild rice (*Oryza rufipogon* L.) were identified. Each of the QTLs contributed to a yield advantage of 18 percent over the high-yielding check hybrid Weiyou 64 (one of the most elite hybrid). By means of molecular marker-assisted backcrosses and field selection, an excellent R line (Q611) carrying one of these QTLs was developed. Its hybrid, J23A/Q611, outyielded the check hybrid by 35 percent in a replicated trial for the second cropping in 2001. Its yield in farmers' field planted as the second cropping is around 10 t/ha, about 2 t/ha higher than that of the CK as mentioned above.

2.Using genomic DNA from barnyard grass (Echinochloa crus-galli) to create a new source of rice

The total DNA of barnyard grass was introduced into a restoring line (R207) by the spike-stalk injection method and variants occurred in the D1. From these variants, new elite stable R lines containing some DNA fragments of barnyard grass have been developed. The most outstanding one is RB207, and its agronomic characters such as number of spikelets per panicle and grain weight are much better than those of the original R207. Particularly, its hybrid, GD S/RB207, has a good plant type and very strong heterosis. The estimated yield was more than 15 t/ha in our experimental plot in 2002. Its yield potential at different ecologic locations is now being evaluated this year. Preliminary results indicate that this hybrid performs very well in higher elevation region. At a mountain location (600 m above sea level), its average yield is 13 t/ha on large scale (6.5 ha) demonstration.

3.C_4 genes from maize have been cloned and successfully transferred into rice plant.

Preliminary test showed that the photosynthetic efficiency of the individual leaves of these transgenic C_4 rice plants is 30 percent higher than that of the CK. By using this transgenic plant as donor to introduce C_4 genes into super hybrid rice is under way. It is expected that the yield potential of C_4 super hybrid rice can be increased by more than 20 percent theoretically.

Because of the above progress, the Phase III super hybrid rice program is proposed, in which the yield target is 13.5 t/ha and will be fulfilled by 2010.

Prospects

The yield standard of Phase II super rice (12 t/ha) can be achieved by 2005. By reaching this target, 2.25 t/ha more rice can be produced, which will increase grain by 30 million tones annually and feed 75 million more people when it is commercialised up to 13 million ha. By reaching the target of Phase III super hybrid rice program, we can increase 3 t/ha, which will increase 39 million tones of rice yearly when it commercialised up to 13 million ha.

Conclusion

Chinese people not only can meet their food demand by themselves, but also can help other developing countries to solve food shortage problem.

Hybrid rice relies greatly on scientific and technological advances and efforts from orgnisations like FAO and IRRI. It will continue to play a key role in ensuring the future food security, not only in China, but even worldwide.

—————————— References ——————————

Longping Yuan, Hybrid Rice for Food Security in the World, FAO Rice Conference, Rome, Italy, 2004.

Oryza Market Report–Philippines, Philippines to Use 4 Hybrid Rice Seeds Commerially, 5 April 2005.

作者：Yuan Longping

注：本文发表于 2005 年 11 月在中国上海召开的亚洲种子大会（2005）。

利用四价病毒基因提高超级杂交水稻的抗性

【摘　要】以籼型超级杂交稻（*Oryza sativa* L.subsp.*indica*）恢复系9311的胚性愈伤组织为受体，采用改良的基因枪轰击法将构建在同一植物表达载体上的四价抗真菌病基因（*RCH10*，*RAC22*，*β-Glu* 和 *B-RIP*）导入9311的基因组中。Southern印迹杂交结果表明，潮霉素抗性再生植株中，*hpt* 标记基因和四价抗真菌病基因是连锁在一起并以孟德尔遗传方式进行传递的。部分转基因 R_1 代和 R_2 代植株分别在烟溪和三亚的典型稻瘟病鉴定圃中表现出对稻瘟病菌［*Magnaporthe grisea*（Hebert）Barr.］的高抗性。在高抗的转基因植株中，多个外源基因均能正常表达。将高抗稻瘟病的9311 R_2 代转基因纯系与培矮 64S 原种进行杂交，杂种 F_1 代除对稻瘟病仍可表现出高抗性外，还同时表现出对稻曲病和黑粉病明显提高的抗性。

【关键词】超级杂交稻；多基因转化；稻瘟病；稻曲病；黑粉病；抗病性

近年来，中国的水稻超高产计划已经取得了举世瞩目的成就，育出了培矮 64S/9311（即两优培九）、P88S/0293（0293 为 9311 的无芒系，两者可通用）等两系亚种间优良杂交组合，完成了中国超级稻第一、第二期的育种目标（即产量分别达到 10.5 t/hm^2 和 12 t/hm^2），被国际水稻界誉为继三系法杂交水稻之后中国育种家新版的"绿色神话"[1-3]。但在实际生产中，水稻的产量和品质仍然面临着多种病虫害的严重威胁。以目前已获得大面积推广的"两优培九"为例，大田中常见的病害就有稻瘟病、纹枯病、稻曲病、黑粉病等[4-6]。

已有的大量研究成果表明，利用基因工程来改良农作物的抗性，可以赋予作物更为全面、广谱、持续、特异的保护作用，而且使用相对较为安全方便，因此已经成为当今农业发展的一个新方向。在植物抗病基因工程研究中，$\beta-1，3-$葡聚糖酶、几丁质酶和核糖体失活蛋白已被证明具有协同的抗真菌作用[7-10]。

本研究拟应用多基因转化策略，将事先构建在同一植物表达载体[11]上的 4 个抗真菌病基因（*RCH10*，*RAC22*，*β-Glu* 和 *B-RIP*）共转化到优质超高产杂交籼稻恢复系 9311 的基因组中，通过分子检测和大田诱发鉴定，筛选到对主要水稻病害具有良好抗性的转基因植株及其后代。然后再经杂交制种得到可高抗、多抗不同病害的新型基因工程改良超级杂交水稻。

1　材料和方法

1.1　材料

超级杂交稻恢复系 9311（*Oryza sativa* L.subsp.*indica*）成熟种子由武汉大学朱英国教授惠赠。单基因质粒载体 pAAG89，pRC24/B-RIP，pARAC2，pARBC6 分别含有苜蓿 *β-*1，3-葡聚糖酶基因（*β-Glu*）、大麦核糖体失活蛋白基因（*B-RIP*）、水稻酸性几丁质酶基因（*RAC22*）和水稻碱性几丁质酶基因（*RCH10*），由美国 Vanderbilt 大学 Salk 研究所许耀教授提供[8-11]。根癌农杆菌双元表达载体 pCAMBIA1300[11]由澳大利亚 CAMBIA 组织提供。植物多基因表达载体 pRAS1300 由本实验室李明等人[11]构建（图 1）。其中 *B-RIP* 基因的启动子为水稻诱导型启动子 pRC24；*RCH10*，*RAC22* 和 *β-Glu* 基因的启动子均为水稻肌动蛋白 Act1 启动子。

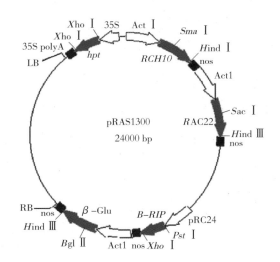

LB 示 T-DNA 左边界；RB 示 T-DNA 右边界；35S 示花椰菜花叶病毒（CaMV）35S 启动子；35S polyA 示 CaMV 35S 终止子；Act1 示水稻肌动蛋白 Act1 启动子；pRC24 示水稻几丁质酶 RC24 启动子；nos 示胭脂碱合成酶 nos 终止子；*hpt* 示潮霉素磷酸转移酶基因；*RCH*10 示水稻碱性几丁质酶基因；*RAC22* 示水稻酸性几丁质酶基因；*B-RIP* 示大麦核糖体失活蛋白基因；*β-Glu* 示苜蓿 *β-*1，3-葡聚糖酶基因。

图 1　pRAS1300 多基因表达载体质粒图

1.2　转基因植株的获得

使用 Bio-Rad 公司生产的 PDS1000/He 型基因枪对 9311 胚性愈伤组织进行轰击。水稻组织培养和基因枪转化的基本程序参照许新萍等人[12]的方法略做修改：选用 NB 培养基为基本培养基；在愈伤组织诱导培养基中加入 5 mg/L　ABA 和 2 mg/L　2,4-D；基因枪的轰击压力为 1100 psi（1 psi=6.895 kPa），真空度 25～27 inHg（1 inHg=3 386.39 Pa），每 3 mg 金粉上包裹 100 μg 纯化的 pRAS1300 质粒。用适量乙醇充分悬浮金粉和 DNA 后，每片载样膜上分数次点上 0.5 mg 金粉。每皿愈伤组织接受 1 次轰击，轰击距离为 6 cm。筛选培养基含 20～25 mg/L 潮霉素（Hygromicin B，购自 Roche 公司），预再生培养基含 30 mg/L 潮霉素，再生培养基含 30～40 mg/L 潮霉素，壮苗培养基含 50 mg/L 潮霉素。幼苗长至 6～8 cm 高、根叶齐全时可移植入土。

1.3　转基因植株的 Southern blot 分析

剪取适量水稻叶子，按 CTAB 法提取 DNA[13]。从单基因质粒载体中酶切分离出所需的目的基因片段作为 Southern 印迹杂交的探针模板[9, 10]，即以 pARBC6/*Hind* Ⅲ +*Sma* Ⅰ 的 0.7 kb 片段为（*RCH10* 基因探针；以 pARAC2/*Hind* Ⅲ +*Sac* Ⅰ 的 0.8 kb 片段为 *RAC22* 基因探针；以 pRC24/B-RIP/*Pst* Ⅰ +*Xho* Ⅰ 的 0.9 kb 片段为 *B-RIP* 基因探针；以 pAAG89/*Hind* Ⅲ +*Bgl* Ⅱ 的 1.1 kb 片段为 *β-Glu* 基因探针；以 pCAMBIA1300/*Xho* Ⅰ 的 1.1 kb 片段为 *hpt* 基因探针。使用 TAKARA 公司的 Ramdom　Primer　DNA Labeling　Kit　Ver.2 试剂盒进行探针的标记，[α-^{32}P]dCTP 购自北京亚辉生物医学工程公司。Southern　blot 的具体方法按文献[14]进行。以含有目的基因的质粒作为阳性对照，以未转基因 9311 原种作为阴性对照。

1.4　转基因植株的 Northern blot 分析

水稻嫩叶总 RNA 的抽提按照 Trizol^{TM} 试剂（Invitrogen 公司）的使用说明书进行。对于诱导型启动子 pRC24 所驱动的 *B-RIP* 基因，检测时应先对水稻植株进行致伤处理：用剪刀在叶片两侧间隔约 1.5 cm 剪开多个伤口，每个伤口深约 1.5 mm。致伤 20 h 后，剪取适量致伤过的叶片提取总 RNA。Northern　blot 的操作方法参见文献[10]。探针的制备方法同第 1.3 小节。

1.5　转基因水稻田间抗病性鉴定

2002 年 9—10 月、2004 年 3—4 月分别在湖南安化烟溪稻瘟病鉴定中心和海南三亚荔枝沟落笔洞典型稻瘟病病圃中对转基因 R_1、R_2 代植株进行了稻瘟病抗性鉴定。其中烟溪病圃四面环山，雨多雾多，日照少，属于典型的稻瘟病山区生态系，菌群的遗传结构多样，具有跨洲际的地理菌源分布，优势小种为 ZB、ZC 群，且有剧毒 ZA 群，故病圃区有很强的鉴测能力和代表性[15]；三亚病圃位于亚热带籼稻区，为国家杂交水稻工程技术研究中心稻瘟病鉴定基地。2002 年共测定 16 个独立株系 102 株 R_1 代转基因水稻的抗性；2004 年共测定 191 个不同 R_1 代转基因植株的后代抗性，每个编号播种 30~50 粒种子，逐株记录叶瘟和穗瘟的发病情况。选取其中抗性较好、分子鉴定结果呈阳性纯合的 12 个 R_2 代 9311 转基因系，与培矮 64S 进行了杂交制种。

2004 年 10 月在湖南长沙国家杂交水稻工程技术研究中心对培矮 64S/ 转基因 9311 R_2 代纯系的杂种 F_1 代进行了稻瘟病、稻曲病和稻粒黑粉病的抗性鉴定。每个杂交组合播种 30~50 粒种子，逐株记录田间发病情况。

田间稻瘟病抗性鉴定以自然诱发为主，人工病株叶接种为辅，供试材料的种子不经任何处理，从播种至收获整个生育期间不施用任何杀菌剂，重施氮肥。种植密度、其他肥水管理同当地一般大田。待鉴定材料全部采用顺序排列，每个供试品系栽种在一个小区中，小区边上插植 2 行稻瘟病高感诱发品种湘矮早 7 号。按国际水稻研究所（IRRI）水稻标准评级系统[16]进行病情调查和记载。

2　结果

2.1　转多基因水稻植株的获得

应用第 1.2 小节所述的改良的基因枪转化方法，将质粒 pRAS1300 导入 9311 的胚性愈伤组织中，共获得 18 个独立的潮霉素抗性再生植株系 62 株抗性再生植株。其中有 1 个株系（D1 株系）不育、1 个株系（E2 株系）表现出极低的育性，其余 60 棵植株的形态、育性未表现出明显异常。

2.2　转基因 9311 R_0 代植株的 Southern blot 分析

从单基因质粒载体 pCAMBIA1300、pAAG89、pRC24/B-RIP、pARAC2 和 pARBC6 中酶切分离出所需的目的基因片段作为 Southern 印迹杂交的探针模板，对 R_0 代

320

所有转基因植株进行 Southern 杂交。结果表明，在 62 个潮霉素抗性再生植株中，*hpt* 基因的导入频率为 100%；pRAS1300 载体上的 4 个外源抗病基因和 *hpt* 基因的共整合率为 100%。也就是说，所有 R_0 代转基因植株都同时含有上述 5 个外源基因。

限于篇幅，本文仅给出了部分 R_0 代植株的 Southern blot 结果（图 2）。由图可见，所有的转基因未酶切样品都仅在高分子量区出现杂交信号，而相应的酶切样品则在低分子量区出现杂交条带，这表明外源转入的基因已整合到受体植物的基因组中。

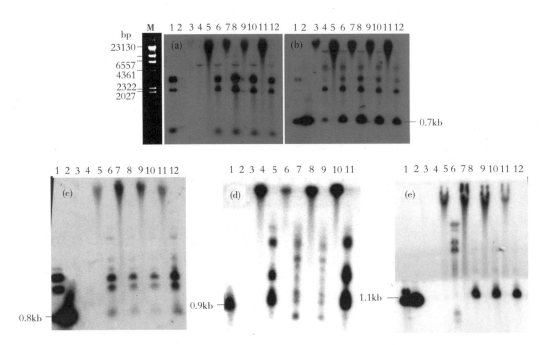

M 示 λ /*Hind* Ⅲ。（a）和（b）为在同一张膜上轮流用不同探针杂交的结果。（a）*RAC22* 基因杂交结果。1 示 pRAS1300/*Hind* Ⅲ +*Sma* Ⅰ；2 示 pARBC6/*Hind* Ⅲ +*Sma* Ⅰ；3 示 9311 原种；4 示 9311 原种 /*Hind* Ⅲ +*Sma* Ⅰ；5 示 A4；6 示 A4/*Hind* Ⅲ +*Sma* Ⅰ；7 示 A5；8 示 A5/*Hind* Ⅲ +*Sma* Ⅰ；9 示 A1；10 示 A1/*Hind* Ⅲ +*Sma* Ⅰ；11 示 A2；12 示 A2/*Hind* Ⅲ +*Sma* Ⅰ.（b）*RCH10* 基因杂交结果。各泳道同（a）。（c）*RAC22* 基因杂交结果。1 示 pRAS1300/*Hind* Ⅲ +*Sac* Ⅰ；2 示 pARAC6/*Hind* Ⅲ +*Sac* Ⅰ；3 示 9311 原种；4 示 9311 原种 /*Hind* Ⅲ +*Sac* Ⅰ；5 示 A1；6 示 A1/*Hind* Ⅲ +*Sac* Ⅰ；7 示 A4；8 示 A4/*Hind* Ⅲ +*Sac* Ⅰ；9 示 A5；10 示 A5/*Hind* Ⅲ +*Sac* Ⅰ；11 示 A2；12 示 A2/*Hind* Ⅲ +*Sac* Ⅰ.（d）*B-RIP* 基因杂交结果。1 示 pRAS1300/*Pst* Ⅰ +*Xho* Ⅰ；2 示 9311 原种；3 示 9311 原种 /*Pst* Ⅰ +*Xho* Ⅰ；4 示 B1；5 示 B1/*Pst* Ⅰ +*Xho* Ⅰ；6 示 B3；7 示 B3/*Pst* Ⅰ +*Xho* Ⅰ；8 示 B3（2）；9 示 B3（2）*Pst* Ⅰ +*Xho* Ⅰ；10 示 E1；11 示 E1/*Pst* Ⅰ +*Xho* Ⅰ.（e）*β-Glu* 基因杂交结果。1 示 pRAS1300/*Hind* Ⅲ +*Bgl* Ⅱ；2 示 pAAG89/*Hind* Ⅲ +*Bgl* Ⅱ；3 示 9311 原种；4 示 9311 原种 /*Hind* Ⅲ +*Bgl* Ⅱ；5 示 A9；6 示 A9/*Hind* Ⅲ +*Bgl* Ⅱ；7 示 A1；8 示 A1/*Hind* Ⅲ + *Bgl* Ⅱ；9 示 A4；10 示 A4/*Hind* Ⅲ + *Bgl* Ⅱ；11 示 A3；12 示 A3/*Hind* Ⅲ + *Bgl* Ⅱ。

图 2　部分 R_0 代转基因 9311 的 Southern blot 分析结果

在图 2（a）和图 2（b）中，分别使用 *RAC22* 和（*RCH10* 基因探针轮流对同一张尼龙膜进行杂交。在图 2（c）中，使用 *RAC22* 基因进行杂交。在上述两个几丁质酶基因的杂交结果中，转基因植株除了出现和 9311 原种一致的较弱的内源性杂交信号外，其酶切样品（泳道 6，8，10，12）还出现了和质粒正对照（泳道 1）片段大小一致的强杂交带。这表明在 9311 原种中，存在着和两个外源几丁质酶基因有一定同源性的序列片段。而通过本研究的遗传操作，两个外源几丁质酶基因（（RCH10 和 RAC22）也已导入到受体细胞中。另外，由于水稻本身存在着多种内源几丁质酶基因，它们的序列间有一定的同源性，而外源（RCH10 和 RAC22 基因之间也具有较高相似性[9, 10, 17]，因此，在杂交结果中出现了多条杂交带并存的现象。

在图 2（d）B-RIP 基因和图 2（e）β-Glu 基因的杂交结果中，非转基因原种完全没有杂交信号。而图 2（d）转基因样品（泳道 5，7，9，11）除了出现和质粒正对照一致的杂交带之外，还出现了许多片段大小不一的杂交信号。在图 2（e）的泳道 6 中也出现了多条与质粒正对照位置不一致的杂交带。这可能是由外源基因的多拷贝整合与重排所导致的。

根据 Southern 杂交的结果，所有的转基因株系都在 hpt、B-RIP 和 β-Glu 这 3 个基因中出现了至少一个外源基因的多拷贝整合与重排（表现为具有两条或两条以上的杂交带）。这表明在利用本方法所得到的转基因群体中，多拷贝和重排现象频繁出现。

2.3　转基因 9311 R_1 代植株的遗传分析

R_0 代植株所收种子种植后得 R_1 代株系。用 β-Glu 和 B-RIP 基因探针对这些株系进行了未酶切总 DNA 的 Southern 杂交分析（图略）。结果发现在同一植株中这两个基因呈现出连锁整合的遗传规律，即在分离的 R_1 代植株中，它们同时存在，或者同时消失。在已检测的 17 个转基因株系中，β-Glu 和 B-RIP 基因杂交阳性和杂交阴性的分离比都符合孟德尔 3∶1 的分离规律。但有部分株系（A3，A6，A7，A9，E2）也不能完全排除 15∶1 分离的可能（表 1）。这表明在 R_0 代转基因植株中，同一外源基因的不同拷贝整合在受体基因组的 1~2 个位点上。

表 1　9311 转基因 R_1 代植株的遗传分析[a]

R_0 代编号	阳性株数	阴性株数	3∶1 分离的 X_i^2	15∶1 分离的 X_i^2
A1	12	5	0.020	14.592
A2	14	5	0.018	12.214
A3	11	1	1.000	0.022
A4	6	4	0.533	18.027
A5	15	5	0.067	11.213

Continued

R$_0$代编号	阳性株数	阴性株数	3∶1分离的 X_i^2	15∶1分离的 X_i^2
A6	15	3	0.296	2.904
A7	14	3	0.176	3.298
A8	14	5	0.018	12.214
A9	18	1	2.965	0.004
B1	14	7	0.397	25.146
B2	8	6	1.524	30.476
B3	12	6	0.296	21.393
B4	6	3	0.037	10.141
B5	15	4	0.018	6.488
B6	12	7	0.860	29.056
E1	28	12	0.300	37.500
E2	4	1	0.000	1.080

注：a）$x_{0.05}^2$=3.841，$x_i^2 < x_{0.05}^2$ 时接受预期分离比的假设。

2.4 转基因 9311 R$_1$，R$_2$ 代植株在稻瘟病病圃中的抗性鉴定

对转基因 9311 R$_1$ 代的稻瘟病抗性鉴定选择在湖南烟溪稻瘟病典型病圃中进行。据统计，102 个 R$_1$ 代转基因植株中，高抗稻瘟病的植株占总植株数的 45.1%。

将中至高抗的 9311R$_1$ 代植株收种，到海南三亚病圃中进行 R$_2$ 代稻瘟病抗性鉴定。在 69 个待测的 R$_1$ 代植株中，共有 19 个 R$_1$ 代植株，其后代中出现了既不感穗瘟又高抗叶瘟的 R$_2$ 代植株；此外，还有 23 个 R$_1$ 代植株，其后代中分离出了完全不感穗瘟的 R$_2$ 代植株。

将另外 122 个未在烟溪进行过抗性鉴定的 9311R$_1$ 代植株收种，其 R$_2$ 代直接在三亚进行稻瘟病抗性鉴定。结果发现共有 13 个 R$_1$ 代植株，其 R$_2$ 代中分离出了既高抗叶瘟又高抗穗瘟的植株；此外，另有 41 个 R$_1$ 代植株，其 R$_2$ 代中出现了完全不感穗瘟的植株。

综上所述，在三亚鉴定的 R$_2$ 代转基因 9311（分别来自 191 个不同的 R$_1$ 代植株）中，共有 96 个 R$_1$ 代植株能产生高抗穗瘟的后代，并可正常结实；在这 96 个 R$_1$ 代植株中，又有 32 个 R$_1$ 代植株，能够产生在整个生育期中都表现出对稻瘟病高抗性的 R$_2$ 代。

需要特别指出的是，在上述 9311R$_2$ 代稻瘟病抗性鉴定的结果中，R$_2$ 代穗瘟发病率低于 10% 的 R$_1$ 代植株共有 4 个。其中编号为 A2-10 的 R$_1$ 代植株，其 R$_2$ 代穗瘟发病率仅为 2.00%，叶瘟发病率为 9.09；编号为 A2-9 的 R$_1$ 代植株，其 R$_2$ 代穗瘟和叶瘟发病率分

别为 9.88% 和 11.11%；编号为 A9-16 的植株，其 R_2 代穗瘟和叶瘟发病率分别为 9.90% 和 10.00%；而在同等栽培条件下，9311 原种的穗瘟和叶瘟发病率都为 100%，病级达到 7~9 级（高感），以致完全颗粒无收（图 3）。

O. 非转基因 9311 原种（感病）；T. 转基因 9311（抗病）。（a）. 转基因 9311 R_1 代植株在湖南烟溪病圃中的抗病情况；（b）. 转基因 9311 R_2 代植株在海南三亚病圃中的抗病情况；（c）. 转基因 9311 R_2 代植株在海南三亚病圃中穗瘟的发病情况。

图 3　转基因 9311 后代植株在稻瘟病病圃中的抗病情况

2.5　转基因 9311 的 Northern blot 分析

为了检验转基因植株中多个外源基因的表达情况，对部分高抗稻瘟病的 9311　R_2 代植株进行了致伤处理和 Northern　blot 分析（图 4），在同一张膜上依次用 *RCH10*，*β-Glu* 和 *B-RIP* 基因探针进行杂交。结果表明：在这些抗性植株中，外源抗病基因多数能够以较高的水平转录。但不同植株中相同基因的表达水平存在差异；同一植株中不同外源基因的表达水平也不完全一致。其中几丁质酶基因的表达水平普遍较高，而 *B-RIP* 基因和 *β-Glu* 基因的表达水平差别较大。如 A2-10（泳道 1）、A9-16（泳道 7）中 3 个基因的转录水平都较高，而 A2-4（泳道 3）、A9-2（泳道 5）、A9-5（泳道 6）中 *B-RIP* 基因和 *β-Glu* 基因的表达水平都偏低，这与它们的田间抗性表现也是基本吻合的。

324

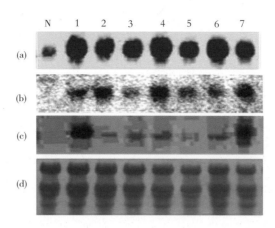

N.9311 原种；1.A2-10；2.A2-13；3.A2-4；4.A5-7；5.A9-2；6.A9-5；7.A9-16。（a）.RCH10 基因杂交结果；（b）.β-Glu 基因杂交结果；（c）.B-RIP 基因杂交结果；（d）.RNA 变性胶电泳结果。

图4　高抗稻瘟病 9311 R$_2$ 代植株的 Northern blot 分析结果

2.6　培矮 64S/ 转基因 9311 杂种 F$_1$ 代的抗病性

选取高抗稻瘟病且 Southern 杂交结果显示为阳性纯合的 R$_2$ 代 9311 转基因系共 12 个，与未转基因培矮 64S 进行杂交。大田隔离条件下，将杂种 F$_1$ 代种植在湖南长沙的田间。2004 年晚造期间，12 个培矮 64S/ 转基因 9311 杂种 F$_1$ 代的稻瘟病田间抗性全部达到 0~2 级（高抗），而原种 F$_1$ 对照的稻瘟病抗性仅为中感到中抗（4~5 级）。

此外，12 个不同品系的转基因 F$_1$ 代中，有 10 个品系完全没有出现稻曲病侵染的症状。另外 2 个品系（04CF069 和 04CF071）出现的稻曲病穗数也分别只占该品系总穗数的 1.2% 和 2.8%。而未转基因培矮 64S/9311 稻曲病穗的出现频率，占原种 F$_1$ 总穗数的 7.7%［图 5（a）］。

从每个杂交组合中随机抽取 6~17 棵植株进行种子实粒数和黑粉病粒数的统计，结果见表 2。可以看到，编号为 04CF077 的转基因 F$_1$ 代杂交组合完全未受黑粉病侵染［图 5（b）］，而其他编号的转基因杂交稻黑粉病发病率也远低于非转基因杂交稻。

因此，从培矮 64S/ 转基因 9311F$_1$ 代田间的表现看来，其对稻瘟病、稻曲病和黑粉病的抗性都比非转基因两优培九有了明显提高，表现出较好的综合抗性。

（a）. 稻曲病发病情况；O. 培矮 64S/ 非转基因 9311 F$_1$ 代；T. 培矮 64S/ 转基因 9311 F$_1$ 代；（b）. 黑粉病发病情况。

图 5　培矮 64S/ 转基因 R$_2$ 代 9311 杂种 F$_1$ 代对稻曲病和黑粉病的抗性

表 2　培矮 64S/ 转基因 R$_2$ 代 9311 杂种 F$_1$ 代黑粉病的发病情况

杂交稻编号	平均发病率 /% （黑粉病粒数 / 实粒数）	每穗平均 发病粒数 / 粒
04CF064	1.87	1.38
04CF065	2.23	1.75
04CF066	5.01	2.92
04CF067	0.81	1.00
04CF068	0.41	0.45
04CF069	5.60	5.29
04CF070	0.60	0.50
04CF071	1.14	1.57
04CF074	5.05	2.29
04CF076	7.10	3.57
04CF077	0.00	0.00
04CF078	6.75	3.25
两优培九原种	17.12	14.50

3 讨论与结论

由于生物体的绝大多数性状和生理功能都是依靠多基因的协调表达而实现的，应用多个基因来定向改变某个或某些性状可以使生物遗传性发生更为显著的改变，因此多基因转化策略必将成为今后基因工程在基础理论与实际应用研究中的主流方向[18, 19]。在植物抗病虫基因工程中，若同时采用不同抗性机制的多个抗性基因，不仅可以扩大转基因作物的抗谱，提高作物的抗性水平，还可以延缓病虫害产生耐受性，从而延长抗性品种的应用年限和适应范围[18-22]。因此，利用多基因转化策略来同时改良水稻及其他农作物对多种病虫害和逆境的抗性，已经成为作物基因工程中的一个新方向[8-10, 18, 19]。

本研究中，我们将来自于禾本科或豆科植物的 4 个抗病基因转化到中籼稻 9311 的基因组中，获得了对稻瘟病抗性有明显提高的转基因水稻植株和后代，并证明以上述转基因高抗病 9311 为父本所配制的两优培九超级杂交稻，能够同时提高对稻瘟病、稻曲病、稻粒黑粉病的抗性。根据已有的文献资料[7-11, 17, 20-24]，我们推测在这些转基因水稻植株中，外源基因协同作用的机制可能为：当病原生物侵入转基因水稻时，首先接触到位于细胞间隙的酸性几丁质酶（组成型表达的 *RAC22* 基因产物），部分真菌细胞壁在几丁质酶的作用下发生裂解，裂解产物作为信号分子进一步诱导植物产生综合的防御机制。同时，由于病菌的入侵，导致创伤诱导表达型启动子 pRC24 驱动下 *B-RIP*（大麦核糖体失活蛋白）基因的表达。当病原菌进一步侵入细胞后，定位于液泡的 β -1，3-葡聚糖酶和碱性几丁质酶（RCH10 蛋白），可联合作用于真菌细胞壁中交织成纤丝状的葡聚糖和几丁质底物上，导致真菌细胞壁彻底瓦解，并使 RIP 蛋白进入真菌细胞质，防止病原生物的蛋白表达和繁殖。

由于本研究所选用的 4 个抗病基因及其启动子（水稻 pRC24 和 Act1 启动子）都来源于可食性的农业作物（水稻、大麦、苜蓿）[8-11]，它们千百年来都被作为人畜的主食而广泛种植，据此推测其环境安全性和食品安全性可能都比较好。另外，从上述基因的作用机制上看，由于哺乳动物的细胞组织结构中不存在几丁质酶、葡聚糖酶的有效作用底物，因此，几丁质酶基因和 β -1，3-葡聚糖酶基因并不会对人畜产生危害[24]；而来自大麦的 I 型 RIP 对完整细胞组织无毒性[24]，因此，转入上述多个基因的植物很可能对人畜是安全的，但对真菌等病原体却有极大抗性。

本研究所选用的基因转化受体品种 9311 为当前水稻超高产育种项目中重要的恢复系，以 9311 及其改良系作为父本配制的两优培九，广两优 6 号，粤优 938，P88S/9311，58S/9311，38S/9311 等不同类型的优质杂交稻组合，目前已在广大中籼稻区推广应用，而 9311 本身也可以作为一个优质高产的常规品种应用于生产实践[25]，因此，本研究中所获得的大量转基因抗性9311植株，为超级杂交稻的抗病育种提供了有用的种质资源，具有广阔的应用前景。今后，通过和常规育种结合，将有望得到综合抗性和农艺性状兼优的新型转多基因超级杂交稻品种（品系）。

─────── R e f e r e n c e s ───────

参考文献

[1] 袁隆平，辛业芸.希望之光—超级杂交稻.世界农业，2001，10（10）：46.

[2] 杨春献，向邦豪，张其茂，等.两系超级杂交稻百亩片平均单产12.26 t/hm² 的栽培技术.杂交水稻，2003，18（2）：42-44.

[3] 王强，卢从明，张其德，等.超高产杂交稻两优培九的光合作用、光抑制和C₄途径酶特性.中国科学，C辑，2002，32（6）：481-487.

[4] 卢开阳，吴和明，卢碧林.优质中稻两优培九和扬稻6号的鉴定.湖北农业科学，2001，（3）：9-12.

[5] 董习华，田学志，汪丽，等.两系杂交中籼稻两优培九的病虫害发生规律及综合防治技术.安徽农业科学，2001，29（3）：342-344.

[6] 马铮，何盛莲，金开美，等.两优培九制种稻粒黑粉病的发生原因及防治方法.信阳农业高等专科学校学报，2001，11（2）：13-15.

[7] Jach G, Gornhardt B, Mundy J, et al. Enhanced quantitative resistanceagainst fungal disease by combinatorial expression of differentbarley antifungal proteins in transgenic tobacco. Plant J, 1995, 8（1）: 97-109.

[8] 冯道荣，卫剑文，许新萍，等.转多个抗真菌蛋白基因水稻植株的获得及其抗稻瘟病菌的初步

研究.中山大学学报（自然科学版），1999，38（4）：62-66.

[9] 冯道荣，许新萍，范钦，等.获得抗稻瘟病和纹枯病的转多基因水稻.作物学报，2001，27（3）：293-301.

[10] 冯道荣，许新萍，邱国华，等.多个抗病抗虫基因在水稻中的遗传和表达.科学通报，2000，45（15）：1593-1599.

[11] 李明，邱国华，许新萍，等.含多种抗真菌病基因植物表达载体的构建.中山大学学报（自然科学版），1999，38（5）：67-71.

[12] 许新萍，卫剑文，范云六，等.基因枪法转化籼稻胚性愈伤组织获得可育的转基因植株.遗传学报，1999，26（3）：219-227.

[13] Doyle J J, Doyle J L. Isolation of plant DNA from fresh tissue.Focus, 1990, 12: 13-15.

[14] 朱华晨，许新萍，李宝健.一种简捷的Southern 印迹杂交方法.中山大学学报（自然科学版），2004，43（4）：128-130.

[15] 肖放华，罗赫荣，贺忠南，等.节水稻（陆稻）资源持久多抗性鉴定及利用研究.湖南农业科学，2001，（5）：39-41.

328

[16] International Rice Research Institute. Standard Evaluation Systemfor Rice (SES). Los Banos, Philippines: IRRI, 2002. 14-18.

[17] Zhu Q, Lamb C J. Isolation and characterization of a rice geneencoding a basic chitinase. Mol Gen Genet, 1991, 226: 289-296.

[18] 李宝健, 朱华晨. 论应用多基因转化策略综合改良生物体遗传性研究方向的前景 I. 多基因转化的基因来源与技术平台. 中山大学学报（自然科学版）, 2004, 43(6): 11-15.

[19] 李宝健, 朱华晨. 论应用多基因转化策略综合改良生物体遗传性研究方向的前景 II. 多基因转化策略中的规律、前景和问题. 中山大学学报（自然科学版）, 2005, 44(4): 79-83.

[20] McDowell J M, Woffenden B J. Plant disease resistance genes: recent insights and potential applications. Trends in Biotechnology, 2003, 21(4): 178-183.

[21] Shah D M. Genetic engineering for fungal and bacterial diseases.Curr Opin Biotechnol, 1997, 8(2): 208-214.

[22] Punja Z K. Genetic engineering of plants to enhance resistance tofungal pathogens-a review of progress and future prospects. CanJ Plant Pathol, 2001, 23: 216-235.

[23] 明小天, 王莉江, 安成才, 等. 利用土壤农杆菌将天花粉蛋白基因转入水稻植株并检测抗稻瘟病活性. 科学通报, 2000, 45(10): 1080-1084.

[24] 单丽波, 贾旭. 核糖体失活蛋白及其在植物抗真菌病基因工程中的应用. 生物工程进展, 2000, 20(6): 74-78.

[25] 白和盛, 詹存钰, 王宝和, 等. 中籼扬稻 6 号及其在杂交稻育种中的应用. 杂交水稻, 2001, (6): 13-15.

作者: 朱华晨　许新萍　肖国樱　袁隆平*　李宝健*
注: 本文发表于《中国科学 C 辑: 生命科学》2006 年第 36 卷第 4 期。

水稻 C815S 及其同源株系的育性光温特征

【摘　要】对 C815S 及其同源株系通过 12 种人控光温组合处理、短日遮光处理以及分期播种试验，就其育性光温特性、发育感光性以及两者之间的关系进行了研究。结果表明，C815S 及其同源株系不育起点温度低，均在 23 ℃以下。在长沙自然长日条件下没有明显不育向可育的转换；在长沙短日条件下，育性出现轻微波动。其育性受光温双因子共同作用，且这种作用的程度与发育感光性强弱似乎存在密切联系。提出了光温敏不育系光温双因子互作量化的光温效应连动假设。这一假设能提供每一光长条件下与之对应的育性转换温度，对光温敏核不育系的选育、制种和繁殖具有实际指导意义。

【关键词】光温敏核不育水稻；育性转换；光敏性；发育感光性

【Abstract】The study of the fertility change mechanism of PTGMS has been a hot topic since 1989, and the prevenient achievements have revealed the fertility alteration models of PTGMS rice in response to photoperiod and temperature condition. There have been mainly four models: the suppose of three developing phase and two photoperiod-sensitive stages, the model of light and temperature in fertility changes, the quantitative model of fertility changes and the three dimensional coordinate model of fertility conversion, and these models and theories had higher utilization values in directing two-line hybrid production. In this paper, fertility photo-thermo characteristics, developmental photo-sensitivity and their relationships in PTGMS rice C815S and its homologous plant lines were studied in 12 controlled combination treatments of three daylength with four lower temperature, treatments of shading, and short photoperiod, and the treatments of different rowing date. The results showed that C815S and its homologous plant lines were a type of typical lower thermo-sensitivity sterile line and their low critical temperature was below 23.0 ℃. In Changsha, there were no obvious fertility fluctuation and alteration from sterility to fertility under the long daylength condition while there was slight fertility fluctuation under the short daylength condition. The different degrees of photoperiod sensitivity sterility in C815S and its homologous plant lines were a result of the interaction of daylength × temperature, and there would be some internal relation between the interaction and developmental photoperiod sensitivity. The interaction of daylength × temperature was further discussed and the interdynamic effect model

330

of daylength and temperature, being of the quantitative fertility response to photoperiod and temperature, was supposed. This hypothesis can offer the corresponding temperature for fertility alteration under the determinate daylength, being used in the selection of PTGMS, hybrid seed production and propagation of PTGMS scientifically. According to this hypothesis, the breeding strategy of selecting the practical two-line genic sterile line can be put forward in the corresponding environment.

【Key words】PTGMS; Fertility alteration ; Photo-period sensitivity; Developmental photo-sensitivity

1989 年，长江中下游地区出现 39 年一遇的盛夏持续低温使育成的一大批两用核不育系表现明显的育性波动，导致两系杂交水稻研究遭受重大挫折。因而，许多学者对水稻光温敏雄性不育系的育性转换机制进行了深入研究，并根据各自的研究材料，提出了水稻两用核不育系的各种光温作用模式。目前主要有以下 4 种光温作用模式来解释光温敏核不育水稻的育性反应，即第二光周期假说[1]；光温作用模型[2]；育性量化模型[3]；光–温–育性三维曲面模型[4]。这些模型对水稻两用核不育系的选育及两系法杂交稻种子的生产具有很好的指导意义。

C815S 是湖南农业大学水稻科学研究所选育的一个具有不育起点温度低（<22℃）、农艺性状优良、异交特性好、品质优、配合力强的中稻籼型两用核不育系。2004 年通过湖南省农作物品种委员会审定。C815S 配制的几个组合在 2005 年湖南省区试或高产示范中表现突出。如 C 两优 396 在超级稻组的省区试中，比对照两优培九增产 12.54%，C 两优 343 在一季晚稻预试中比对照汕优 63 增产 15%，C 两优 87 在一季晚稻区试中比对照汕优 63 增产 7.3%，3 个组合的区试产量均列各供试组第一位。C 两优 87 还创造了湖南省单产的最高纪录（13.55 t/hm²）。C815S 及其同源株系以及培矮 64S 雄性育性均表现不同程度的光敏特性。在相同遗传背景下，通过研究这些材料育性光温特性，探明在不同感光背景下光温双因子对这些材料育性转换的一些基本规律，不仅可以增进人们对水稻发育感光性与育性光敏性的认识，而且也为这种类型不育系尤其是 C815S 的生产利用提供理论依据。

1 材料与方法

1.1 供试材料

C815S 及 4 个同源株系 3MS135、3MS136、3MS138、3MS148 与培矮 64S（CK）。

1.2　研究方法

1.2.1　人工控制光温条件下育性转换的光温反应

试验于 2004 年在湖南农业大学实习基地（位于长沙市芙蓉区马坡岭）进行。3 月 28 日开始播种，共分 10 期，每期相隔 10 d，单本植，常规田间管理，剥检幼穗了解各期材料的发育进度。在第一播期幼穗分化的第二次枝梗原基分化期开始移入日光型人工气候室，用 12.0 h、13.0 h 和 14.0 h 3 种光长处理 7 d，然后移入 SANYO Gallenkamp PLG 型人工气候箱（英国）进行 5 d 光温组合处理，每份材料每种光温组合处理 9 株。人工气候箱光温处理参照中国水稻所两系不育系的鉴定方式[5]，设 3 档光长（12.0 h，13.0 h，14.0 h）×4 档温度（21.0℃，22.0℃，23.0℃，24.0℃）共 12 种处理，温差 ±0.3℃。光长处理与日光型人工气候室的光长处理对应。光强 200 $\mu mol \cdot m^{-2} \cdot s^{-1}$，相对湿度 75%。

处理后，标记剑叶与倒二叶叶枕距 ±1.0 cm 的茎蘖，并移至自然条件下抽穗。于标记茎蘖开始抽穗时起，用 2%I_2-KI 溶液染色镜检当天上部 5 朵颖花的花粉育性，连续镜检 4 d，尽量使各处理材料在相同的光温条件下有花粉育性资料可供比较。可染花粉率（%）=（染色花粉数 / 总花粉粒数）×100。

1.2.2　感光性试验

试验分自然条件下种植和遮光处理两部分，于 2004 年夏季在长沙（28°12′N）湖南农业大学水稻科学研究所网室进行。5 月 16 日播种，6 月 13 日移栽 40 株，单本植，其中 20 株从 5 叶期开始遮光处理，每天光照时数 10 h，连续处理 30 d，分别记载抽穗期，并参照文献[6]计算短日出穗促进率和划分感光性类别。

1.2.3　供试材料分期播种育性观察

2004 年在长沙对分期播种的 C815S、3MS148 和培矮 64S 进行育性观察，育性检查从第 1 期播种材料抽穗开始至 9 月下旬结束，每隔 2 d 镜检 1 次，当发现有材料育性波动时，则每 1 d 镜检 1 次。各供试材料每次取 5 株，每株取 1 穗，以花粉可染率作为育性表现的直观分析。并与人工光温处理资料进行对比。光长引自参考文献[7]的旬平均日长，温度资料来源于长沙马坡岭气象观测站 2004 年气象资料。

1.2.4　光温育性效应计算

参照马国辉[8]的方法。光敏性的计算采用同一温度条件下不同光长对花粉育性的影响程度，即光长影响率 =（12.0 h 处理花粉可染率 - 14.0 h 处理花粉可染率）×100/12.0 h 处理花粉可染率。本文采用 21℃ 温度不同光长的花粉可染率作为计算依据。

2 结果与分析

2.1 12种光温组合处理的花粉可育程度

供试材料用21.0℃、22.0℃、23.0℃、24.0℃四档温度和12 h、13 h、14 h 三档光长12种光温组合处理后的可育程度见表1。

表1 供试材料12种光温组合处理的花粉可育程度

材料	光长	温度			
		21℃	22℃	23℃	24℃
C815S	12	38.85±5.28	23.08±2.65	0.50±0.12	0
	13	29.88±7.71	7.77±1.05	0	0
	14	14.27±2.29	0.58±1.07	0	0
3MS135	12	42.33±4.57	6.75±0.35	0.15±0.07	0
	13	33.50±8.50	3.05±1.34	0.05±0.04	0
	14	24.20±3.63	1.00±0.71	0	0
3MS136	12	32.98±7.40	16.26±7.81	0	0
	13	41.68±2.40	8.57±3.07	0	0
	14	7.10±5.14	2.68±0.90	0	0
3MS138	12	48.80±2.43	12.03±10.38	0.13±0.06	0
	13	34.50±7.03	5.05±4.74	0	0
	14	11.77±9.29	0.3±0.00	0	0
3MS148	12	42.65±9.86	4.25±0.29	1.45±0.13	0.1±0.05
	13	10.50±0.42	2.00±0.00	0	0
	14	4.46±3.66	0	0	0
Pei'ai 64S（CK）	12	31.83±3.81	52.14±9.73	13.83±7.30	0
	13	22.33±5.31	27.75±8.23	8.40±0.00	0
	14	16.25±4.99	15.00±0.00	3.57±2.98	0

由表1可知，在同一温度条件下，6个材料均存在长光条件下花粉可染率下降；短光条件下花粉可染率升高的趋势。3MS148在21℃时，这种趋势最明显，染色花粉率由14 h的4.46%上升到12 h的42.65%，3MS135、对照培矮64S也存在这种趋势；C815S、3MS138、对照培矮64S在22℃时，这种趋势表现明显，染色花粉率分别由14 h的0.58%、1%、15.0%上升到12 h的23.08%、12.03%、52.14%；在23℃时，C815S

及其同源株系短光照条件下，有很低的花粉可育度，而在长光照条件下，完全表现为败育；
3MS148在24℃、12 h 光长条件下，存在极少数可染花粉。

在同一光长条件下，不同温度对各供试材料花粉育性的效应存在差异。在光长12 h的
条件下，21℃处理所有的供试材料花粉可染率都较高；22℃处理，除对照培矮64S外，
C815S及其同源株系的花粉可染率均有不同程度的下降，其中3MS135和3MS148 下
降的幅度最大，分别达84.1%和90.0%；23℃处理，C815S及其同源株系的花粉可染
率均低于1.5%。在光长14 h 条件下，21℃处理各供试材料之间花粉可染率差异相当
大，3MS135、培矮64S、C815S、3MS138、3MS136、3MS148的花粉可染率分别为
24.20%、16.25%、14.27%、11.77%、7.10%和4.46%；22℃时，C815S及其同
源株系花粉可染率普遍下降，均低于4.0%，其中C815S、3MS138、3MS148败育彻底；
23℃时，C815S及其同源株系均彻底败育，而对照培矮64S仍有3.57%的可染花粉。

参照前人分析方法[9]，通过图像可以很直观地了解光温双因子对供试材料育性的影响（图
1～图3）。

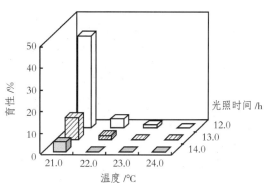

图1　光温对3MS148育性转换的影响

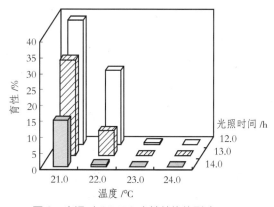

图2　光温对C815S育性转换的影响

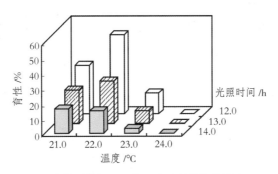

图3　光温对培矮64S（CK）育性转换的影响

2.2　供试材料分期播种的育性表现

从 2004 年分期播种育性观察结果（表2）可以看出，C815S 及其同源株系在长沙 7 月 6 日—9 月 21 日处于完全不育阶段，短日的较低温度没有使育性产生波动。在 9 月中旬的低温导致 9 月下旬抽穗的 C815S 和 3MS148 出现较轻度的育性恢复，表明这种育性恢复只有在供试材料育性敏感期同时处于短光照和低温条件下才可能发生。

表2　供试材料分期播种的育性表现

日期	光长/h	日均温/℃	可染花粉率 /%			日期	光长/h	日均温/℃	可染花粉率 /%		
			3MS148	C815S	CK				3MS148	C815S	CK
6 月 11 日 *	13.86	27.4				8 月 21 日	12.81	29.1	0	0	0
6 月 14 日		23.0				8 月 23 日		24.7	0	0	0
6 月 15 日		20.7				8 月 25 日		27.6	0	0	0.5
6 月 16 日		22.1				8 月 26 日		26.6	0	0	0.1
6 月 17 日		24.1				8 月 27 日		26.6	0	0	0.5
6 月 18 日		24.8				8 月 28 日		26.4	0	0	0.1
6 月 19 日		25.9				8 月 29 日		26.3	0	0	0.7
6 月 20 日		27.2				8 月 30 日		23.3	0	0	15.3
6 月 21 日	13.90	28.5				8 月 31 日		25.7	0	0	18.7
6 月 26 日		27.3				9 月 01 日	12.52	27.3	0	0	7.3
6 月 30 日		32.3				9 月 02 日		26.0	0	0	17.9
7 月 01 日	13.83	31.2		0	15.2	9 月 03 日		28.2	0	0	4.5
7 月 03 日		29.8		0	3.7	9 月 04 日		28.9	0	0	5.0

续表

日期	光长 /h	日均温 /℃	可染花粉率 /%			日期	光长 /h	日均温 /℃	可染花粉率 /%		
			3MS148	C815S	CK				3MS148	C815S	CK
7 月 04 日		30.5	0	0	1.0	9 月 05 日		25.4	0	0	9.3
7 月 06 日		25.5	0	0	0	9 月 06 日		24.5	0	0	3.0
7 月 09 日		30.3	0	0	0	9 月 07 日		20.9	0	0	1.5
7 月 11 日	13.71	25.5	0	0	0	9 月 09 日		20.5	0	0	5.0
7 月 14 日		26.3	0	0	0	9 月 10 日		23.3	0	0	1.0
7 月 16 日		30.3	0	0	0	9 月 11 日	12.23	24.4	0	0	0
7 月 18 日		29.2	0	0	0	9 月 12 日		23.7	0	0	0.2
7 月 21 日	13.54	29.7	0	0	0	9 月 13 日		23.7	0	0	0.3
7 月 23 日		30.5	0	0	0	9 月 14 日		23.4	0	0	0.1
7 月 25 日		32.1	0	0	0	9 月 15 日		23.5	0	0	0.1
7 月 28 日		32.4	0	0	0	9 月 16 日		25.5	0	0	1.3
7 月 30 日		30.3	0	0	0	9 月 17 日		27.0	0	0	4.7
8 月 02 日	13.32	32.2	0	0	0	9 月 18 日		28.3	0	0	1.3
8 月 04 日		30.1	0	0	0	9 月 19 日		28.1	0	0	0.5
8 月 06 日		27.6	0	0	0	9 月 21 日	11.94	20.9	0	0	7.5
8 月 08 日		31.0	0	0	0	9 月 22 日			0	1.2	48.9
8 月 10 日		32.4	0	0	0	9 月 23 日			1.4	6.3	65.3
8 月 12 日	13.08	28.7	0	0	0	9 月 24 日			11.1	0.1	53.7
8 月 14 日		23.5	0	0	0	9 月 25 日			0	0	53.1
8 月 15 日		22.0	0	0	0	9 月 26 日			0	0	87.3
8 月 17 日		23.0	0	0	0	9 月 27 日			0.1	0.5	89.5
8 月 19 日		25.6	0	0	0.2	9 月 28 日			0	0	84.1

注：＊表示供试材料已进入幼穗分化。

2.3　供试材料人控光温处理与分期播种的育性表现对比

　　分期播种的温度、光照、湿度比 12 种人控光温组合处理条件要复杂得多，通过对供试材料的分期播种与人控光温处理的育性表现的拟合，可以更准确判断各供试材料的育性光温特点。

6月中旬（光照时数 13.86 h），长沙出现了连续 3 d 日平均温低于 23.0 ℃、连续 7 d 日最低温小于 22.0 ℃的低温天气，在这种光温条件下，3MS148、C815S 育性没有恢复，而对照培矮 64S 可染花粉率高的达 15.2%，育性出现了波动。这与 23.0 ℃、14.0 h 光温组合处理结果吻合较好。到 8 月中旬又出现连续 4 d 日平均温低于 23.1 ℃，连续 6 d 日最低温低于 22.0 ℃的盛夏低温天气，此时长沙的日照时数为 13.08 h，在这种光温条件下，3MS148、C815S 仍没有出现育性恢复，而对照培矮 64S 可染花粉率达 18.7%，这与 23.0 ℃、13.0 h 的光温组合处理结果吻合；到 9 月上中旬长沙地区再次出现连续 3 d 日平均温低于 21.9 ℃、连续 10 d 最低温低于 22.0 ℃的天气，此时的日照时数为 12.23 h 左右，3MS148、C815S 和对照培矮 64S 分别出现了 1.4%、6.3% 和 65.3% 的可染花粉率，这与 22.0 ℃、12.0 h 的光温组合处理结果较吻合，上述对比表明，严格的人控光温处理结果完全能作为两用核不育系选育和种子生产的育性光温特性依据。图 4 是上述 3 个时段分期播种的光温条件对供试材料花粉育性影响与相应人工光温处理结果的对比结果。

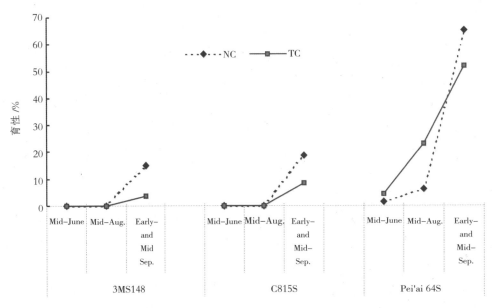

图 4　人工光温处理与自然光温对供试材料育性影响的对比

注：实线表示人工光温处理下的育性表现；虚线表示分期播种育性表现。

2.4　发育感光性与光温敏不育性的关系

对表 1 中具有代表性的 3 份供试光温敏核不育系 3MS148、C815S、CK 的花粉可染率经反正弦转换所作的光长、温度两因子方差分析和育性光敏性分析，以及利用感光性试验数据进行感光性分析表明（表3），各材料育性转换的光长效应、温度效应及光温双因子互作效应明显不同，感光性中等的 3MS148 的光长效应、温度效应及光温互作效应均极显著，感光性弱的 C815S 光长效应和光温互作效应显著，温度效应极显著，感光性弱的 CK 的温度效应极显著，光温互作效应显著，而光长效应不显著。反映出各材料感光性的强弱对育性转换的光长效应、光温互作效应存在一定的关联性。

表 3　供试材料育性的光、温效应方差分析及发育感光性和育性光敏性

材料	方差分析			短日出穗促进率 /%	感光性类别	育性光敏性 /%
	光长效应	温度效应	光 × 温			
3MS148	90.79**	97.48**	3.52**	25.6	中	89.54
C815S	76.21*	89.40**	1.26*	8.4	弱	63.27
CK	64.39	80.71**	1.08*	5.7	弱	31.83

注：*，** 分别表示 F 值在 5% 和 1% 水平上显著。

3　讨论

根据表 1 的结果，在同一温度条件下，短光照能显著促进 C815S 及其同源株系由不育向可育转变，表明其育性表达具有明显的光敏特性。C815S 及其同源株系间光敏性存在显著差异，据此推断，控制 C815S 及其同源株系光敏性的遗传基础相当复杂，可能属于数量性状，受微效多基因控制。利用 C815S 及其同源株系具有光敏性的特点，在每一光照条件下，可以初步确定 C815S 及其同源株系的育性转换温度。

陈立云把两用核不育系分成 4 种类型[10]，即长光高温不育型、高温不育型、短光低温不育型和低温不育型。本文将其中第一和第三种类型的光温不育特性作如下假设描述，即光温关系好像一个装了水的 "U" 型管（图 5）。当在 "U" 型管的一边施压时，另一边的水就会上升，施压大小决定另一边水的上升程度，水稻的光温敏特性正是这种状况。当光照加长时，光温敏不育系的不育起点温度就会降低；光照逐渐变短时，不育系的不育起点温度会慢慢升高。作者把这种现象称之为 "光温效应连动"。据此假设，光温敏或反光温敏不育系的不育临界光长和育性转换温度不是固定不变的，而是动态的。但不同不育系之间这种 "光温效应连动" 的程度是不同的，其大小决定于不育系的发育感光性，发育感光性强时，意味着施压力量大，不育临界温度的变幅大；发育感光性弱时，施压力量小，不育临界温度的变化则小。这一模型的另一依据是植株育性转换是一个可逆过程，育性变化是一个连续过程[3]。这一假设与光温互作模型相比，进一步量化了光温双因子互作程度，从 "U 模型" 中能看出，随临界光长的升降，不育临界温度也随之降低或升高。因而能提供每一光温敏核不育系光温连动指标，即每一

光长条件下与之对应的育性转换温度。这种指标能更确切地反映每一光温敏核不育系育性的光温效应模式。针对光温敏核不育系的选育、制种和繁殖，这种量化的指标具有很好的实际指导意义。

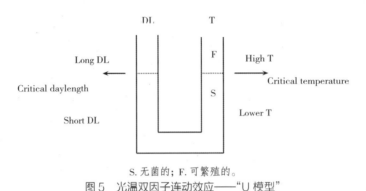

S. 无菌的；F. 可繁殖的。

图5　光温双因子连动效应——"U 模型"

　　依据这一假设，作者提出如下选育实用两用核不育系的育种策略。一是选育在长光照条件下（13.5 h以上）不育起点温度低（21 ℃以下）；在短光照条件下（12 h左右）不育起点温度高（26 ℃以上）的两用核不育系，这种不育系既制种安全，又易繁殖，是最理想的两用核不育系。但这种不育系往往只能从发育感光性强的品种中获得，且只适应于华南稻区配制两系杂交组合，因感光性强的亲本所配杂种生育期容易超亲，不能在我国华南以外的地区种植。二是选育在长光照条件下（13.5 h以上）不育起点温度低（22 ℃以下）；在短光照条件下（12 h左右）不育起点温度较低（23 ℃左右）的两用核不育系，这种不育系既制种安全，又能配制出生育期不超长的杂交组合。但需在短光照条件下进行冷水串灌才易繁种成功。C815S、3MS138就是这种类型的典型代表。光敏核不育系的发育感光性与育性光敏性一直是困扰育种家的难点和热点。有的学者认为发育感光性和不育性的光敏特性由2个独立的遗传系统决定[11-12]，也有的认为发育感光性强弱不是育性光敏性强弱的决定因子，育性光敏性强弱受遗传背景多因子综合作用[13-14]。本研究采用的供试材料都表现出不同程度的光敏性，而且感光性强的3MS148表现的光敏性强，培矮64S感光性弱，其育性光敏性也弱，而C815S的感光性介于两者之间，其育性光敏性也介于两者之间。这一结果似乎表明育性光敏性与发育感光性存在某种内在联系，虽然这一现象的遗传机制还不清楚，但可以把这一重要研究结果运用到不育的选育、鉴定及两系法杂交稻种子生产的实践中去。

4　结论

　　水稻两用核不育系的育性光温效应作用模式对光温敏核不育系的选育、制种和繁殖具有很好的实际指导意义。通过对 C815S 及其同源株系光温育性特性的研究，提出了有别于前人研究成果的光温双因子连动效应——"U 模型"假设，这一假设能量化每一光长条件下与之对应的育性转换温度，并根据这一假设提出了选育实用两用核不育系的育种策略。

────────────── R e f e r e n c e s ──────────────

参考文献

［1］Yuan SC，Zhang ZG，Xu CZ.Studies on the critical stage of fertility change induced by light and its phase developmental inHPGMR.*Acta Agron Sin*，1988，14（1）：7-13（in Chinese with English abstract）

［2］Zhang ZG，Zeng HL，Yuan SC，Zhang DP，Wang BX，Li YZ.Studies on the response to photoperiod and temperature of photoperiod sensitive genic male sterile rice（PGMR）. *J Huazhong Agric Univ*，1992，11（1）：1 6（in Chinese with English abstract）

［3］Yao KM，ChuCS，Yang YX，Sun RL. A preliminary study of the fertility change mechanism of the photoperiod（temperature period）sensitive genic male sterile rice（PSGM R）.*Acta Agron Sin*，1995，21（2）：187 297（in Chinese with English abstract）

［4］Xue GX，Shen YZ.Studies on the fertility conversion model of photoperiod sensitive genic male sterile（PGMS）rice in response to photoperiod and temperature conditions. *Acta Agron Sin*，1995，21（2）：198 203（in Chinese with English abstract）

［5］Cheng SH，Sun ZX，Si HM. Uniformity of transformation of photo-thermoperiod sensitive genic male sterile rice lines in controlled photo and thermoperiod condition. *Zhejiang J Agric Sci*，1993，5（3）：133-137（in Chinese with English abstract）

［6］Diao CQ. CropCultivation Science（special）. Beijing: China Agriculture Press，1994. p51（in Chinese）

［7］Lu XG，Yuan QH，Yao KM. Study of the Ecological Adaptabi lity of the Photo-Thermo-Sensitive Genic Male Sterile Lines in Rice. Beijing: Meteorologic Press，2001. pp 6 7（in Chinese）

［8］Ma G H. Effect of temperature and light on the ferti lity rate of two-line hybrids（F$_1$）based on different kinds of TGMS. *Acta Agron Sin*，1999，25（6）：742-746（in Chinese with English abstract）

［9］Cheng SH，Sun ZX，Si HM，Zhuo LS. Classification of fertility response to photoperiod and temperature in dual-purpose genic male sterile lines（*Oryza sativa L.*）. *Sci Agric Sin*，1996，29（4）：11-16（in Chinese with English abstract）

［10］Chen LY. The Principi les and Techniques of Two-Line Hybrid rice. Shanghai: Shanghai Sci &Tech Press，2004.pp 106 109（in Chinese）

［11］Zhang ZG，Yuan SC，Xu CZ. The inf luence of photoperiod on the fertility changes of Hubei photoperiodsensitive genic male-sterile rice（HPGM R）. *Chin J Rice Sci*，1987，1（3）：137-143（in Chinese with English abstract）

340

［12］Yuan SC, Zhang ZG, Xu C-Z. Studies on the critical stage of fertility change induced by light and its phase development inHPGMR. *Acta Agron Sin*, 1988, 14（1）: 7−13（in Chinese with English abstract）

［13］Zeng HL, Zhang DP, Zhang ZG, Zhang ZY, Yang J. Photothermol Responses of Development and male sterility in male fertile alterant plants from the offspring of w6154SNongken 58. *Acta Agron Sin*, 1997, 23（6）: 693−698（in Chinese with English abstract）

［14］Zeng HL, Zhang ZG, Zhang DP, Yang J, Zhang ZY. Studies on the expression characterization of thermo-sensitive male sterility（TGMS）of Annong S-1 in the strong-photoperiodic-sensitive genetic background.*Chin High Technol Lett*, 1996, 6（5）: 12−15（in Chinese with English abstract）

作者：何强　陈立云＊　邓华凤　唐文邦　肖应辉　袁隆平＊

注：本文发表于《作物学报》2007 年第 33 卷第 2 期。

超级杂交水稻"种三产四"丰产工程

　　粮食安全始终是关系到国计民生的头等大事。针对我国人增地减的严峻形势和近期超级杂交稻取得的重大进展，作者于 2006 年提出超级杂交稻"种三产四"丰产工程的设想和建议，希望通过这一工程的实施，大幅度提高现有水稻的单产和总产，提高农民种粮的经济效益，确保国家粮食安全。在此，特别建议，应把"种三产四"丰产工程作为重点项目，纳入"国家粮食丰产科技工程"，加大力度实施，以期 5 年内实现在全国种 6 000 万亩（等于 400 万 hm²）第二期超级杂交稻，产出 8 000 万亩（约 533.3 万 hm²）粮食的奋斗目标（图 1）。

图 1　湖南省湘阴县白泥胡"种三产四"丰产工程示范现场

　　水稻超高产研究始于 20 世纪 80 年代初期。日本曾率先于 1981 年启动水稻超高产育种，计划 15 年内将水稻的产量提高 50%，即由当时单产 6 150~7 800 kg/hm² 提高到 9 300~12 300 kg/hm²。1989 年国际水稻研究所提出超级稻育种计划：到 2000 年要把水稻的产量潜力提高 20%~25%，即由单产 9 900 kg/hm² 提高到 12 000~12 450 kg/hm²。我国农业部跟踪国际前沿，于 1996 年制定了"中国超级稻育种计划"，分两个阶段实施：1996—2000 年为第一期，其中一季稻的产量指标是连续两年大面积示范产量 10 500 kg/hm²；2001—2005 年为第二期，产量指标是 12 000 kg/hm²。

迄今，以利用水稻亚种间杂种优势为主的超级杂交稻研究已获得成功，育成具有强大杂种优势的亚种间超级杂交稻品种，并分别于 2000 年实现农业部超级稻第一期大面积示范产量 10 500 kg/m²、2004 年（提前一年）实现第二期产量 12 000 kg/hm² 的目标。这些结果表明，我国在超级稻育种方面处于世界领先水平。

超级杂交稻的研究与开发，有助于促进农业结构调整，提高稻作产出与投入比。近几年在大面积生产上应用的第一期超级杂交中稻的平均产量为 8 250 kg/hm² 左右，年推广面积约 133.3 万 hm²，主要品种是两优培九，该组合已成为我国继"汕优 63"之后推广面积最大的水稻品种。第二期产量 12 000 kg/hm² 超级杂交稻已于 2006 年开始推广，种植面积达 10.67 万 hm²，其产量又上了一个新台阶，千亩（约 66.7 hm²）和万亩（约 666.7 hm²）示范片平均单产达 9 750～10 500 kg/hm²，在大面积生产上比第一期超级稻高 750 kg/hm² 以上。如浙江金华市的千亩（约 66.7hm²）示范片平均产量为 9 855 kg/hm²，湖南溆浦县 800 hm² 产量超过 10 500 kg/hm²，自然条件较差的贵州黔东南自治州 0.67 万 hm² 平均产量在 9 000 kg/hm² 以上。这些高产实例表明，"种三产四"在技术上已经成熟（图 2）。

图 2　袁隆平超级杂交稻"种三产四"示范点（安徽省芜湖县六郎镇）

所谓"种三产四"就是运用超级杂交稻的技术成果，为争用三亩地（等于 0.20 万 hm²，下同）产出现有四亩地（约 0.27 万 hm²，下同）的粮食，节余 1/4 也就是等于增加 1/4 的粮食耕地，即种三亩超级杂交稻，产出现有四亩地的粮食总产。建议到 2011 年全国推广 6 000 万亩（等于 400 万 hm²），产出现有 8 000 万亩（约 533.3 万 hm²）的粮食，就等于增加了 2 000 万亩（约 133.3 万 hm²）粮食耕地，可多养活 3 000 多万人。

这一项目受到湖南省领导和市、县有关部门、有关人员的高度重视和积极响应。2007 年，湖南省率先在 20 个县启动实施，其中有双季早、晚稻，一季中稻和一季晚稻。2007 年 7 月 9 日和 10 日，湖南省农业厅组织专家对湘阴县（中产区）和醴陵市（高产区）千亩（约 66.7 hm²）示范片的双季超级早稻进行验收，前者平均产量 7 590 kg/hm²，后者 8 415 kg/hm²，均比示范基地前三年的平均单产高三分之一，达到了"种三产四"的指标。当前，超级杂交中稻的长势更加喜人，达标在望。

实施超级杂交稻"种三产四"丰产工程，不仅具有重大的现实意义，而且能推进我国超级稻研究向更深层次发展。

作者：袁隆平

注：本文发表于《作物研究》2007 年第 3 期。

实施超级杂交稻"种三产四"丰产工程的建议

粮食安全始终是关系到国计民生的头等大事。针对我国人增地减的严峻形势和当前超级杂交稻取得的重大进展，本人于 2006 年特提出"超级杂交稻'种三产四'丰产工程"（以下简称"种三产四"）的设想和建议，希望通过这一工程的实施，大幅度提高现有水稻的单产和总产，提高农民种粮的经济效益，确保国家粮食安全。

所谓"种三产四"，就是运用超级杂交稻的技术成果，用 3 hm^2 地产出现有 4 hm^2 地的粮食，节余 1/4 的面积也就是等于增加 1/4 的粮食耕地。建议用 5 年的时间，到 2011 年全国推广超级杂交稻 400 万 hm^2，产出现有 533 万 hm^2 的粮食，等于增加 133 万 hm^2 粮食耕地，可多养活 3 000 多万人。

近年来，我国的水稻平均单产为 6.3 t/hm^2 左右，其中杂交稻 7.05 t/hm^2；日本的水稻平均单产为 6.6 t/hm^2。我国的超级稻育种计划，分 3 个阶段进行：第 1 期（1996—2000 年）的产量指标是大面积示范单产 10.5 t/hm^2，已于 2000 年实现；第 2 期（2001—2005 年）是单产 12.0 t/hm^2，已提前一年于 2004 年达标；第 3 期（2006—2015 年）的指标是单产 13.5 t/hm^2，目前正在攻关中。

近几年在大面积生产上应用（约 133 万 hm^2）的第 1 期超级杂交中稻的平均单产为 8.25 t/hm^2 左右，2006 年第 2 期超级杂交中稻开始在生产上推广，其产量又上了一个新台阶。如浙江金华市的"千亩示范片"平均单产为 9.86 t/hm^2，湖南溆浦县 800 hm^2 单产过了 10.5 t/hm^2，自然条件较差的贵州黔东南自治州 6 667 hm^2 平均单产在 9.0 t/hm^2 以上。这些高产实例表明，"种三产四"在技术上已经成熟。因此我向湖南省领导建议实施该项目，得到他们的高度重视和支持，采纳了我的建议，并于 2007 年率先在 20 个县启动实施，其中有双季早、晚稻，一季中稻和一季晚稻。2007 年 7 月 9 日和 10 日，湖南省农业厅组织专家对湘阴县（中产区）和醴陵市（高产区）"千亩示范片"的双季超级早稻进行验收，前者平均单产 7.59 t/hm²，后者 8.42 t/hm²，均比示范基地前三年的平均单产高三分之一，达到了"种三产四"的指标。这个结果，使我对实现该工程更加充满了信心。

　　现在我特别地建议，应把"种三产四"作为重点项目，纳入国家"粮食丰产科技工程"，加大力度实施，以期在 5 年内实现全国种植 400 万 hm² 超级杂交稻产出现有 533 万 hm² 粮食的奋斗目标。

作者：袁隆平

注：本文发表于《杂交水稻》2007 年第 22 卷第 4 期。

Relationship Between Grain Yield and Leaf Photosynthetic Rate in Super Hybrid Rice

【Abstract】In order to explore the relationship between grain yield and photosynthesis, the yield composition and leaf photosynthetic rate in some super hybrid rices and ordinary hybrid rice Shanyou 63 as control were measured in 2000−2005. The results were as follows. (1) The yield levels of the four super hybrid rices, Pei'ai 64S/E32, P88S/0293, Jin23A/611 and GD−1S/RB207, were significantly higher, being 108%−120% of Shanyou 63. (2) These super hybrid rices had a better plant type with more erect upper layer leaves and bigger panicles or more spikelets per panicle, being 125%−177% of Shanyou 63. (3) Net photosynthetic rates of these super hybrid rices were significantly higher in the second leaf but not necessarily in the first leaf or flag leaf than those of Shanyou 63. (4) The removal of half flag leaf led to a decline in the seed-setting rate, while the removal of half panicle induced its increase in GD−1S/RB207. Hence, higher yield in these super hybrid rices can be atributed to their bigger panicles, better plant type and higher light use efficiency of their canopies. Raising the photosynthetic capacity of each leaf, especially flag leaf, is the key to overcome the photosynthate-source restriction on grain yield and to make a new breakthrough of yield potential in future breeding of super hybrid rice.

【Key words】erect leaf; grain yield; net photosyntheic rate; plant type; super hybrid rice; yield composition

It was estimated that rice production must increase substantially, by 50% by 2050, to meet the demand of the rising population of Asia (Mitchell and Sheehy 2000). Rice is the most important grain crop in China. Increasing rice yield per unit area by development of science and technology is an inevitable choice to resolve food problem in China with more people and smaller fields. The improvement through breeding plays a key role in increasing rice yield per unit area (Qing and Wang, 2001).

To meet the food requirement of all the Chinese people in the 21st century, the Ministry of Agriculture of the People's Republic of China formally set up the super rice research program in 1996. Yuan (1997)

suggested a morphological model of super high-yielding rice and some strategies using intersubspecific heterosis, favorable genes from wild rice and IRRI's new plant type rice. In 1998 the super hybrid rice-breeding program presided by Yuan got the support of the Premier Fund. The phase aims of grain yield in Chinese super-rice breeding research program were 10.5 t/ha, 12.0 t/ha, and 13.5 t/ha in a rather large area more than 100 mu (6.7 ha), respectively in 2000, 2005, and 2010.

Now the first and second aims have been achieved. Owing to unremitting efforts of rice scientists some hybrid rices with high yield potential were developed. For example, the hybrid rice Pei'ai 63S/E32 planted at four locations (total area 2.35 ha) of Jiangsu and Hunan Provinces in 1998 reached a yield level higher than 12 t/ha, and it created a record yield of 17.1 t/ha in an experimental plot (0.07 ha) of Yongsheng County, Yunnan Province in 1999 (Yuan, 2001). Another hybrid rice P88S/0293 also achieved grain yields higher than 12 t/ha at five locations in 2003 and eight locations in 2004 (Yuan et al., 2004). There is a bright prospect.

However, what is the main reason for the high yield in these hybrid rices? What is the relationship between the levels of grain yield and of leaf photosynthetic rate? How should people further raise the yield of hybrid rice by breeding new hybrid rice? To answer these questions has undoubtedly great significance in guiding to improving rice breeding and achieving the third phase target (13.5 t/ha) of the super rice research program. In 2000 - 2005, therefore, a cooperative research of comparing the new hybrid rices with ordinary hybrid rice Shanyou 63 was made in the experimental fields of National Hybrid Rice Research and Development Center in Changsha (112° 9′ E, 28° 2′ N), Hunan Province to answer these questions.

1　Materials and Methods

1.1　Plant materials

Super hybrid rices used in this study were Pei'ai 64S/E32 (2000, 2001, 2005), Jin 23A/611 (2002), P88S/0293 (2002, 2003, 2004), GD-1S/RB207 (2003 and 2004), and Y58S/9311 (2005). The hybrid rice Shanyou 63 was used as control in all years of 2000 - 2005.

Twenty-to-twenty-five-days-old seedlings were transplanted from seedling bed to experimental plots. The area of each plot was 33.3 m^2, and the planting density was 16.7 cm×26.7 cm. Each plot had 750 hills with two seedlings per hill. In the experimental field there were three, four, and two plots in 2000-2001, 2002-2004, and 2005, respectively.

1.2　Fertilizer and water managements

The fertilization of rice plants in the experimental field was performed in the ways of typical agronomic management for high-yield rice. Pig manure (15 t/ha) was used as base fertilizer, while calcium super-phosphate (750 kg/ha) and compound fertilizer containing N, P_2O_5 and K_2O (15% each) (300 kg/ha) were applied as top dressing. Top-dressing was done three times 5-7 d after transplanting, 15 d before heading, and at the full heading time. The dosages for both the first and second times were urea (NH_2CONH_2) 150 kg/ha and KCl 150 kg/ha, while that for the third time was urea 37.5-75 kg/ha, dependent on plant growth situation. The levels of fertilization in the whole crop season were N 225 kg/ha, P_2O_5 135 kg/ha, and K_2O 225 kg/ha.

Water management of the experimental field was conducted by regulating the depth of water layer above the soil surface. It was kept at about 5 cm during the days from transplanting to turning green, 2—3 cm at the tillering stage, 0 cm in the middle of the growth season, 3—5 cm in the period from booting to full heading, and intermittent irrigation was carried out after flowering.

1.3 Growth observation

During rice growth and development the growth periods and tillering dynamics were recorded, and before grain maturation the effective panicle number per unit ground area was examined on the basis of five points per plot and each point included ten hills.

1.4 Leaf area measurement

In the period from booting to full heading the leaf areas of all leaves except the yellow ones of two plants in a hill of each plot were measured using a portable leaf area analyzer (CID, USA) and the leaf area index was calculated using the data of leaf area measurements.

1.5 Grain yield examination

The number of spikelets per panicle, seed-setting rate, and 1000-grain weight were examined on plants of five hills per plot. Grain yields per hectare was calculated from the weights of grains harvested from each plot, and harvest index was calculated on the basis of grain and shoot dry weights of plants of five hills per plot after maturity.

1.6 Leaf photosynthetic gas exchange measurement

During grain filling net photosynthetic rates of rice leaves were measured at a photosynthetic photon flux density (PPFD) of 1200 μmol $\cdot$ m $\cdot^{-2} \cdot$ s^{-1} and a CO_2 concentration of 350 μmol/mol using a portable gas analysis system LI—6400 (LI—COR Inc., Lincoln, Nebraska, USA). Photosynthetic measurements were made during 10:00—12:00 AM and 14:00 —16:00 PM on clear days. CO_2 concentration of the air used in the measurement was controlled with a LI—COR CO_2 injection system, and the measuring light was supplied by a LI—COR LED light source. The air temperature of leaf chamber was kept at about 30 ℃.

1.7 Removal of half flag leaf or half panicle

The removal of half flag leaf or half panicle was done on twenty plants of the super hybrid rice GD—1S/RB207 at the early filling stage.

2 Results

2.1 Some characteristics of plant type and canopy in the super hybrid rices

The super hybrid rices, Pei'ai 64S/E32, P88S/0293, Jin23A/611, GD—1S/RB207 and Y58S/9311, had some common characteristics of plant type and canopy. They had thicker canopies and lower panicle positions. Their uppermost three leaves were long, erect, narrow, concave, and thick (Fig.1). The long leaves had large leaf area and the erect leaves could intercept solar irradiation from both sides and did not shade each other. The narrow and concave leaves were easy to keep themselves erect and form a better light environment within the canopy (Fig.2). The thick leaves had frequently higher photosynthetic capacity and trended towards late senescence. Their panicles were below the flag leaves and bigger than those of Shanyou 63. Also, their stems were shorter and sturdier, more resisting

Fig.1　Difference in morphology of the flag leaves and canopies between super hybrid rices and ordinary hybnid rice Shanyou 63' at the flling stage

A and B : Changsha，2000. C and D : Yunnan，1999 and Changsha，2003，respectively.

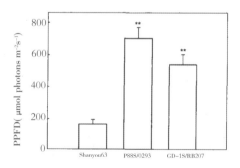

Fig.2 Photosynthetic photon flux density（PPFD）within the canopy of super hybrid rices and ordinary hybrid rice Shanyou 63（2003）

PPFD was measured with the digital luxmeter ZDS-10（Xuelian Instrument Factory，Jiading，Shanghai，China）placed on the base of the leaf below flag leaf，namely the second leaf，at noon on a clear day. The height of cach column is the mean±SE expressed by vertical bar of 4 plots（6 spots in each plot）. The significant levels of difference between super hybrid rice and Shanyou 63 are shown by asterisks for $P<0.01$.

lodging than Shanyou 63. For example, after the experimental field suffered storm at the beginning of September (the metaphase of grain filling) in 2000 all plants of Pei'ai 64S/E32 still stood erectly, while those of Shanyou 63 were all lodged (Fig.3). Such features of these super hybrid rices led to good light and aeration conditions within their canopies, benefiting canopy photosynthesis.

Fig.3 Difference in resistance to lodging of super hybrid rice Pei'ai 64S/E32 and ordinary hybrid rice Shanyou 63
All plants of Pei'ai 64S/E32 still stood erectly , while those of Shanyou 63 were all fallen after a storm at the beginning of September in 2000.

2.2 Yield composition in the super hybrid rices

From Table 1 it can be seen that grain yields of the four super hybrid rices were 8%−20% higher than that of Shanyou 63, with differences significant at the $P<0.05$ level, though their effective panicle numbers per hectare were not higher. The thousand-grain weights of all these super hybrid rices except GD−1S/RB207 were significantly lower than that of Shanyou 63. The differences in the full grain percentages between these super hybrid rices and Shanyou 63 were not consistent, for example, the full grain percentage of GD−1S/RB207 was lower but the full grain percentage of Pei'ai 64S/E32 was higher than that of Shanyou 63. However, spikelets per panicle of these super hybrid rices were significantly higher, being 125%−177% of Shanyou 63.

Table 1　The yield compositions in super hybrid rices and ordinary hybrid rice Shanyou 63 as control
(2001—2004)

	No. of panicle /million ha	No. of spikelets per panicle	Seed-setting Rate/%	Grain weight / (g/1000)	Grain yield / (t/ha)	HI	LAI
2001							
Shanyou 63	265.50±2.25	132.10±7.20	75.60±1.20	28.15±0.05	9.75±0.00		8.19
Pei'ai 64S/ E32	223.95± 1.05**	234.20± 12.00*	82.65± 1.75	22.70± 0.10**	11.03± 0.23*		8.04

Continued

	No. of panicle /million ha	No. of spikelets per panicle	Seed-setting Rate/%	Grain weight / (g/1000)	Grain yield / (t/ha)	HI	LAI
2002							
Shanyou 63	261.00± 1.35	134.75± 1.96	84.60± 0.77	27.33± 0.12	8.99± 0.27	0.50± 0.01	9.60± 0.57
P88S/0293	263.55± 4.35	169.08± 2.57**	81.58±1.68	25.75± 0.19**	10.75± 0.2**	0.47± 0.02	8.38± 0.75
Jin23A/611	265.20± 6.00	185.27± 4.70**	67.93± 2.60**	23.08± 0.21**	10.61± 0.36*	0.53± 0.01*	8.87± 0.38
2003							
Shanyou 63	290.40± 5.40	100.18± 2.50	69.33± 2.22	27.98± 0.23	9.03± 0.35		8.54± 0.24
P88S/0293	266.85± 2.40**	145.88± 1.93**	71.16± 0.55	25.60± 0.14**	9.81± 0.25		7.57± 0.34
GD−1S/ RB207	231.30± 3.90**	165.70± 2.43**	68.48± 0.81	29.55± 0.13**	10.76± 0.17**		7.19± 0.29*
2004							
Shanyou 63	287.70± 2.85	122.48± 1.03	77.48 ± 1.90	27.63± 0.19	8.95± 0.17	0.47± 0.01	6.24± 0.31
P88S/0293	277.50± 2.40*	158.90± 1.56**	82.28± 0.69	25.05± 0.22**	10.64± 0.10**	0.51± 0.01**	
GD−1S/ RB207	243.75± 4.50**	161.60± 2.39**	72.03± 2.10	30.10± 0.24**	10.81± 0.19**	0.54± 0.01**	6.73± 0.46

Note: The significant levels of difference between super hybrid rice and Shanyou 63 are shown by asterisks * for $P<0.05$ and ** for $P<0.01$. Each value in the table is the mean±SE of 2 plots (2001) or 4 plots (2002, 2003, and 2004). HI: Harvest index; LAI: Leaf area index.

2.3　Leaf photosynthesis in the super hybrid rices

The super hybrid rices had a superiority in photosynthesis, as reported by Zhang et al. (1996), Wang et al. (2000, 2002), and Zhai et al. (2002). In comparison with those of Shanyou 63 the flag leaves of some super hybrid rices i.e., Pei′ai 64S/E32 and Y58S/9311 had remarkably higher net photosynthetic rates (P_n), while the second leaves of all the super hybrid rices examined had significantly higher P_n (Fig.4). These results indicate that the superiority in photosynthesis is more frequently observed in the second leaves but not in the flag leaves of these super hybrid rices. Noteworthily, the flag leaves of Pei′ai 64S/E32 and Y58S/9311 had significantly higher values of stomatal conductance (g_s) and intercellular CO_2 concentration (C_i) than Shanyou 63 (Table 2).

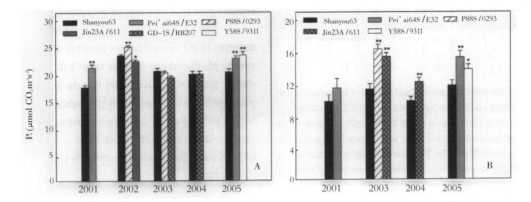

Fig.4 Net photosynthetic rate (P_n) in flag leaf (A) and leaf below flag leaf (the second leaf) (B) of super hybrid rices and ordinary hybrid rice Shanyou 63 (2001—2005)

Each value in the figure is the mean±SE expressed by vertical bar of 10 leaves. The significant levels of difference between super hybrid rice and Shanyou 63 are shown by asterisks * for $P<0.05$ and ** for $P<0.01$.

Table 2　Net photosynthetic rate (P_n) , stomatal conductance (g_s) and intercellular CO_2 concentration (C_i) in flag leaves of super hybrid rices Pei′ai 64S/E32 and Y58S/9311 as well as ordinary hybrid rice Shanyou 63

Date		Shanyou 63			Pei′ai 64S/E32 or Y58S/9311		
		P_n (μmol $CO_2m^{-2}s^{-1}$)	g_s (mol $H_2Om^{-2}s^{-1}$)	C_i (μmol CO_2/mol)	P_n (μmol $CO_2m^{-2}s^{-1}$)	g_s (mol H_2O $m^{-2}s^{-1}$)	C_i (μmol CO_2/mol)
Aug.28, 2001	AM	19.02±0.40	0.46±0.03	237.46±3.10	21.50±0.42**	0.56±0.03*	252.33±2.71**
	PM	17.25±0.41	0.36±0.02	228.88±3.12	20.79±0.61**	0.55±0.02**	253.76±1.66*
Sep.1, 2001	AM	18.37±0.31	0.31±0.01	211.86±3.51	22.06±0.65**	0.55±0.03**	251.47±2.77**
	PM	17.94±0.31	0.29±0.01	208.86±3.98	22.54±0.75**	0.53±0.03**	247.32±3.28**
Aug.28, 2005	PM	20.06±0.57	1.28±0.08	290.82±1.57	22.88±0.64**	2.37±0.17**	299.38±0.78**

Note: The significant levels of difference between Pei′ai 64S/E32 or Y58S/9311 and Shanyou 63 are shown by asterisks * for $P<0.05$ and ** for $P<0.01$. Each value in the table is the mean±SE of 10 leaves. AM: In the morning; PM: In the afternoon. Pei′ai 64S/E32 and Y58S/9311 were measured in 2001 and 2005, respectively.

2.4　Effects of removal of half flag leaf or half panicle on the seed-setting rate

To explore the source-sink relation in photosynthate supply-demand an experiment was performed in which the source-sink relation was modified by removal of part of flag leaf or panicle in GD−1S/

RB207. The results showed that the removal of half flag leaf led to a decline in the seed-setting rate, but the difference between treatment and control was not significant. And the removal of half panicle resulted in a significantly higher the seed-setting rate (Fig.5).

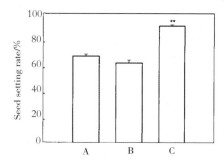

Fig.5　The effect of removal of half flag leaf or half panicle on the seed-setting rate in super hybrid rice 'GD-1S/RB207' (2004) (A): Control;(B): Removal of half flag leaf;(C): Removal of half panicle. Each value in this figure is the mean±SE expressed by vertical bar of 20 panicles. The removal of half leaf or half panicle was made at the early filling stage. The significant levels of difference are shown by asterisks **for $P < 0.01$.

3　Discussion

Our study has shown that the four super hybrid rices examined have significantly higher grain yields than Shanyou 63, being 108%—120% of it. This is mainly due to the significantly larger number of grains per panicle, being 125%—177% of Shanyou 63 (Table 1). The results are similar to those of super high-yield hybrid rice Xieyou 9308 with bigger panicles than control (Zhai et al., 2002).

Photosynthetic productivity is the most important determinant of crop yield, while its magnitude depends mainly on photosynthetic area, rate, and duration of leaves. In the aspects of photosynthetic area and photosynthetic duration, there is no significant difference between these super hybrid rices and Shanyou 63. These super hybrid rices have similar maximal leaf area indexes (Table 1), growth durations, grain-filling periods and maturity time to those of Shanyou 63, but significantly higher net photosynthetic rates than Shanyou 63. This superiority of photosynthesis is more common in the second leaves than in the first or flag leaves (Fig.4). The photosynthetic superiority of the second leaves may be due to that the erect character of flag leaves in these super hybrid rices (Fig. 1) which improves the light penetration and aeration of middle and low layer leaves within canopy (Fig.2, Liu et al., 2005), leading to higher photosynthetic capacity of these leaves. The higher photosynthetic capacity in the second leaves possibly makes higher canopy photosynthesis of these super hybrid rices than that of Shanyou 63. It appears that the higher canopy photosynthesis is an important physiological basis for higher yield in these super hybrid rices. The higher photosynthetic rate in the second leaf is likely to be a reliable index for breeding super high-yielding hybrid rices in future.

Among all leaves of a super hybrid rice plant the flag leaf has the biggest area, the highest photosynthetic capacity (Zhang et al., 2005), and the best illumination and aeration conditions, thus making the greatest contribution to grain filling. It has been reported that the daily yield per plant is positively correlated with flag leaf area, but negatively correlated with flag leaf angle (Liu et al., 2005). Nevertheless, compared with the Shanyou 63, the superiority in photosynthetic rate of the flag leaves is not universal in these super hybrid rices. For example, net photosynthetic rate in the flag leaf of GD-1S/RB207 is not higher than that of Shanyou 63, and the rate of the flag leaves of Jin 23A/611 sometimes is even significantly lower than that of Shanyou 63 (Fig. 4A). These facts mean that the contribution of flag leaf photosynthesis to the higher grain yield is very limited in these super hybrid rices. It is reasonable to predict that grain yield must increase substantially if the photosynthetic capacity of the flag leaves is raised remarkably.

In the first Green Revolution, the grain yield of some crops was raised greatly by the improved plant types with higher canopy light utilization efficiency and larger harvest index. Now, the high-yield breeds of rice, wheat and some other crops have almost optimal canopy architecture, and the room of plant type improvement is quite limited. Hence, increasing single-leaf photosyn-thesis could be the only way to substantially increase yield potential (Peng, 2000; Horton, 2000). Improving photo-synthetic efficiency of crops has become a central object of the new or second Green Revolution (Xu and Shen, 2002). In future a great effort increasing the photo-synthetic capacity in leaf, especially the flag leaf, should be made for breeding new super rice.

Current situation of the photosynthate source-sink relation in these super hybrid rices can also illustrate the necessity of increasing leaf photosynthetic capacity in breeding of super hybrid rice in the future. The super hybrid rices mentioned above have some defects such as lower seed-setting rate and lower grain weight though they had the advantage of larger panicle. Of course, there were some exceptions. For example, the seed-setting rate of Pei′ai 64S/E32 and the grain weight of GD-1S/RB207 were higher than those of Shanyou 63 (Table 1). In comparison with Shanyou 63 the number of spikelets per panicle is 25%−77% higher, while the grain yield is only 8%−20% higher in these super hybrid rices (Table 1), indicating that their potential of being higher yield due to larger panicles is not in full play, and implying that production ability of leaves as the main photosynthate source cannot meet the need of panicles as the main photosynthate sink, as shown by the higher percentages of unfertilized spikelets and unfilled grains in some super hybrid rices (Table 1). The experimental results that the removal of half of the flag leaf led to a decline in the filled spikelet percentage, while the removal of half panicle resulted in a significant increase in the seed-setting rate (Fig.5) support this deduction. Therefore, the problem of source ability inadequacy must be resolved if one wants to give full play to the potential of increasing yield by big panicle of super hybrid rice. There are three possible ways raising the photosynthate supply ability of the source: increasing leaf area, prolonging leaf photo synthetic duration or life, and enhancing leaf photosynthetic capacity. The maximal leaf area index of these hopeful super hybrid rices have reached 8−9 now (Table 1), leaving no room for further increase. So, increasing leaf photosynthetic capacity and avoiding premature leaf senescence will be the only two possible ways to overcome the photosynthate-source limitation and to achieve a new breakthrough in grain yield level of super hybrid rice.

It should be emphatically pointed out that the higher net photosynthetic rates of super rices Pei′ai 64S/E32 and Y58S/9311 than that of Shanyou 63 (Fig.4), we observed, are associated with the higher stomatal conductance and intercellular CO_2 concentration (Table 2). This fact means that the higher leaf photosynthetic rate of the super rices is due to the bigger stomatal conductance and whereby more CO_2 supply rather than an increase in photosynthetic activity of the mesophyll cells. Therefore, on the basis of keeping the higher stomatal conductance, efforts enbancing photosynthetic capacity, especially carboxylation capacity of the mesophyll cells, should be made to overcome the photosynthate-source restriction of super hybrid rice. The amounts of ribulose-1, 5-bisphosphate carboxylase/oxygenase (Rubisco) have been shown to limit maximal photosynthetic rate in rice leaves (Makino et al., 1985). Furthermore, a study by Zhang et al. (2005) has shown that the carboxylation capacity of Rubisco is the determinant of the photosynthetic capacity in rice leaves. Surprisingly, it was reported that the lower P_n was accompanied by a higher Rubisco content in rice variety japonica (new plant type, NPT) compared with variety indica (IR72) (Murchie et al., 2002). Nevertheless, this does not necessarily mean that Rubisco is excessive or its amount is not related to photosynthetic capacity. Under saturating light and ambient CO_2 concentration P_n is probably higher in NPT than IR72 if all of Rubisco are activated. So to increase the photosynthetic capacity in rice leaves, the amount of Rubisco per unit leaf area should be increased. On the other hand, the carboxylation efficiency of Rubisco should also be improved, because Rubisco is the key enzyme in photosynthetic carbon assimilation and its low efficiency is the biggest barrier of improving leaf photosynthetic capacity. Some scientists have tried to transform Rubisco through genetic engineering to enhance crop photosynthesis (Mann, 1999). In addition, creating C_4 rice by genetic engineering may be the best approach for increasing rice production in the next half-century (Sage, 2000). And the key enzyme in photosynthetic C_4 pathway, phosphoenolpyruvate carboxylase (PEPC), has been increased many times in rice leaves by genetic engineering (Ku et al., 1999). However, there has been no formal report suggesting that both leaf photosynthetic rate and grain yield are higher significantly in the transgenic rice. It appears that increasing photosynthesis is unlikely to be brought about by altering the expression of a single photosynthetic gene (Horton, 2000).

In conclusion, the higher yield in the present super hybrid rices can be attributed mainly to their larger panicles, better plant type and higher light use efficiency of their canopies. Although the yields are significantly higher than that of normal hybrid rice Shanyou 63, there is a source restriction on grain yield in some super hybrid rices. So enhancing leaf photosynthetic efficiency is an urgent affair. By summarizing breeding experience over 40 years Yuan (1997) pointed out that the key for breeding super high-yielding rice is to get a huge source of photosynthate on the basis of present large sink. Hereafter, in breeding new hybrid rice, photosynthetic efficiency of single-leaf, mainly the uppermost two leaves of the plant, should be enhanced studiously by using traditional and modern methods including heterosis and gene engineering under presupposition of keeping good plant type characters such as short stalk and big panicle. This is the most important way to further raise super rice yield potential, and may also be the most important objective for breeding high-yielding varieties.

Acknowledgments

JIANG Hua, YUAN Lin, CAI Shi-Qing, ZHANG Hai-Bo, LIAO Yi and XIAO Yuan-Zhen partially took part in the work of photosynthetic measurements in the fields. We thank Academician SHEN Yun-Kang for critically reading the manuscript and giving useful comments.

356

References

Horton P (2000). Prospects for crop improvement through the genetic manipulation of photosynthesis: morphological and biochemical aspects of light capture. J Exp Bot 51: 475-485

Ku MSB, Agarie S, Nomura M, Fukayama H, Tsuchida H, Ono K, Hirose S, Toki S, Miyao M, Matsuoka M (1999). High-level expression of maize phosphoenolpyruvate carboxylase in transgenic rice plants. Nat Biotechnol 17: 76-80

LiuJ-F, Yuan L P, Deng Q-Y, Chen L-Y, Cai Y-D (2005). A study. on characteristics of photosynthesis in super high-yielding hybrid rice. Sci Agr Sin 38: 258-264 (in Chinese)

Makino A, Mae T, Ohira K (1985). Photosynthesis and ribulose-1, 5-bisphosphate carboxylase/oxygenase in rice leaves from emergence through senescence. Quantitative analysis by carboxylation/oxygenation and regeneration of ribulose-1, 5-bisphosphate. Planta 166: 414-420

Mann CC (1999). Genetic engineers aim to soup up crop photosynthesis. Science 283: 314-316

Mitchell PL, Sheehy JE (2000). Performance of a potential C_4 rice: overview from quantum yield to grain yield. In: Sheehy JE, Mitchell PL, Hardy B (eds). Redesigning Rice Photosynthesis to Increase Yield. Amsterdam, the Netherlands: Elsevier 145-163

Murchie EH, Hubbart S, Chen Y, Peng S, Horton P (2002). Acclimation of rice photosynthesis to irradiance under field conditions. Plant Physiol 130: 1999-2010

Peng S (2000). Single-leaf and canopy photosynthesis of rice. In: Sheehy JE, Mitchell PL, Hardy B (eds). Redesigning Rice Photosynthesis to Increased Yield.

Amsterdam, the Netherlands: Elsevier 213-228

Qing X-G, Wang X-H (2001). View of super-paddy's research: background and progress. Res Agr Modern 22: 99-102 (in Chinese)

Sage RF (2000). C_3 versus C_4 photosynthesis in rice: ecophysiological perspectives. In: Sheehy JE, Mitchell PL, Hardy B (eds). Redesigning Rice Photosynthesis to Increased Yield. Amsterdam, the Netherlands: Elsevier 13-35

Wang Q, Zhang Q-D, Jiang G-M, Lu C-M, Kuang T-Y, Wu S, Li C-Q, Jiao D-M (2000). Photosynthetic characteristics of two super high-yield hybrid rice. Acta Bot Sin 42 (12): 1285-1288

Wang Q, Zhang Q-D, Lu C-M, Kuang T-Y, Li C-Q (2002). Pigments content, net photosynthesis rate and water use efficiency of two super high-yield rice hybrid at different developmental stages. Acta Phytoecol Sin 26 (6): 647-651 (in Chinese)

Xu D-Q, Shen Y-K (2002). Photosynthetic efficiency and crop yield. In: Pessarakli M (ed). Handbook of Plant and Crop Physiology (New Ed). New York: Marcel Dekker, Inc 821-834

Yuan L-P (1997). Hybrid rice breeding for super high yield. Hybrid Rice 12: 1-6 (in Chinese)

Yuan L-P (2001). Breeding of super hybrid rice. In: Peng S, Hardy B (eds). Rice Research fr Food Security and Poverty Alleviation. Los Baños, Philippines: International Rice Research Institute 143-149

Yuan L-P, Deng Q-Y, Liao C-M (2004). Status of commercialization and technologies on hybrid rice. In: the High Technique Industry Department of Chinese Development and Reform Comitee, the Chinese Society of Biological Engineering (eds). The Report on Chinese Biological Technique Industry Development. Beijing: Chemical Industry Press, the Publishing Center of Modern Biological Technique and Medicine 119–130 (in Chinese)

Zhai H-Q, Cao S-Q, Wan J-M, Zhang R-X, Lu W, Li-L-B, Kuang T-Y, Min S-K, Zhu D-F, Cheng S-H (2002). Relationship between leaf photosynthetic function at grain filling stage and yield in super high-yielding hybrid rice (Oryza sativa L.). Sci Chin Ser C45 (6): 79–88

Zhang Q-D, Lu C-M, Zhang Q-F, Zhang S-P (1996). The comparative studies of photosynthetic characteristics among hybrid rice and their parents of some hybrid combinations. Acta Biophys Sin 12 (3): 511–516 (in Chinese)

Zhang D-Y, Wang X-H, Chen Y, Xu D-Q (2005). The Determinant of the photosynthetic capacity in rice leaves. Photosynthetica 43: 273–276

作者: Chen Yue　Yuan Long-Ping[*]　Wang Xue-Hua　Zhang Dao-Yun

Chen Juan　Deng Qi-Yun　Zhao Bing-Ran　Xu Da-Quan[*]

注: 本文发表于《植物生理与分子生物学学报》2007 年第 33 卷第 3 期。

Enhancing Disease Resistances of Super Hybrid Rice With Four Antifungal Genes

【Abstract】A plant expression vector harboring four antifungal genes was delivered into the embryogenic calli of '9311', an indica restorer line of Super Hybrid Rice, via modified biolistic particle bombardment. Southern blot analysis indicated that in the regenerated hygromycin-resistant plants, all the four antifungal genes, including $RCH10$, $RAC22$, β-Glu and B-RIP, were integrated into the genome of '9311', co-transmitted altogether with the marker gene hpt in a Mendelian pattern. Some transgenic R_1 and R_2 progenies, with all transgenes displaying a normal expression level in the Northern blot analysis, showed high resistance to $Magnaporthe\ grisea$ when tested in the typical blast nurseries located in Yanxi and Sanya respectively. Furthermore, transgenic F_1 plants, resulting from a cross of R_2 homozygous lines with high resistance to rice blast with the non-transgenic male sterile line Peiai 64S, showed not only high resistance to $M.\ grisea$ but also enhanced resistance to rice false smut (a disease caused by $Ustilaginoidea\ virens$) and rice kernel smut (another disease caused by $Tilletia\ barclayana$).

【Key words】Super Hybrid Rice; multi-gene transformation; rice blast; rice false smut; rice kernel smut; anti-fungi; disease resistance

In recent years, China has made remarkable achievements and new breakthroughs in breeding new types of hybrid rice. The development of a series of elite hybrid combinations with a two-line approach has attracted global attention, putting China on the cutting edge of the world's Super Hybrid Rice Plan (SHRP). Among these most promising two-line hybrid varieties, Peiai 64S/9311 (also known as Liang-you-pei 9 or LYP9) and P88S/0293 ('0293' is a non-arista line derived from '9311', which could be an alternative of '9311' in practice) have reached a super high-yield of 10 500 kg and 12 000 kg per hectare respectively, fulfilling the first-stage and second-stage goal set by the 'China Super Hybrid Rice Project'. Subsequent to the success in three-line hybrid rice breeding, this progress was reputed as a new version of the 'Great Green Myth' by the international rice research community[1-3].

Although the Super Hybrid Rice had such great yield potential, and some varieties had been cultivated in large scale in south China, their quality and yield were often compromised by the infestation of plant pathogens and insect pests. As an example, rice false smut (caused by *Ustilaginoidea virens*), rice kernel smut (caused by *Tilletia barclayana*), rice sheath blight (caused by *Rhizoctonia solani*) and rice panicle blast (caused by *Magnaporthe grisea*) were the most prevalent diseases in fields planted with LYP9[4-6].

As a novel and powerful tool, genetic modification of crops provides a viable strategy to enhance plant resistances to insects and diseases. Compared with the conventional methods, this technology can endow crops with more comprehensive, more specific and more durative tolerance with wider spectrum, which results in a more environmentally friendly protection and more convenience in use. In plant genetic engineering for antifungal breeding, it was well documented that β-1, 3-glucanases, chitinases and ribosome inactivating proteins (RIPs) had a synergistic protective interaction and could coordinately inhibit the growth of fungi[7-10].

In this research, we applied the Multi-gene Transformation Strategy (MTS) to enhance disease resistance of the elite indica restorer line '9311' of the Super Hybrid Rice. Four antifungal genes (*RCH10*, *RAC22*, β-*Glu* and *B-RIP*) carried on a plant expression vector[11] were co-introduced into the genome of '9311' by the biolistic method. A batch of transgenic plants and progenies with improved resistance to the main fungal diseases of rice were selected by molecular detection and field tests. By hybridizing with other suitable male sterile lines, we are expecting to obtain novel types of genetically modified Super Hybrid Rice with high, multiple resistances to different diseases.

1　Materials and methods

1.1　Materials

Mature seeds of indica (*Oryza sativa* L.subsp. indica) rice variety '9311', a restorer line of Super Hybrid Rice, were kindly provided by Prof. Zhu Yingguo from Wuhan University. Vectors with a single antifungal gene including pAAG89, pRC24/B-RIP, pARAC2 and pARBC6 were gifts from Prof. Xu Yao of the Salk Institute for Biological Studies, Vanderbilt University, harboring the alfalfa glucanase gene (β-*Glu*), barley ribosome-inactivating protein gene (*B-RIP*), rice acidic chitinase gene (*RAC22*) and the rice basic chitinase gene (*RCH10*), respectively[8-11]. *Agrobacterium tumefaciens* binary vector pCAMBIA1300[11] was provided by Dr. Richard Jefferson of CAMBIA Center, Australia. Plant multi-gene expression vector pRAS1300 was constructed by our colleagues, Li Ming et al.[11] (Figure 1), harboring four antifungal genes (*RCH10*, *RAC22*, *B-RIP*, β-*Glu*) and the hygromycin phosphotransferase marker gene (*hpt*). *B-RIP* gene was driven by a rice inducible promoter, pRC24, *hpt* by CaMV35S, and the other 3 genes were all under the control of rice actin1 promoter (Act1).

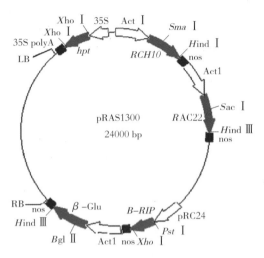

Figure 1 Plasmid map of the plant multigenic expression vector , pRAS1300[11]. LB : T-DNA left border ; RB : T-DNA right border ; 35S : cauliflower mosaic virus (CaMV) 35S promoter ; 35S polyA : terminator of CaMV35S ; Act1 : promoter of rice actin1 gene ; pRC24 : promoter of ricechitinase gene RC24 ; nos : promoter of the nopaline synthase gene ; hpt : hygromycin phosphotransferase gene ;(RCH10 : rice basic chitinase gene ; RAC22 : rice acidic chitinase gene ; B−RIP : barley ribosome-inactivating protein gene ; β −Glu : alfalfa glucanase gene.

1.2　Generation of rice transformants

Bio-Rad Biolistic® PDS1000/He Particle Delivery System was used to transfer plasmid pRAS1300 into the '9311' embryogenic calli. The rice tissue culture process and the method for biolistic transformation was as described in ref. [12] with some modifications: Using NB medium as the basic culture medium for '9311', 5 mg/L ABA and 2 mg/L 2, 4-D were added into the calli-induction medium as the additive phytohormones; bombardments were conducted using helium as a carrier gas at 1 100 psi (1 psi=6.895 kPa) in a vacuum of 25-27 inHg (1 inHg=3.386 kPa); per 3 mg gold particles were coated with 100 μg purified pRAS1300 plasmid DNA. After the DNA-coated microcarriers were fully suspended with plenty of ethanol, pipet them onto the center of each macrocarrier till the gold particles reached 0.5 mg/macrocarriers. Each sample was bombarded at a distance of 6 cm only once. The selection medium, pre-regeneration medium, regeneration medium and rooting medium should contain 20-25 mg/L, 30 mg/L, 30-40 mg/L and 50 mg/L Hygromicin B (Roche) respectively. Plantlets with vigorous roots of about 6-8 cm in height were suitable to sow in pots and grow in the greenhouse.

1.3　Southern blot analysis of the transgenic plants

DNA samples were extracted from the fresh rice leaf tissue using the CTAB method [13]. Probes were labeled using TAKARA Random Primer DNA Labeling Kit Ver.2, and [α -³²P] dCTP was purchased from Yahui Biomedical Engineering Corp of Beijing. Labeled Hind Ⅲ /Sma Ⅰ fragment (about 0.7 kb) from plasmid pARBC6 was used to probe RCH10 gene; pARAC2/Hind Ⅲ +Sac Ⅰ fragment (about 0.8 kb) for RAC22 gene; pRC24/B-RIP/Pst Ⅰ +Xho Ⅰ fragment (about 0.9 kb) for B-RIP gene; pAAG89/Hind Ⅲ +Bgl Ⅱ fragment (about 1.1 kb) for β -Glu gene; and a pCAMBIA1300/Xho Ⅰ fragment of 1.1 kb for *hpt* gene[9, 10]. Details of Southern blot analysis were as described in ref. [14]. Plasmid harboring the corresponding target gene was used as a positive control, while the untransformed '9311' plant as a negative control.

1.4　Northern blot analysis of the transgenic plants

Rice total RNA was extracted from the fresh leaves according to the user's manual of the Trizol™ reagent (Invitrogen) .As for detection of the *B-RIP* gene which was driven by the pRC24 inducible promoter, a wound treatment was needed beforehand: making small cuts along both edges of the leaf blade at an interval of 1.5 cm without damaging the middle vein. Each cut might be around 1.5 mm in depth. After 20 h, this wounded leaf was excised and total RNA was extracted from it.Northern blot analysis was carried out as in a previous report[10], and the preparation and labeling of probes were as described in sec. 1.3.

1.5　Disease resistance field tests of the rice transformants

Blast resistance appraisal of the transgenic '9311' R_1 plants from 16 lines was performed in Yanxi Blast Nursery, Hunan Province during September to October, 2002. Yanxi Nursery is located in a moist valley with little sunshine throughout a year, where the hereditary constitution of the local *Magnaporthe grisea* strains is in great diversity. As it was reported, the dominant isolates there were ZB and ZC groups, and the extremely virulent ZA group showed up frequently as well, thus making this region a very typical mountain ecosystem for blast[15].

Evaluation of '9311' R_2 plants' resistance to blast was conducted in Sanya Nursery, Hainan Province from March to April in 2004. Sanya Nursery was in a subtropical indica rice region and was a blast assessment base of the National Hybrid Rice R&D Center. All the R_2 plants to be tested derived from 191 R_1 plants. For each code of R_1 plants, 30-50 seeds were sown. The scales of the panicle blast (PB) and the leaf blast (Bl) were evaluated and scored by every single plant[15, 16].

Later on, with the evidence from the molecular detection and the field test, 12 homozygous lines with high resistance to rice blast were chosen accordingly to mate with non-transgenic Peiai 64S, giving rise to the transgenic F_1 generation.

In October 2004, field tolerance of the transgenic F_1 hybrid rice was tested against rice blast, rice false smut and kernel smut by the National Hybrid Rice R&D Center in Changsha, Hunan Province. Non-transgenic '9311' and LYP9 were used as a control when biological resistance characterization was carried out. For every hybrid combination, 30-50 seeds were sown. The incidence of fungal disease was recorded by each individual plant according to the Standard Evaluation System for Rice (SES) of IRRI[16].

As a supplement to the natural induction of rice disease, artificial induction and inoculation with infected leaves were necessary as well. Besides, all seeds for test were grown without any treatments beforehand. No disinfectants were applied to the tested plants throughout their whole life cycle. Excessive nitrogenous fertilizer was used to favor the blast. Sensitive rice varieties (Xiang'aizao 7) were planted closely around the objective rice sample in 2 lines per region. Irrigation or density of plants was the same to common field management.

2 Results and analysis

2.1 Generation of the multi-gene transformed rice plants

Plasmid pRAS1300 was introduced into the '9311' embryogenic calli with a modified biolistic process as described in sec. 1.2. In total, 62 hygromycin-resistant regenerated plants of 18 independent transgenic lines were generated in this research. Most of the transgenic plants were normal in morphology, except the sterile D1 line and the less-reproductive E2 line.

2.2 Southern blot analysis of the transgenic R_0 plants of '9311'

Monogenic vectors including pCAMBIA1300, pAAG89, pRC24/B-RIP, pARAC2 and pARBC6 were digested with proper restriction enzymes (see sec. 1.3) and subjected to electrophoresis to isolate the corresponding target gene fragments. Then these fragments were labeled as gene-specific probes for the Southern blot analysis of all the transgenic R_0 plants. The results showed that among the 62 hygromycin-resistant R_0 plants, the introduction frequency of *hpt* gene was 100%. Moreover, the co-integration frequencies of *hpt* gene and the other 4 antifungal genes lined up in pRAS1300 vector were also 100%, which meant that all these foreign genes were successfully co-transferred into the '9311' regenerated plants.

The Southern blot analysis of '9311' R_0 plants is shown in Figure 2. It was clear that all the undigested DNA samples of the transgenic plants gave a hybridization signal in the high-molecular-weight (MW) region, while the corresponding digested ones gave bands in the low-MW region. Therefore, it was concluded that the foreign transgenes had integrated into the genome of the recipient plants.

As shown in Figure 2 (a) and (b), when the nylon membrane was probed by labeled *RAC22* and (*RCH10* fragments in turn, transgenic samples upon digestion (lanes 6, 8, 10, 12) gave hybridization bands in consistence with both the positive control in lane 1 and the negative control in lane 4. Similar phenomenon occurred in Figure 2 (c) when *RAC22* probe was used to detect another membrane, indicating that the two foreign chitinase genes both had been integrated into the genome of '9311' after the pRAS1300 plasmid was bombarded into the target calli. Besides, all the '9311' plants, including the transformants and the untransgenic samples, carried endogenous sequences homologous to the foreign (*RCH10* and *RAC22* genes so that there were background hybridization bands in both pictures. Since there were many different rice chitinase genes with high homology to each other, and the (*RCH10* and *RAC22* genes shared high similarity[9, 10, 17], more than one bands appeared in the Southern blot assay of the chitinase genes.

In Figure 2 (d) for a *B-RIP* gene assay, transgenic samples upon digestion (lanes 5, 7, 9, 11) gave hybridization bands not only at about 0.9 kb just as the plasmid control did, but also in the other regions with different sizes, whereas the negative control showed no hybridization signal at all. Similar circumstance could be found in Figure 2 (e) for a β-*Glu* gene assay, where digested DNA sample in lane 6 gave several extra bands when compared with positive control in lane 1 and lane 2. It was speculated that bands not corresponding to the plasmid control might be due to multicopy integration and rearrangements of the target genes.

According to the Southern blot analysis of the *hpt*, *B-RIP* and β-*Glu* genes, all the transgenic lines showed a multicopy integration and rearrangement mode at least in one of these 3 genes. They

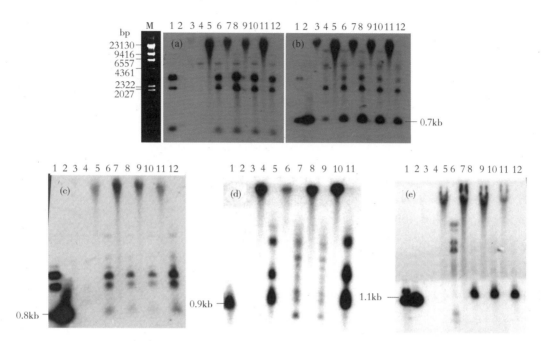

Figure 2　Southern blot analysis in part of the transgenic '9311' R₀ plants. M : λ /Hind Ⅲ. In (a) and (b) , the same nylon membrane was hybridized with different probes in turn. (a) A labeled Hind Ⅲ +Sac Ⅰ fragment from plasmid pARAC2 (0.8 kb) was used as the RAC22 gene probe. 1 , pRAS1 300/ Hind Ⅲ +Sma Ⅰ ; 2 , pARBC6/Hind Ⅲ +Sma Ⅰ ; 3 , '9311' non-transformed control (NT) ; 4 , '9311' NT/ Hind Ⅲ +Sma Ⅰ ; 5 , A4 ; 6 , A4/Hind Ⅲ +Sma Ⅰ ; 7 , A5 ; 8 , A5/ Hind Ⅲ +Sma Ⅰ ; 9 , A1 ; 10 , A1/ Hind Ⅲ +Sma Ⅰ ; 11 , A2 ; 12 , A2/ Hind Ⅲ +Sma Ⅰ . (b) A labeled pARBC6/ Hind Ⅲ +Sma Ⅰ fragment of about 0.7 kb was used as the RCH10 gene probe to hybridize with the same membrane as in (a). (c) A labeled pARAC2/ Hind Ⅲ +Sac Ⅰ fragment of 0.8 kb was used as the RAC22 geneprobe. 1 , pRAS1 300/ Hind Ⅲ +Sac Ⅰ ; 2 , pARAC6/ Hind Ⅲ +Sac Ⅰ ; 3 , '9311' NT ; 4 , '9311' NT / Hind Ⅲ +Sac Ⅰ ; 5 , A1 ; 6 , A1/ Hind Ⅲ +Sac Ⅰ ; 7 , A4 ; 8 , A4/ Hind Ⅲ +Sac Ⅰ ; 9 , A5 ; 10 , A5/ Hind Ⅲ +Sac Ⅰ ; 11 , A2 ; 12 , A2/ Hind Ⅲ +Sac Ⅰ . (d) A labeled pRC24-B-RIP/ Pst Ⅰ +Xho Ⅰ fragment of 0.9 kb was used as the B-RIP gene probe. 1 , pRAS1 300/ Pst Ⅰ +Xho Ⅰ ; 2 , '9311' NT ; 3 , '9311' NT / Pst Ⅰ +Xho Ⅰ ; 4 , B1 ; 5 , B1/ Pst Ⅰ +Xh o Ⅰ ; 6 , B3 ; 7 , B3/ Pst Ⅰ +Xho Ⅰ ; 8 , B3 (2) ; 9 , B3 (2) Pst Ⅰ +Xho Ⅰ ; 10 , E1 ; 11 , E1/ Pst Ⅰ +Xho Ⅰ . (e) A labeled pAAG89 / Hind Ⅲ +Bgl Ⅱ fragment of 1.1 kb was used as the β -Glu gene probe. 1 , pRAS1 300/ Hind Ⅲ +Bgl Ⅱ ; 2 , pAAG89/ Hind Ⅲ +Bgl Ⅱ ; 3 , '9311' NT ; 4 , '9311' NT/ Hind Ⅲ +Bgl Ⅱ ; 5 , A9 ; 6 , A9/ Hind Ⅲ + Bgl Ⅱ ; 7 , A1 ; 8 , A1/ Hind Ⅲ + Bgl Ⅱ ; 9 , A4 ; 10 , A4/ Hind Ⅲ + Bgl Ⅱ ; 11 , A3 ; 12 , A3/ Hind Ⅲ + Bgl Ⅱ .

gave more than one hybridization bands in these assays. It was concluded that transgenic plants obtained by the above-mentioned method had high frequency in multicopy integration and rearrangement of the transgenes.

2.3　Genetic and molecular analysis of the transgenic R1 plants of '9311'

Seeds set by the R₀ transgenic plants were sown and grown up as the R₁ plants. Total DNA was then prepared from the fresh leaves of the R₁ plants, and without digestion by any restriction enzymes,

Southern blots of these uncut DNA were performed using β-*Glu* and *B-RIP* genes as probes in turn (Figures are not shown). Among the tested R_1 population, β-*Glu* and *B-RIP* genes were always co-transmitted, suggesting that they were genetically linked. As shown in Table 1, segregation of these 2 genes in each of the 17 transgenic lines conforms to Mendal's 3 : 1 ratio, but in some of these lines (such as A3, A6, A7, A9 and E2), the possibility of a 15 : 1 segregation ratio could not be completely excluded. This result indicated that the multiple copies of the foreign genes might be inserted at the same or adjacent chromosome site, or integrated into 2 separate chromosomes.

2.4 Disease resistance tests of the transgenic '9311' R_1, R_2 plants in the blast nursery

Disease resistance field test of the '9311' R_1 transformants was carried out in Yanxi Blast Nursery, Hunan Province. According to the calculations, 45.1% of the 102 R_1 plants showed high resistance level to rice blast, though this percentage differed from lines to lines.

Seeds from the resistant R_1 plants were collected and the R_2 plants were grown in Sanya Blast Nursery, Hainan Province. Among the 69 R_1 plants tested, 19 gave R_2 offspring with no symptom of either panicle blast or leave blast. Another 23 R_1 plants had R_2 progenies showing high resistance to panicle blast and slight infection symptom of leave blast.

Seeds from 122 R_1 plants which had not yet been tested for blast tolerance were planted in Sanya as well. Among them, 54 produced offspring plants without any panicle infections, and 13 out of these 54 even gave rise to R_2 plants with no symptom of blast.

Table 1　Genetic analysis of the transgenic R_1 plants of '9311'

R_0 transgenic line	No. of Southern blot positive plant	No. of Southern blot negative plant	Chi-squared (X_i^2) of the expected 3 : 1 segregation ratio	X_i^2 of the expected 15 : 1 segregation ratio
A1	12	5	0.020	14.592
A2	14	5	0.018	12.214
A3	11	1	1.000	0.022
A4	6	4	0.533	18.027
A5	15	5	0.067	11.213
A6	15	3	0.296	2.904
A7	14	3	0.176	3.298
A8	14	5	0.018	12.214
A9	18	1	2.965	0.004
B1	14	7	0.397	25.146
B2	8	6	1.524	30.476
B3	12	6	0.296	21.393
B4	6	3	0.037	10.141
B5	15	4	0.018	6.488

Continued

R_0 transgenic line	No. of Southern blot positive plant	No.of Southern blot negative plant	Chi-squared (X_i^2) of the expected 3 : 1 segregation ratio	X_i^2 of the expected 15 : 1 segregation ratio
B6	12	7	0.860	29.056
E1	28	12	0.300	37.500
E2	4	1	0.000	1.080

Note：$X_{0.05}^2 = 3.841$. If $X_i^2 < X_{0.05}^2$, the expected segregation ratio was accepted.

In total, there were 96 R_1 plants out of the 191 samples tested in Sanya showing a potential to produce offspring with high resistance to panicle blast. Among them, there were 32 R_1 plants whose R_2 progenies had exhibited a persistently high resistance to blast throughout their whole growth periods.

In particular, it was worthy of notice that of all the transgenic plants tested for blast resistance, there were 4R_1 plants, whose R_2 progenies had a panicle blast incidence lower than 10%. Impressively enough, the most supereminent line was an R_1 plant coded as A2-10. Only 2.00% of its R_2 plants showed panicle blast lesions and 9.09% showed small necrotic spots on the leaves. As for A2-9, the corresponding percentage came to 9.88% and 11.11%,; and for A9-10, 9.90% and 10.00%, respectively.Except this small percentage, the progenies of these lines were immune from rice blast, while the untransgenic '9311' and some lines of the transformants revealed an incidence of either panicle blast or leaf blast of 100%, scale 7-9 (highly sensitive), without a single grain for harvest (Figure 3).

(a)　　　　(b)　　　　(c)

Figure 3　Disease resistance of the transgenic offspring of '9311' in the blast nurseries. O : Untransgenic '9311' (highly sensitive) ; T : transgenic '9311' (highly resistant). (a) '9311' transgenic R_1 plants in the Yanxi Blast Nursery, Hunan Province ;(b) '9311' transgenic R_2 plants in the Sanya Blast Nursery, Hainan Province ;(c) panicles of the '9311' transgenic R_2 plants in the Sanya Blast Nursery, Hainan Province.

2.5　Northern blot analysis of the '9311' transgenic plants

In order to detect the expression of the multiple transgenes at mRNA level, Northern blot was performed on some of the high resistance R_2 plants. 20 h before the RNA was extracted, the plants were wounded according to the method described in sec. 1.4. Then (*RCH10*, *β*-*Glu* and *B-RIP* gene probes were hybridized with the RNA on the same membrane in turn (Figure 4). Results indicated that most of the foreign genes could be expressed at high levels in these high resistance plants, but the 3 genes might not be expressed at a consistent level. Chitinase gene seemed to have a higher expression level in general, whereas *β*-*Glu* and *B-RIP* genes appeared to vary in its mRNA level. For example, all the 3 genes detected in A2-10 (lane 1) and A9-16 (lane 7) revealed a high expression level, whereas A2-4 (lane 3), A9-1 (lane 5), A9-5 (lane 6) showed a much lower level in the expression of *β*-*Glu* and *B-RIP* genes. Such differences in the mRNA expression levels between plant lines correlated with that in the resistance levels in the fields.

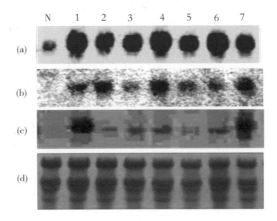

Figure 4   Northern blot analysis of the '9311' R_2 plants with high resistance to blast. N, Untransgenic '9311'; 1, A2-10; 2, A2-13; 3, A2-4; 4, A5-7; 5, A9-2; 6, A9-5; 7, A9-16. (a) Hybridized with an (*RCH10* gene probe; (b) hybridized with a *β*-*Glu* gene probe; (c) hybridized with a *B-RIP* gene probe; (d) RNA stained in the denaturing gel.

2.6　Disease resistance of the F_1 hybrid rice plants produced by the transgenic 9311/Peiai 64S

Southern blot was performed in the transgenic R_2 populations to screen for the homozygous lines. Each of the 12 presumptive pure lines with high resistance to blast was then crossed with the non-transgenic male sterile line Peiai 64S, and the F_1 hybrid rice plants were grown in an isolated field in Changsha, Hunan.

During the late-season in 2004, all the 12 transgenic F_1 lines showed a high resistance against rice blast (scale 0-2), whereas the untransgenic F_1 of 9311/Peiai 64S (LYP9) had an intermediate resistance or intermediate sensitive level in the fields (scale 4-5).

When 12 F_1 populations of the transgenic 9311/Peiai 64S combinations were tested for rice false smut, 10 of them appeared to have no infected symptom. The other two lines of the transgenic F_1 population (coded as 04CF069 and 04CF071) each had one panicle affected, thus Line 04CF069 gave an incidence of infected panicles of 1.2%, incidence of infected plants of 5.9%, while Line 04CF071 gave the corresponding incidence of 2.8% and 14.3%. As a control, untransgenic 9311/Peiai 64S hybrid rice showed a panicle infection incidence of about 7.7% [Figure 5 (a)].

From each F_1 population, 6-17 plants were randomly selected to calculate the number of filled grains and the amount of infected seeds with kernel smut. As listed in Table 2, transgenic F_1 population with a code of 04CF077exhibited no visible lesions [Figure 5 (b)], and the other lines of transgenic hybrid F_1 showed a much lower incidence of kernel smut, compared with the untransgenic ones.

Figure 5　Field resistance of the F_1 hybrids of the transgenic '9311' R_2 plants/Peiai 64S against rice false smut and kernel smut. O : Untransgenic control (F_1 of the non-transformed 9311/Peiai 64S) ; T : F_1 of the transgenic 9311/Peiai 64S. (a) Rice false smut ;(b) rice kernel smut.

Table 2　Incidence of rice kernel smut in F_1 hybrids of the transgenic '9311' R_2 plants/Peiai 64S

Line code of the hybrids	Average incidence (No. of seeds with kernels mut/total No. of filled grains×100%)	Average No. of infected seeds per panicle
04CF064	1.87	1.38
04CF065	2.23	1.75
04CF066	5.01	2.92
04CF067	0.81	1.00
04CF068	0.41	0.45
04CF069	5.60	5.29
04CF070	0.60	0.50
04CF071	1.14	1.57
04CF074	5.05	2.29
04CF076	7.10	3.57
04CF077	0.00	0.00

Continued

Line code of the hybrids	Average incidence (No. of seeds with kernels mut/total No. of filled grains×100%)	Average No. of infected seeds per panicle
04CF078	6.75	3.25
Untransgenic 9311/Peiai 64S	17.12	14.50

According to the above data, transgenic F_1 plants of 9311/Peiai 64S have showed a more comprehensive resistance to multiple fungal diseases in the fields, with significantly enhanced resistances to *M.grisea*, *Ustilaginoidea virens* and *Tilletia barclayana*, when compared with the untransformed LYP9 hybrid rice.

3 Discussion and conclusion

As most biological traits and physiological functions in an organism are manifested through the cooperative expression of multiple genes, the inherited characteristics may be changed and modified to a greater extent if Multi-gene Transformation Strategy (MTS) is adopted. It is foreseeable that MTS will certainly become the mainstream in the future research of gene engineering, either in basic theoretical research or practical application area[18, 19]. With regards to plant genetic manipulation for crop protection, multiple transgenes with different resistant mechanisms could broaden the resistance spectrum of transgenic crops, improve their resistance level, delay the development of tolerance of the pathogens and insects, and prolong the usefulness of the existing cultivars[18-22].Hence, applying multiple genes to simultaneously improve various kinds of biotic or abiotic stress resistance in rice and other crops has already become a new trend for developing genetically modified crops[8-10, 18, 19].

In this research, 4 antifungal genes from rice, barley and alfalfa were co-delivered into the genome of '9311'.Up to date, we have obtained a batch of transgenic plants and progenies with a significant increase in their resistance to rice blast. Subsequent research proved that these high resistance transgenic '9311' plants could serve as a paternal line in the breeding of LYP9, conferring simultaneous resistances against rice false smut, kernel smut and blast to their F_1 progenies. On the basis of the previous reports[7-11, 17, 20-24], we inferred that the possible synergistic mechanism of the 4 anti-fungal transgenes in the resistant rice plants might be as follows: when pathogens invaded the transgenic plants, they first encountered the constitutively expressed acidic chitinase RAC22 (driven by the Act1 promoter) in the intercellular space. Chitinase could lyse the cell wall of most fungi, and the lysate would serve as a signaling molecule to further induce the systemic defense mechanism against the intruders. Meanwhile, the invasion of pathogens might trigger the expression of *B-RIP* gene as it was under the control of an inducible promoter, pRC24. Once the remainder of the pathogens entered the cell, the β-1, 3-glucanase and the basic chitinase of RCH10 released from the vacuoles would align themselves to act on their substrates—the glucans and chitins, which have been interlaced in the fungal cell wall in a filamentous form.This joint action might result in a thorough disorganization or lysis of the cell wall, which would facilitate the entry of RIP into the cytoplasm as well. As a consequence, the protein synthesis and propagation of the fungi would eventually be completely inhibited.

Since the 4 antifungal genes and their promoters (rice Act1 promoter and rice pRC24 promoter) used in this research were all derived from the edible crops[8-11], including rice, barley and alfalfa, which had served as the staple food for human or livestock, and had been widely planted and spread over thousands of years, they were supposed to present less problems in the food and environment safety evaluation. On the other hand, when the action mechanisms of these genes were concerned, there were no active substrates for β-1, 3-glucanase and chitinase in the mammalian cell[24], and the B-RIP was a type-I RIP which was avirulent to intact tissues or cells[24]. Therefore, all these genes expressed in a genetically modified plant were theoretically safe to domestic animals and human beings, but deleterious enough to eliminate the fungal pathogens.

As an elite indica restorer line involved frequently in the breeding of Super Hybrid Rice, '9311' itself could be used as a high-quality and high-yielding conventional variety in agricultural practice as well. By now, multiple hybrid combinations such as Liangyoupeijiu (LYP9), Guangliangyou 6, Yueyou 938, P88S/9311, 58S/9311 and 38S/9311 have been popularized in country-wide indica rice-growing region, using '9311' and its derivatives as the restorer line[25]. For this reason, we are convinced that the transgenic '9311' plants attained in this research will provide a useful germplasm resource for the breeding of super high-yielding rice with high resistance to multiple fungal diseases. By integrating the conventional breeding methods into the succeeding cultivation, it is highly possible that new types of multigene-transgenic Super Hybrid Rice with good agricultural traits and comprehensive resistances will be widely recognized in the near future.

We thank Prof.Y.G., Zhu Prof.Y.Xu and Dr.R.Jefferson for their kind gifts of rice seeds and plasmids. We are also grateful to Dr.Qiuyun Liu and Dr.Zengfu Xu for helpful discussions in the preparation of this manuscript.

References

[1] Yuan L P, Xin Y Y. Light of hope: Super Hybrid Rice. World Agr (in Chinese), 2001, 10 (10): 46

[2] Yang C X, Xiang B H, Zhang Q M, et al. Cultural techniques of achieving a yield of 12.26 t/hm² in the demonstrative production of two-line Super Hybrid Rice under an area over 6.7 hectares. Hybrid Rice (in Chinese), 2003, 18 (2): 42-44

[3] Wang Q, Lu C M, Zhang Q D, et al. Characterization of photosynthesis, photoinhibition and the activities of C4 pathway enzymes in a superhigh-yield rice, Liangyoupeijiu, Sci China Ser C-Life Sci, 2002, 45 (5): 468-476

[4] Lu K Y, Wu H M, Lu B L. Identification of the elite medium rice variety Liangyoupeijiu and Yangdao 6. Hubei Agr Sci (in Chinese), 2001, (3): 9-12

[5] Dong X H, Tian X Z, Wang L, et al. The occurrence rule and its IPM technique of the diseases and pests of two-line hybrid medium indica rice-Liangyou Peijiu. J Anhui Agr Sci (in Chinese), 2001, 29 (3): 342-344

[6] Ma Z, He S L, Jin K M, et al. The reason of the occurrence of rice kernel smut in the hybrid seed production of Liangyou Peijiu and its prevention and cure. J Xinyang Agr College (in Chinese), 2001, 11 (2): 13-15

[7] Jach G, Gornhardt B, Mundy J, et al. Enhanced quantitative resistance against fungal disease by combinatorial expression of different barley antifungal proteins in transgenic tobacco. Plant J, 1995, 8 (1): 97-109

[8] Feng D R, Wei J W, Xu X P, et al. Introduction of multiple antifungal protein genes into rice and preliminary study on resistance to *Pyricularia oryzae* of transgenic rice. Acta Sci Nat Univ Sunyatseni (in Chinese), 1999, 38 (4): 62-66

[9] Feng D R, Xu X P, Fan Q, et al. Rice plants of multiple transgenes for resistance to rice blast and sheath blight diseases. Acta Agron Sin (in Chinese), 2001, 27 (3): 293-301

[10] Feng D R, Xu X P, Qiu G H, et al. Inheritance and expression of multiple disease and insect resistance genes in transgenic rice. Chin Sci Bull, 2001, 46: 101-106

[11] Li M, Qiu G H, Xu X P, et al. The construction of plant expression vector containing multi-antifungal genes. Acta Sci Nat Univ Sunyatseni (in Chinese), 1999, 38 (5): 67-71

[12] Xu X P, Wei J W, Fan Y L, et al. Fertile transgenic indica rice from microprojectile bombardment of embryogenic callus. Acta Genet Sin (in Chinese), 1999, 26 (3): 219-227

[13] Doyle J J, Doyle J L. Isolation of plant DNA from fresh tissue. Focus, 1990, 12: 13-15

[14] Zhu H C, Xu X P, Li B J. A simple, rapid method of Southern blot analysis. Acta Sci Nat Univ Sunyatseni (in Chinese), 2004, 43 (4): 128-130

[15] Xiao F H, Luo H R, He Z N, et al. Identification and application of the multiple durable resistance of the dryland rice. Hunan Agr Sci (in Chinese), 2001, (5): 39-41

[16] International Rice Research Institute. Standard Evaluation System for Rice (SES). Los Banos: IRRI, 2002, 14-18

[17] Zhu Q, Lamb C J. Isolation and characterization of a rice gene encoding a basic chitinase. Mol Gen Genet, 1991, 226: 289-296

[18] Li B J, Zhu H C. On the prospects of applying the Multi-gene Transformation Strategy to modify the inheritance of organisms I. The gene resources and technical platform for multi-gene transformation. Acta Sci Nat Univ Sunyatseni (in Chinese), 2004, 43 (6): 11-15

[19] Li B J, Zhu H C. On the prospects of applying the Multi-gene Transformation Strategy (MTS) to modify the inheritance of organisms II. General principles, possible problems and prospects of the MTS. Acta Sci Nat Univ Sunyatseni (in Chinese), 2005, 44 (4): 79-83

[20] McDowell J M, Woffenden B J. Plant disease resistance genes: recent insights and potential applications. Trends in biotechnol, 2003, 21 (4): 178-183

[21] Shah D M. Genetic engineering for fungal and bacterial diseases. Curr Opin Biotechnol, 1997, 8 (2): 208-214

[22] Punja Z K. Genetic engineering of plants to

enhance resistance to fungal pathogens-A review of progress and future prospects. Can J Plant Pathol, 2001, 23: 216-235

[23] Ming X T, Wang L J, An C C, et al. Resistance to rice blast (*Pyricularia oryzae*) caused by the expression of trichosanthin gene in transgenic rice plants transferred through agrobacterium method. Chin Sci Bull, 2000, 19: 1774-1778

[24] Shan L B, Jia X. Ribosome inactivating protein and its application in plant anti-fungal disease genetic engineering. Progr Biotechnol (in Chinese), 2000, 20 (6): 74-78

[25] Bai H S, Zhan C Y, Wang B H, et al. Medium indica rice variety Yangdao 6 and its application in hybrid rice breeding. Hybrid Rice (in Chinese), 2001, 16 (6): 13-15

作者: Zhu Huachen　　Xu Xinping
Xiao Guoying　　Yuan Longping[*]　　LI Baojian[*]

注: 本文发表于 *Science in China Serc: Life Sciences* 2007 年第 50 卷第 1 期。

超级杂交水稻育种研究的进展

【摘　要】本文回顾了我国超级杂交水稻育种研究取得的成绩和进展，总结提出了形态改良、提高杂种优势水平、借助分子技术等提高水稻产量的技术路线，并对超级稻的发展应用前景进行了展望。

【关键词】超级稻；研究进展；技术路线；应用前景

1　前言

目前中国人口有 13 亿，人均可耕地仅 934 m^2，预计 2030 年人口将增至 16 亿，人均可耕地会减少到 667 m^2 左右，面对人口增长压力和耕地减少的严峻形势，为在 21 世纪让所有中国人吃饱吃好，中国农业部 1996 年立项了超级水稻育种计划，其中超级杂交稻的产量指标见表 1。

表 1　中国超级杂交水稻产量指标　　　　（kg/ 亩）

阶段	杂交水稻			增长率 /%
	早季	晚季	单季	
1996 年水平	500	500	550	0
第一阶段 1996—2000	650	650	700	>20
第二阶段 2001—2005	750	750	800	>40

注：连续两年在两个示范点，每点面积 6.67 hm^2（100 亩）的平均单产。

2　超级稻育种研究取得的成绩与进展

通过形态改良及利用籼粳亚种间杂种优势，至 2000 年，已育成几个超级杂交稻先锋组合，并达到了第一阶段单季水稻产量指标。其中，以两优培九（P64S/9311）表现最好，2000 年有 20 多个示范点，每点 100 亩或 1 000 亩，其平均产量超过 700 kg/ 亩。该组合近几年的推广面积在 120 万～200 万 hm^2，平均每亩产量达 550 kg，而同期全国水稻的平均产量是 420 kg/ 亩，杂交稻为 470 kg/ 亩。

1999 年在云南永胜县的实验田（720 m²），苗头组合 P64S/E32 的产量达 1 139 kg/ 亩，创造了当时的水稻超高产纪录。

从 2001 年起开展了第二阶段的超级杂交水稻选育工作，并取得了突破，且提前 1 年实现了第二阶段超级杂交水稻的产量目标。

根据 2001 年的进展，2002 年在多个示范点（每点 100～120 亩）安排了一些有希望的新组合示范，其中最好的是 P88S/0293，2002 年在湖南龙山县平均产量达 820 kg/ 亩；2003 年该组合在海南省三亚市、澄迈县和湖南省汝城、隆回、中方、湘潭四县共 6 个百亩示范片，平均单产超过 800 kg/ 亩。2004 年在湖南、安徽和贵州有 2 个组合共 12 个点，单产在 800 kg/ 亩以上，其中湖南的汝城、隆回、中方三县的百亩片是连续两年达标，即提前 1 年实现了第二阶段超级杂交稻的产量指标。2005 年，在湖南溆浦县更有一个千亩（1 240 亩）示范片单产超过 800 kg/ 亩。同时，在选育超级杂交晚稻上也有重大进展，一新育成生长期短的三系杂交稻组合金 23A/Q611，作双季晚稻栽培，2003 年在浏阳市、2004 年在洪江市分别安排一个百亩示范片，其平均产量在 660 kg/ 亩左右，比对照 V 优 6 号增产近 30%，达到了双季晚稻第一阶段的产量指标。2006 年该组合在桂阳县的百亩示范片，平均单产高达 712 kg/ 亩。

3　技术路线

育种实践表明，迄今为止，通过育种提高作物产量，只有两条有效途径：一是形态改良，二是杂种优势利用。单纯的形态改良，潜力有限；杂种优势不与形态改良结合，效果必差。其它育种途径和技术，包括基因工程在内的高技术，最终都必须落实到优良的形态和强大的杂种优势上，否则，就不会对提高产量有贡献。但是，另一方面，育种要进一步向更高层次的发展，又必须依靠生物技术的进步。

3.1　形态改良

优良株型是高产的基础。自从 1968 年 Dr Donald 提出理想株型概念后，很多水稻育种家特别注意这一重要课题并设想了多种高产水稻模型。其中著名的是国际水稻研究所 Dr Khush 提出的 "新株型" 稻，其主要特征是：①大穗，每穗 250 粒；②分蘖少，每株 3～4 个有效分蘖；③短而壮实的秆。这种模型是否高产，还有待实践证明。

根据我们的研究，特别是受超高产组合 P64S/E32（其单产高达 1 139 kg/ 亩）的显著特征所启发，我们发现超高产品种有如下形态特征。

3.1.1　高冠层

上三叶叶片应长、直、窄、凹、厚。长而直的叶子不仅叶面积大而且能两面受光又互不遮荫，因此能更有效地利用光能；窄叶所占的空间面积小，能增加有效的叶面积指数；凹字形可使叶片坚挺不披；厚叶光合功能强且不易早衰。总之，具有这种形态特征的水稻品种，才能有最大的有效叶面积指数和光合功能，为超高产提供充足的光合产物即有机源。

3.1.2　矮穗层

成熟期稻穗顶部离地面仅 60~70 cm，这种结构由于重心下降，可使植株高度抗倒伏。抗倒是培育超高产水稻必备的特性。

3.1.3　中大穗

每穗谷重约 5 g，每平方米约 300 穗。理论上，其产量潜力为 1 000 kg/ 亩。

稻谷产量 = 生物学产量 × 收获指数。现行的矮秆品种，收获指数（HI）已很高（>0.5），进一步提高收获指数难度很高，因此，主要应依赖提高生物学产量才能进一步提高稻谷产量。

从形态学观点来看，提高植株高度是提高生物学产量有效而可行的方法。然而这种方法会引起倒伏。为解决这个问题，不少育种家正试图使茎秆更粗壮，但此举会导致收获指数下降，因此，很难达到超高产。上述由叶片组成的高叶冠层植株模型能同时将高生物学产量、高收获指数和高度抗倒伏三者较好地统一起来，从而能实现超高产。

3.2　提高杂种优势水平

水稻的杂种优势强弱有如下的总趋势：籼粳交 > 籼爪交 > 粳爪交 > 籼籼交 > 粳粳交。依据我们研究，籼粳杂交稻库大源足，其产量潜力比现行在生产上应用的品种间杂交稻可高 30% 以上。但是，要利用籼粳杂种优势的难度很大，最主要的是籼稻和粳稻为不同亚种，亲缘关系较远，二者之间存在不亲和性，致使籼粳杂种的受精结实不正常，一般结实率仅 30% 左右，因此，其实际产量不高。现举例 1986 年的一个试验例子，即足以说明（表 2）。

表 2　籼粳 F_1 杂种的产量潜力

组合	株高/cm	每穗颖花数 / 个	每株颖花数 / 个	株数/（株 / 亩）	结实率/%	实产/（kg/ 亩）
城特 232（粳）×26 窄早（籼）	120.0	269.4	1 779.4	25 000	54.0	550
威优 35（对照，籼 × 籼）	89.0	102.6	800.3	25 000	92.9	575
优势 /%	34.8	162.8	122.4		−41.9	−4.3

注：* 小区面积 6.67 m²，7 月 29 日插秧，株行距 13.32 cm×19.98 cm。

从表 2 可知，城特 232×26 窄早的每亩颖花数比对照多 122%，可达 4 400 万个以上，尽管结实率仅 54%，但产量几乎与品种间强优组合威优 35 相当。如果能将结实率提高到 80%，作双季晚稻栽培，其产量潜力每亩可超过 800 kg。

经过近 10 年的努力，我们在利用籼粳杂种优势育种上终于取得成功，育成一批结实率正常的具有超高产潜力的籼粳杂交稻组合。主要经验，一是利用广亲和基因以克服籼粳稻之间的不亲和性；二是用具有混合亲缘的中间型材料而不是用典型的籼、粳品种作亲本，以协调其他方面的矛盾。

3.3　借助分子技术

常规育种与生物技术相结合是今后作物育种的发展方向，这也是具有巨大潜力的超级杂交稻的选育途径。

3.3.1　利用野生稻中的有利基因

基于分子技术和田间试验，已鉴别出两个源于普通野生稻（*O. rufipogon*）的增产 QTL 位点，每个 QTL 位点具有比高产的对照杂交组合（V 优 64）增产 18% 的效应。通过分子标记辅助选择和田间选择，已育成一个带有一个上述 QTL 位点的优良的恢复系（Q611）。它的杂交种 J23A/Q611，作晚稻试验示范，比杂交稻对照增产极显著，示范田的平均单产达 660 kg/ 亩左右。

3.3.2　利用稗草的 DNA 创造水稻新资源

通过穗颈注射法将稗草的总 DNA 导入恢复系（R207），后代产生变异，从这些变异株中选育出新的优良、稳定的恢复系 RB207-1，它具有如下农艺特征：①穗大粒多；②粒重比原始 R207 大大提高。尤其是其杂交种 GDS/RB207-1，株型良好，杂种优势强，在海拔较高的山区（400～800 m）表现特别好，2005 年有 3 个试验点，小面积（2～3 亩）单产都在 900 kg/ 亩以上。经分子检测，RB207-1 含有稗草 DNA 的片断。

3.3.3　选育 C_4 型超级杂交稻

我们与香港中文大学合作，将源于玉米的 C_4 基因成功克隆并正在导入超级杂交稻的亲本。理论上讲，C_4 基因的光合效率比水稻的 C_3 基因高 30%，初步测定，有个别含 C_4 基因的亲本植株，其叶片的光合效率比对照高 10%～30%，但还不稳定，目前仍在继续选育中。

4　前景

现在，已有两个第二阶段超级杂交稻组合通过了省级品种委员会的审定，2006 年开始投入生产应用。大面积的生产显示，第二阶段超级杂交稻的增产效果很好，如浙江金华万亩以上的两优 293 平均单产为 657 kg/ 亩，湖南溆浦 800 hm^2（1.2 万亩）单产过 700 kg/ 亩，贵州黔东南自治州 0.67 万 hm^2（10 万亩）准两优 527 单产接近 700 kg/ 亩。基于这些成绩，我们特提出一个超级杂交稻"种三产四"的丰产工程计划，即种三亩超级杂交稻产出现有四亩地所生产的粮食。计划用 5 年时间在全国发展 400 万 hm^2，产 534 万 hm^2 的粮食。该工程既能保证我国的粮食安全，又能节约 134 万 hm^2 耕地，以发展其他经济效益更高的项目，为农民致富创造条件。由于这项计划具有十分重要的现实意义，因而受到了各方面的高度重视和支持。2007 年已在湖南省率先实施。

科技进步永无止境，水稻还有很大的产量潜力。根据以上各项研究进展，2006 年农业部已立项和启动了第三阶段超级杂交稻育种计划，即到 2015 年第三期单季超级杂交稻大面积示范的产量达到 900 kg/ 亩。

作者：袁隆平

注：本文发表于《中国稻米》2008 年第 1 期。

发展杂交水稻　保障粮食安全（A）

选用两个在遗传上有一定差异，同时优良性又有互补的水稻品种进行杂交，生产具有杂交优势的第一代杂种用于生产，这就是杂交水稻。杂交水稻的优势很明显，它穗大粒多，抗逆性强，产量高。我国用十年时间就培育出第一代杂交种子，现在全国每年杂交稻的种植面积有 2.4 亿亩，每年增产的粮食可多养活 7 000 多万人。

为了满足新世纪全球居民对粮食的需求，很多国家开展了超级杂交水稻的研究。日本在 1981 年启动一个水稻超高产的计划，要求在 15 年之内把水稻的年产量提高到每公顷 10~12.5 吨。这个指标非常高大，到现在日本还没有实现。我国为满足全国人民的粮食需要，也开展了杂交水稻研究。农业部规划，分两个阶段达到两大增产目标：第一阶段是 1996—2000 年，指标是每公顷产水稻 10 吨；第二阶段是 2001—2005 年，每公顷产水稻 12 吨。事实上，我国杂交水稻已连续两年超过了亩产 800 公斤的产量。对于湖南省来讲，到 2000 年，杂交水稻产量已达到了农业部制定的第一期指标，2000 年以后，开始推广应用杂交水稻新型品种。目前，杂交水稻新型品种已在全国推广种植 3 000 万亩左右。这就是我国自行研究生产的第一期超级水稻。

另外，在 20 世纪 90 年代末，我国的杂交水稻曾在小面积田里创造出了亩产 1 139 公斤的生产记录。这个记录现又已被打破了，最高亩产达 1 200 公斤了。这让世界各国很惊讶，中国杂交水稻有效地缓解了中国这个世界第一人口大国吃粮的压力。

目前，我国杂交水稻第三期研究已经开始，到 2010 年大面积的杂交水稻亩产将达到 900 公斤。这个研究已经有了一定的进展，我们有信心按时完成这个任务。根据超级稻研究的重大进展，我们提出了"种三产四"工程的建议，即：种三亩超级杂交稻，产量要有现行种植的四倍，湖南省去年就开始实施了超级水稻的工程，其中有 7 个县很成功。这说明此工程可行，增产潜力大。

在世界粮食供给矛盾凸显的形势下，中国杂交水稻正大步地走向世界。目前，在国际上已有 7~8 个国家大面积种植杂交水稻，其中面积最大的是印度、越南、孟加拉、菲律宾、美国等国家，2006 年美国杂交水稻产量占稻谷总产的 25%，今年估计要占美国水稻面积的 1/3。可见，杂交水稻在新世纪对保证世界粮食安全发挥了重大的作用，如果全世界 50% 的粮田种植杂交水稻，每年可增产 1.5 亿吨粮食，可多养活 4 亿人。

杂交水稻的产业化就是种子生产，但近年全国杂交水稻种子生产形势不好。2005—2006 年，杂交水稻种子总量可满足生产需要，但 2007 年锐减了很多，今年预计还要进一步的下降。这将影响我国杂交水稻的进一步推广种植，进而威胁到国家的粮食安全。分析杂交水稻种子产量上不去的原因，主要是现在农民种粮效率不高，收入低。因此，我要大声呼吁：国家和各级地方政府要出台更加优厚的惠农政策，采取更加有效的措施，调动广大农民种粮的积极性，以确保我们国家的粮食安全。

作者：袁隆平

注：本文发表于《农村工作通讯》2008 年第 18 期。

Development of Hybrid Rice for Food Security in the World

The current world population is over 6 billion and will reach 8 billion in 2030. Meanwhile, the annual loss of land to other use is 10 to 35 million hectares, with half of this lost land coming from cropland. Facing such severe situation of population growth pressure plus cropland reduction, it is obvious that the only way to solve food shortage problem is to greatly enhance the yield level of food crops per unit land area through advance of science and technology.

Rice is a main food crop. It feeds more than half of world population. It has been estimated that the world will have to produce 60% more rice by 2030 than what it produced in 1995. Therefore, to increase production of rice plays a very important role in food security and poverty alleviation. Theoretically, rice still has great yield potential to be tapped and there are many ways to raise rice yield, such as building of irrigation works, improvement of soil conditions, cultural techniques and breeding of high yielding varieties. Among them, it seems at present that the most effective and economic way available is to develop hybrid varieties based on the successful experience in China.

It has proved practically for many years that hybrid rice has more than 20% yield advantage over improved inbred varieties. In recent years, hybrid rice covers 57% or 16 million ha of the total rice area in China. The nationwide average yield of hybrid rice is 7.2 t/ha, about 1.4 t/ha higher than that of inbred varieties (5.8 t/ha). The yearly increased paddy in China due to growing hybrid rice can feed 70 million people each year. Therefore, hybrid rice has been playing a critical role in solving the food problem of China thus making China the largest food self-sufficient country.

China makes increasing progress in development of hybrid rice technology. Following the success of three-line hybrid rice in 1970s, two-line hybrid rice was successfully commercialized in 1995. The extension of two-line hybrid rice has been very fast in these years. The area of two-line hybrid rice was 2.6 million ha, about 18% of total hybrid rice area in 2002. The yield advantage of two-line hybrid rice is 5% – 10% higher than that of the existing three-line hybrid rice.

More encouragingly, good results have been achieved in developing super hybrid rice varieties since the initiation of the super rice research program in 1996. Several pioneer super hybrids have a yield advantage of around 20% over current three-line hybrids on commercial scale. In recent years the area planted to super hybrid rice is 2 million ha with an average yield of around 8.5 t/ha. In addition, a two-line super hybrid P64S/E32 and a three-line super hybrid II－32A/Ming 86 created a record yield of 17.1 t/ha and 17.95 t/ha, respectively. In the meantime, grain quality of super hybrid rice varieties is very good. Based on successful development of the first generation super hybrid rice (yield level at 10.5 t/ha), efforts were focused on breeding second generation super hybrid rice (yield target is 12 t/ha) and good results are obtained. For example, in 2003, the second super hybrid rice varieties yielded over 12 t/ha at five 6.7 ha locations in Hunan province. In 2004, twelve 6.7 ha or 67 ha locations in the southern provinces reached the yield level of 12 t/ha These hybrids were released for commercial production since 2006.The area under which was 200 000 ha and the average yield was over 9 t/ha. The above fact indicate that the super hybrid rice shows a very bright future. If super hybrid rice covers an annual area of 13 million ha in China and calculating by a yield increase of 2.25 t/ha, it is expected that the annual increased grains will reach 30 million tons, which means 75 million people more can be fed every year.

Hybrid rice has been proved to be a very effective approach to greatly increase yield not only in China, but also outside China. Vietnam and India have commercialized hybrid rice for years. In 2007 about 667 000 hectares were covered with rice hybrids in Vietnam. On average, the yield of rice hybrids is 6.3 t/ha while that of the inbred varieties is 4.5 t/ha. Because of planting hybrid rice on large-scale commercial production, Vietnam becomes the second largest rice export country. Besides, many other countries, such as the Philippines, Bangladesh, Indonesia, Pakistan, Ecuador, Guinea and the USA, have also achieved great progress in extending hybrid rice technology. A lot of experimental trials and large-scale demonstrations in farmers' field conducted in these countries have shown that hybrid rice can significantly outyield their local CK varieties. For example, in the Philippines, under technical assistance from FAO, IRRI and China National Hybrid Rice R&D Center, hybrid rice has been commercialized since 2002.Especially, a super hybrid rice variety called SL－8 has been developed by my assistant in the Philippines, it was planted to about 3 000 ha in 2003 and the average yield was 8.5 t/ha, more than doubled the country's average yield. In 2007, the area under rice hybrids was expanded rapidly to 233 000 ha in the Philippines. Based on this achievement, the Philippines government has made an ambitious plan. The target is to expand the number of hectares under hybrid rice cultivation to one million ha by 2010.

These facts clearly show that hybrid rice technology developed by China is also effective to greatly increase rice yield worldwide. It is estimated if 50% of conventional rice were replaced by hybrid rice, and calculated on 2 t/ha yield advantage of hybrid rice, the total rice production in the world could be increased by another 150 million tons of rice which can feed 400 million people each year.

Therefore, I firmly believe that hybrid rice, relying on scientific and technological advances and the efforts from all other aspects, including governments, private sector, NGO and particularly from FAO and IRRI, will have a very good prospect for commercial production and continue to play a key role in ensuring the future food security worldwide in the new century.

作者：Yuan Longping

注：本文发表于 2009 年 9 月 11—12 日在长沙召开的国家杂交水稻合作论坛。

长江流域超级杂交稻产量稳定性研究

【摘 要】以农业部认定的 6 个超级杂交稻组合、3 个省级主推超级杂交稻组合、2 个苗头超级杂交稻组合和普通高产杂交稻汕优 63（对照）为材料，研究了超级杂交稻在长江流域 7 个生态试验点的产量稳定性。结果表明，超级杂交稻准两优 527、红莲优 6 号、C 两优 87 和 Y 两优 1 号 4 个组合表现超高产且在各试验点产量稳定性较高，结实率高而稳定，在 80% 以上，表现出良好的生态适应性；其他组合则在特定的生态条件下表现较高的产量和结实率。对超级杂交稻的生态适应性及其分类进行了讨论。

【关键词】超级杂交稻；产量稳定性；结实率；生态适应性；长江流域

中国的粮食安全问题几度成为国内外关注的话题，解决中国粮食安全问题的唯一出路在于提高粮食作物单位面积产量。因此，中国在水稻矮化育种和杂种优势利用之后，立项了超级稻育种计划并成功选育了一批超级杂交稻组合。目前超级稻已进入大面积推广应用阶段，产生了巨大的经济效益和社会效益。到 2008 年止，在中国主推超级稻新品种中，超级杂交稻组合占 70% 以上，快速发展超级杂交稻是中国农业发展的需要。但是同一超级杂交稻组合在不同的生态和稻作条件下表现不一，在某一区域增产显著深受农民欢迎，在另一区域则表现较差，甚至减产；有的组合在低海拔高产区表现较好，而在海拔较高的山区表现较差；有的地方适宜于大穗型组合推广，而有的区域则适宜于多穗型组合种植。说明不同的生态稻作区需要不同类型的超级杂交稻组合。因此，开展长江流域超级杂交稻生态适应性研究，培育和推广产量潜力更高、稳产性更强、适应区域更广的广适型超级杂交稻新组合，对于进一步提高中国水稻的平均单产，确保国家粮食安全具有重要意义[1]。

1 材料与方法

1.1 供试材料

本研究选用以下超级杂交稻组合为试验材料：两优培九（超级杂交稻对照）、准两优527、Ⅱ优明86、Y两优1号、协优9308和国稻1号为农业部认定的超级杂交稻[2-3]，两优293、C两优87和红莲优6号为省级主推超级杂交稻，P88S/747和GD-1S/RB207为超级杂交稻苗头组合，汕优63为普通高产杂交稻对照。其中协优9308和国稻1号由中国水稻所供种，C两优87由湖南农业大学供种，Ⅱ优明86由福建省农科院供种，红莲优6号由武汉大学供种，其他均由国家杂交水稻工程技术研究中心供种。所有试验材料都为一季中籼品种。

1.2 试验方法

1.2.1 试验地点选择

选择了长江上、中、下游稻区共7个生态试验点，分别为江苏南京、安徽安庆、江西南昌、湖南长沙、湖南桂东、湖北武汉和重庆北碚。

1.2.2 田间试验设计

2006—2007年连续2年试验，各生态点根据当地一季中稻播种时间播种，按照当地种植习惯育秧，随机区组试验设计，3次重复，每小区300株，插单本（1粒谷秧），移栽规格19 cm×26 cm。按超高产栽培管理，每公顷施纯N 225 kg、P_2O_5 120 kg、K_2O 240 kg（具体施肥量根据当地施肥水平酌情调整），其中N、P以基肥为主，田间肥水管理、病虫害防治和栽培技术与超级杂交稻大田栽培相同。

1.2.3 考查性状

每小区取20株调查有效穗数，并取平均有效穗数的单株5株室内考种，考查单株平均结实率。成熟期每小区取中间50株单收单晒，称重测产。

1.3 数据分析

试验资料利用Excel 2003数据分析工具和DPSV3.01专业版数据处理系统进行分析。采用温振民等[4]提出的高稳系数法计算各组合产量和结实率的高稳系数（HSC）：

$$HSCi = [(Xi\text{-}Si) / (1.10 \times X_{CK})] \times 100\%$$

式中，$HSCi$为第i个参试组合性状的高稳系数，其值越大，表明该性状稳定性越好；Xi为第i个参试组合的多点性状平均值，Si为第i个参试组合性状变异的标准差，X_{CK}为对照汕优63性状平均值。

2　结果与分析

2.1　产量表现

对长江流域 7 个生态试验点 12 个组合在 2006—2007 年试验中的产量表现进行方差分析，结果表明，供试组合产量在不同年份、不同生态点间的差异及其互作均达到显著或极显著水平，表明 12 个组合在不同年份、不同生态区域间的产量表现差异较大，可进一步分析评价超级杂交稻的生态适应性及其适宜种植的稻作区域。

从供试超级杂交稻组合 2006—2007 年连续 2 年在长江流域 7 个不同生态试验点的产量表现可以看出，与对照汕优 63 相比，增产幅度达到 8% 以上（超级稻区试增产标准）的组合有准两优 527、红莲优 6 号、C 两优 87、Y 两优 1 号和 Ⅱ 优明 86，其增产幅度分别是 13.6%，12.1%，11.9%，11.2% 和 8.1%，均达到极显著水平；与超级杂交稻对照两优培九相比，这 5 个组合的增产幅度也达到了极显著水平（表 1）。从表 1 还可看出，超级杂交稻国稻 1 号产量与对照汕优 63 相当，协优 9308 比汕优 63 极显著减产。

表 1　供试超级杂交稻组合在长江流域的产量及其稳定性表现

组合	产量 /(t·hm^{-2})	标准差 /(t·hm^{-2})	变异系数 /%	比对照增产 /%		高稳系数 /%
				CK$_1$	CK$_2$	
准两优 527	9.99 aA	1.78	17.86	13.6	7.3	84.85
红莲优 6 号	9.86 abA	1.86	18.85	12.1	5.9	82.66
C 两优 87	9.84 bA	2.00	20.36	11.9	5.7	81.03
Y 两优 1 号	9.78 bAB	1.61	16.42	11.2	5.0	84.44
Ⅱ 优明 86	9.51 cB	1.55	16.25	8.1	2.1	82.27
两优 293	9.43 cdBC	1.39	14.78	7.2	1.3	83.12
两优培九（CK$_2$）	9.31 dC	1.36	14.56	5.9		82.19
P88S/747	9.31 dC	1.57	16.88	5.8	−0.1	79.95
GD-1S/ RB207	9.09 eD	1.34	14.76	3.3	−2.4	80.04
汕优 63（CK$_1$）	8.80 fE	1.38	15.68		−5.5	76.65
国稻 1 号	8.80 fE	1.38	15.64	0.0	−5.5	76.65
协优 9308	8.57 gF	0.36	4.19	−2.6	−8.0	84.83

注：同列数据后带相同大、小写字母者分别表示在 1% 和 5% 水平上差异不显著。下同。

2.2 稳定性分析

2.2.1 产量稳定性

从表1可知，准两优527、红莲优6号、C两优87和Y两优1号4个组合的产量显著或极显著高于其他组合，从其高稳系数（HSC）可以看出，这4个组合在不同生态试验点的产量表现出较高的稳定性，与汕优63相比，这4个组合的HSC高出4~8个百分点，属于适宜种植区域较广的组合。两优293、Ⅱ优明86、两优培九、GD-1S/RB207和P88S/747的HSC也高于汕优63。在7个生态试验点产量稳定性表现较差的是汕优63和国稻1号，其HSC最小，属于适宜种植区域相对较窄的组合。就协优9308而言，尽管表现出很高的产量稳定性，但其产量却较低。

2.2.2 结实率稳定性

供试组合结实率的标准差、变异系数和高稳系数见表2。准两优527、Y两优1号平均结实率高，且其标准差、变异系数均比对照汕优63小，高稳系数比对照大，其结实率在不同生态区表现很稳定，表明这类超级杂交稻适应性广。而其他超级杂交稻组合结实率的标准差、变异系数均比对照汕优63大，高稳系数比对照小，其结实率在不同生态区的稳定性较差。

表2　超级杂交稻在长江流域的结实率稳定性表现　　　　　　　　单位：%

组合	结实率	标准差	变异系数	高稳系数
准两优527	90.00	3.52	3.91	93.30
Y两优1号	86.94	3.13	3.60	90.42
汕优63（CK$_1$）	84.26	4.84	5.74	85.69
红莲优6号	82.62	6.90	8.35	81.70
Ⅱ优明86	82.48	6.22	7.55	82.28
C两优87	82.45	5.77	7.00	82.73
协优9308	81.75	7.59	9.29	80.01
两优293	80.94	5.32	6.57	81.59
两优培九（CK$_2$）	79.32	6.68	8.42	78.37
国稻1号	77.26	6.66	8.62	76.17
GD-1S/RB207	73.26	6.17	8.42	72.38
P88S/747	70.08	8.56	12.22	66.37

2.3　超级杂交稻在长江流域各生态点的具体产量表现

2006—2007 年 12 个供试组合在长江流域各生态试验点的具体产量表现见表 3，每个生态试验点组合间产量差异均达显著或极显著水平。

江苏南京点产量表现最好的是 C 两优 87、准两优 527 和 II 优明 86，分别比对照汕优 63 增产 12.9%，12.4% 和 10.1%，比两优培九增产 14.6%，14.1% 和 11.7%，均达极显著水平。其次是国稻 1 号、两优 293 和 Y 两优 1 号，分别比对照汕优 63 增产 5.3%，5.3% 和 4.2%，比两优培九增产 6.9%，6.9% 和 5.8%。表现较差的为红莲优 6 号和 P88S/747，分别比汕优 63 减产 4.8% 和 4.3%。

安徽安庆点产量最高的为红莲优 6 号，比汕优 63 增产 31.8%，比两优培九增产 11.7%，均达极显著水平；P88S/747、两优 293、II 优明 86、两优培九、C 两优 87 和准两优 527 产量较高，分别比对照汕优 63 增产 24.7%，22.3%，18.5%，18.1%，15.2% 和 15.2%，均达极显著水平；汕优 63 产量最低。

湖北武汉点产量表现最佳的为 II 优明 86、P88S/747，分别比汕优 63 增产 8.9% 和 7.4%，比两优培九增产 14.2% 和 12.7%；其次为国稻 1 号、红莲优 6 号和 Y 两优 1 号，分别比对照两优培九增产 10.2%，2.9% 和 2.0%；C 两优 87 和两优 293 产量较低，分别比汕优 63 减产 9.7% 和 9.3%，比两优培九减产 5.3% 和 4.9%。

重庆北碚点 C 两优 87、红莲优 6 号、准两优 527 和 Y 两优 1 号产量表现较好，分别比对照汕优 63 增产 26.5%，24.1%，23.7% 和 22.1%，比两优培九增产 20.1%，17.8%，17.4% 和 15.9%，均达极显著水平。协优 9308、国稻 1 号表现较差，分别比汕优 63 减产 19.6% 和 18.4%，比两优培九减产 23.7% 和 22.5%。

江西南昌点 Y 两优 1 号、准两优 527 产量表现较好，分别比汕优 63 增产 16.6% 和 14.3%，达极显著水平，比两优培九增产 6.4% 和 4.4%。两优培九、红莲优 6 号和两优 293 次之，分别比汕优 63 增产 9.5%，9.5% 和 9.0%；国稻 1 号、GD-1S/RB207 和汕优 63 产量最低，分别比两优培九减产 15.5%，9.2% 和 8.7%。

湖南长沙点协优 9308 产量最高，比汕优 63 增产 29.0%，比两优培九增产 12.6%，均达极显著水平。其次是红莲优 6 号、准两优 527 和 Y 两优 1 号，分别比汕优 63 增产 25.7%，25.7% 和 21.2%，均达极显著水平；比两优培九增产 9.7%，9.7% 和 5.7%，达显著水平。汕优 63、II 优明 86、P88S/747 和国稻 1 号产量表现较差，分别比两优培九减产 12.7%，8.1%，6.3% 和 5.7%。

湖南桂东点产量表现好的是准两优527和C两优87，分别比汕优63增产14.7%和13.0%，分别比两优培九增产11.0%和9.4%，均达极显著水平；Y两优1号、国稻1号、红莲优6号、两优培九、Ⅱ优明86和GD-1S/RB207的产量较好，分别比汕优63增产6.9%，4.9%，3.3%，3.3%，2.9%和2.5%。产量表现较差的协优9308比汕优63减产16.6%，比两优培九减产19.2%。

表3　供试超级杂交稻组合在各生态试验点的产量表现　　　单位：t/hm²

组合	南京	安庆	武汉	北碚	南昌	长沙	桂东
准两优527	9.82 a AB	10.21 cd DEF	8.23 de CD	12.73 a A	9.03 ab AB	8.02 ab AB	11.89 a A
红莲优6号	8.32 e E	11.68 a A	8.86 bcd BCD	12.77 a A	8.65 bc ABC	8.02 ab AB	10.71 bcd BC
C两优87	9.87 a A	10.21 cd DEF	8.15 e D	13.02 a A	8.40 c BCD	7.52 cd BC	11.72 a A
Y两优1号	9.11 c CD	10.00 de EF	8.78 cde BCD	12.56 a A	9.20 a A	7.73 bc ABC	11.09 b B
Ⅱ优明86	9.62 ab ABC	10.50 c CD	9.83 a A	11.05 b B	8.19 cd CD	6.72 fg EF	10.67 bcd BC
两优293	9.20 bc BCD	10.84 b BC	8.19 e D	11.21 b B	8.61 bc ABC	7.60 cd BC	10.37 cd C
P88S/747	8.36 e E	11.05 b B	9.70 a A	11.26 b B	8.40 c BCD	6.85 f DEF	9.53 e D
两优培九（CK₂）	8.61 de DE	10.46 c CDE	8.61 cde CD	10.84 bc B	8.65 bc ABC	7.31 de CD	10.71 bcd BC
GD-1S/RB207	8.74 cde DE	9.83 e FG	8.57 cde CD	10.71 bc B	7.85 d DE	7.27 de CDE	10.63 bcd BC
汕优63（CK₁）	8.74 cde DE	8.86 g I	9.03 bc ABC	10.29 c B	7.90 d DE	6.38 g F	10.37 d C
协优9308	8.86 cd DE	9.20 f HI	8.32 de CD	8.27 d C	8.44 c BCD	8.23 a A	8.65 f E
国稻1号	9.20 bc BCD	9.41 f GH	9.49 ab AB	8.40 d C	7.31 e E	6.89 ef DEF	10.88 bc BC

3　讨论

长江流域超级杂交稻的生态适应性主要表现为抽穗扬花期对夏季异常高温的耐受性，其次是抗病性和灌浆结实后期的耐低温能力，其最直接的指标就是结实率。据报道，2003年湖北种植的两优培九由于在抽穗扬花期受到高温影响，空壳率在53.1%～73.3%之间，平均达到67.8%[5]。本研究中，供试组合在7个生态试验点之间结实率的标准差、变异系数和高稳系数均存在差异，超级杂交稻国稻1号和两优培九的结实率低于80%，变异系数较大，结实率稳定性相对较差，可能是其抽穗扬花期在多数试验点受到高温影响导致灌浆结实不良而结实率下降。

对7个生态试验点的产量分析结果表明，12个组合在不同的生态条件下产量各异，如C两优87在重庆、南京点产量排第1位，而在武汉点则为末位；协优9308在长沙点产量最高，而在桂东、重庆点则最低；表明不同组合对不同生态稻作区的适应能力存在明显的差异。本研究中，农业部认定的6个超级杂交稻中，准两优527、Y两优1号生态适应范围较广，国稻1号的生态适应区域较狭窄，Ⅱ优明86、两优培九的生态适应性则为中等，协优9308虽然在各试点间产量表现稳定（高稳系数大），但产量偏低，极显著低于对照汕优63。因此，在超级杂交稻大面积推广时，在考虑超高产的前提下，有必要对其生态适应性进行科学的评价，以确定其适宜推广种植的区域。

根据超级杂交稻在各生态试验点的产量表现、产量和结实率的稳定性，本研究将超级杂交稻分为两种类型：一种是能适应不同生态环境的广适型超级杂交稻，这是当前超级杂交稻育种面临的重要课题，也是难点、热点问题。这类超级杂交稻，在区试产量比对照增产8%的基础上，在不同的生态稻作区均表现高而稳定的结实率，产量潜力能得到充分而稳定的发挥。在供试材料中，准两优527、红莲优6号、C两优87和Y两优1号可以列为这一类型。另一种是适应特定生态稻作区种植的超级杂交稻，暂称为区域型超级杂交稻。这类超级杂交稻在一定的生态种植区域能表现很高的产量和结实率，而在其他的稻作区其超高产潜力不能充分发挥。Ⅱ优明86、两优293和两优培九就属于这一类。本研究中，协优9308和国稻1号因整体产量水平与普通高产杂交稻汕优63相当，不参与超级杂交稻的生态适应性评价。

References
参考文献

[1]邓华凤.长江流域超级杂交稻目标性状研究[D].长沙：湖南农业大学，2008.

[2]农业部推荐的28个符合超级稻标准的水稻品种[J].中国农技推广，2005，（5）：24.

[3]农业部发布的2006年21个超级稻示范推广品种[J].中国农技推广，2006，（11）：21.

388

［4］温振民，张永科.高稳系数法估算玉米杂交种高产稳产性的探讨 [J]. 作物学报，1994，20（4）：508-512.

［5］王前和，潘俊辉，李晏斌，等.武汉地区中稻大面积空壳形成的原因及防止途径 [J]. 湖北农业科学，2004，（1）：27-30.

作者：邓华凤　何强　陈立云　张武汉　舒服　袁隆平

注：本文发表于《杂交水稻》2009 年第 24 卷第 5 期。

A Transcriptomic Analysis of Super Hybrid Rice *LYP9* and its Parents

【Abstract】By using a whole-genome oligonucleotide microarray, designed based on known and predicted *indica* rice genes, we investigated transcriptome profiles in developing leaves and panicles of super hybrid rice *LYP9* and its parental cultivars *93-11* and *PA64s*. We detected 22,266 expressed genes out of 36,926 total genes set collectively from 7 tissues, including leaves at seedling and tillering stages, flag leaves at booting, heading, flowering, and filling stages, and panicles at filling stage.Clustering results showed that the F_1 hybrid's expression profiles resembled those of its parental lines more than that which lies between the 2 parental lines.Out of the total gene set, 7,078 genes are shared by all sampled tissues and 3,926 genes (10.6% of the total gene set) are differentially expressed genes (DG).As we divided DG into those between the parents (DG_{PP}) and between the hybrid and its parents (DG_{HP}), the comparative results showed that genes in the categories of energy metabolism and transport are enriched in DG_{HP} rather than in DG_{PP}.In addition, we correlated the concurrence of DG and yield-related quantitative trait loci, providing a potential group of heterosisrelated genes.

【Key words】heterosis; hybrid rice; transcriptome; quantitative trait loci; differentially expressed genes.

Extensive sequence diversity at the microstructural level has been demonstrated in a number of plant species (1), and such diversity can extend even to allelic regions (2).These intraspecific allelic variations should have impacts on gene expressions that lead to phenotypic variation, perhaps including hybrid vigor as a beneficial trait used in crop breeding.In a hybrid, in which 2 different alleles of a gene are often brought together, the combined allelic expression may deviate from that of either parent or the midparent predictions (3).In maize, both allelic diversity and expression variation were found between inbred parents and their hybrid (4). In maize hybrids, not only the allelic variation in gene expression but also different responses to extrinsic stimuli supported the presence of allelic expression variation in the same genetic context

(5) .Large-scale transcriptome profiling has been used for heterosis studies in maize (6), *Arabidopsis* (7), and wheat (8) .In rice, an investigation of a yield-related quantitative trait locus (QTL) resulted in a discovery of allelic variation that affected the expression of a leucinerich repeat receptor kinase gene cluster (9) .Another survey with a cDNA microarray concerning 9, 188 expressed sequence tags on expression polymorphism between an elite rice hybrid and its parental varieties revealed significant heterotic expression for 141 expressed sequences (10) .

We have recently focused our heterosis research on *Liang-You-Pei-Jiu* (*LYP9*), a super hybrid rice strain from a cross between the maternal inbred *PA64s*, a photothermosensitive male sterile line, and the paternal inbred *93-11*, an elite *indica* variety, after we sequenced the 2 parental genomes (11, 12) .Two-dimensional electrophoresis analysis among *93-11*, *PA64s*, and *LYP9* revealed significant numbers of different embryo protein spots, many of which were shown to display mirrored relationships between parents and the first filial generations (13) .Further analysis on mature embryos of this hybrid triad identified 54 differentially expressed proteins involved in major biological processes including nutrient reservoir, response to stress, and metabolism.Among these embryos, most of the storage proteins exhibit overdominance and stress-induced proteins display additivity (14) .We also carried out transcriptome profiling for the hybrid and its parents using both sequencingbased (15-17) and hybridization-based methods (18) .We now report a rather large-scale comparative transcriptome analysis of the triad, concerning 7 tissues sampled across developmental times and different tissues.We expect this genome-wide transcriptome comparison to be an initial step forward in understanding the causative mechanism of the altered gene expression in the hybrid and the molecular mechanism underlying heterosis.

Results

The Rice Whole-Genome Microarrays Are of Satisfactory Quality. Our 70-mer oligonucleotide microarray, with 36, 926 unique features identified, was designed based on known and predicted gene models of the indica rice *93-11* genome (18) .We calibrated our microarray by doing 4 preliminary tests. First, a selfhybridization experiment was conducted, detecting only 9 false differentially expressed genes (DG) with marginal intensity above the background.Second, we conducted hybridizations between the seedling shoot and the filling panicle and discovered>5,000 DG with correlation coefficients of 0.85 between duplication and correlation coefficients of 0.81 in dye-swapping experiments.Third, to better define the background and fold changes we introduced a polyubiquitin gene as positive control, the fold changes of which are both consistent and always below the threshold (Fig.S1) .We acquired at least 3 independent replicates for each sample pair in general and a total of 48 raw datasets (96 slides) for 7 tissues from the triad (collective correlation coefficient among all replicates>0.8) .Finally, we validated our microarray results with semiquantitative RT-PCR, and out of 25 primer pairs with amplification products, 20 (80%) DG showed consistent results compared with those obtained from the microarray data（Fig.S2）. Collectively, these results demonstrated the satisfactory quality of our experimental procedures and data.

Transcriptome Profiles of LYP9 and Its Parents Revealed Consistent Trends with Phenotypic Observations. Our data were derived from 7 tissues of the LYP9 hybrid triads, including seedling shoot, leaf at tillering stage, flag leaf at booting stage, flag leaf at heading stage, flag leaf at flowering stage, flag leaf at filling stage, and panicle at filling stage, out of which we identified 11,448-14,592 genes expressed in each pairwise comparison (Table S1) .Our analysis revealed 7,078 genes expressed in all studies tissues and 22,266 genes expressed collectively.

We used a cluster analysis method to investigate correlations among transcriptome profiles.The results revealed that tissues from different cultivars at the same developmental stage always formed the primary groups (Fig.1) .In a broader spectrum, the transcriptome profiles of *LYP9* are similar to *PA64s* (maternal) at the early developmental stages but closer to *93-11* (paternal) at the later stages.Both are consistent with the morphological appearances or characteristics of the hybrid plant at corresponding stages, observed empirically in the field as either *93-11*-like or *PA64s−like*.A distinct result was found in the cluster of the panicle at filling stage, where the profile of *LYP9* is more similar to that of *93-11* because *PA64s* is a photothermosensitive male sterile rice line (19), and many of its genes may not express appropriately or at levels comparable to those of *93-11* and *LYP9*.

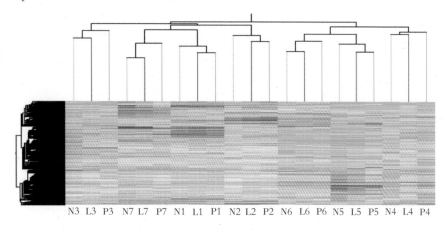

Fig.1　Hierarchical clustering analysis of all gene models based on expression data.Normalized expression values for the microarray (37K) clustered with Genespring (Silicon Genetics) .Each horizontal line refers to a gene.The color represents the logarithmic intensity of the expressed genes.N, L, and P stand for *93-11*, *LYP9*, and *PA64s*, respectively.Numbers 1-7 denote samples from the following tissues in order : seedling shoot, leaf at tillering stage, flag leaf at booting stage, flag leaf at heading stage, flag leaf at flowering stage, flag leaf at filling stage, and panicle at filling stage.

When we looked at universally expressed genes, some are undoubtedly housekeeping genes whereas the molecular category of structure was found overrepresented (Fig.S3) .We also noticed that among structure molecules, genes encoding cytoplasmic (60S/40S) protein and plastid ribosomal (50S/30S) protein have a synergistic expression profile except in the filling stage panicle where the former are up-regulated and the latter are down-regulated (Fig.S4) as compared with those in other tissues.This result is consistent with the fact that the number of chloroplasts in panicles is significantly lower than that found in leaf tissues.

DG and Their Functional Analysis. We defined DG between the parental lines as DG_{PP} and those between the hybrid and its parents as DG_{HP}. DG_{PP} only denote the differences between 2 inbred lines, but DG_{HP} may underlie heterosis because differences in expression between hybrid and parents should underlie their phenotypic differences. DG_{HP} can be divided into 2 classes—i.e., those shared by DG_{PP} and DG_{HP} (DG_O) and those uniquely belonging to DG_{HP} (DG_{HPU}). We found 3,926 (10.6%) DG observed at least once among all sample pairs (Dataset S1), and the numbers of DG_{HPU} are larger than DG_O in all 7 tissues investigated (Table 1). By comparing DG between the hybrid and its parents, we found that the great majority of DG are close to either their maternal or their paternal parent and that a minority of them are close to neither parent. To further understand the function of DG, we classified these genes according to their functional categories and relatedness. For instance, DG_{PP} are enriched in 16 out of 161 categories as compared with DG_{HP}, which are enriched in 25 function categories (Table S2 and S3). Since DG_{HP} are composed of DG_O and DG_{HPU}, we expected that heterosis-related genes may be enriched in DG_{HPU} rather than in DG_O. Indeed, the DG_{HPU} identified in this study are enriched mostly in the categories of energy metabolism and transport (Table 2).

Table 1　Number and classification of DG

Sample	DG_{PP}	DG_{HP}								
		L/N	L/P	DG_{HPU}	DG_O	H2P	CHP	B2P	CLP	L2P
S1	305	243	167	215	161	19	190	21	142	4
S2	312	309	266	328	201	17	247	46	208	11
S3	472	323	412	424	272	14	465	42	174	1
S4	389	345	313	447	180	36	324	17	235	15
S5	342	337	333	401	208	57	315	40	182	15
S6	331	313	323	347	203	36	222	53	199	40
S7	383	405	451	505	289	24	321	11	400	38
Total	2132	1913	1898	2260	1280	196	1851	198	1316	108

Note: N, L, and P refer to *93-11*, *LYP9*, and *PA64s*, respectively. DG_{PP} refers to DG between both parents, DG_{HP} refersto DG between the hybrid and parent. DG_{HPU} denotes the unique portion of DG_{HP}, and DG_O denotes the overlapbetween DG_{PP} and DG_{HP}. H2P, CHP, B2P, CLP, and L2P represent higher than both parents, close to higher parent, between both parents, close to lower parent, and lower than both parents, respectively.

Table 2　Functional classification of unique portion of DG_{HPU}

Functional categories	S1	S2	S3	S4	S5	S6	S7
Metabolism							
Amino acid metabolism	7	14[*]	14	10	16	19[**]	29[**]
Biosynthesis of polyketides and nonribosomal peptides	0	1	0	1	0	1	1

Continued

Functional categories	S1	S2	S3	S4	S5	S6	S7
Biosynthesis of secondary metabolites	7	15	23	15	19	19	20
Carbohydrate metabolism	19	28	39	37	47**	37*	40
Energy metabolism	9**	14**	17**	11	11	29**	39**
Glycan biosynthesis and metabolism	0	0	6	3	3	6	9*
Lipid metabolism	5	7	8	7	8	12*	6
Metabolism of cofactors and vitamins	10	19	21	14	24	17	16
Metabolism of other amino acids	4	8*	4	4	8*	3	7
Nucleotide metabolism	6	5	5	8	8	2	10
Xenobiotics biodegradation and metabolism	9	19	24	20	21	19	12
Genetic Information Processing							
DNA metabolism	6	1	3	3	9	2	0
RNA metabolism	4	15	26	26	26	13	24
Cellular protein metabolism	17	34	61**	50	44	22	41
Environmental Information Processing							
Signal transduction	1	6	12	17*	17**	5	3
Transport	11	27	43**	39*	42**	33**	38
Cellular Processes							
Cell motility	0	1	1	1	1	0	0
Cell cycle	2	6	4	2	4	2	7
Cell-cell signaling	0	0	2	1	1	1	1
Cell death	5	4	1	2	5	5	3
Cell growth	0	1	1	1	0	0	0
Other	28*	32	59**	59**	52**	50**	75**
Unknown	123	181	193	229	155	152	249
Total	215	328	424	447	401	347	505

Note: * and ** denote significant enrichment of DG among function category with $P<0.05$ and $P<0.01$, respectively.

Since the most important trait of hybrid rice is grain yield, we analyzed the genes involved in carbohydrate biosynthesis (20, 21)—such as starch biosynthesis—and noticed that genes involved in starch synthesis have much higher expression in the panicle of *LYP9* than *PA64s* at filling stage, including the key enzymes in starch biosynthesis such as sucrose synthase, ADP-glucose pyrophosphorylase, and starch synthase. The result is in agreement with the fact that starch biosynthesis cannot take place in the panicles of *PA64s*. In addition, rubisco, a key protein in the pathway, showed a lower expression level in *LYP9* than in *PA64s*, thus supporting the fact that the panicle of *PA64s* remained green long

394

after flowering was observed in the field.It is interesting that the genes taking part in sucrose and starch metabolism, such as ADP-glucose pyrophosphorylase, sucrose-P synthase, invertase, and branching enzyme, tend to be highly expressed in the hybrid (Fig.2).

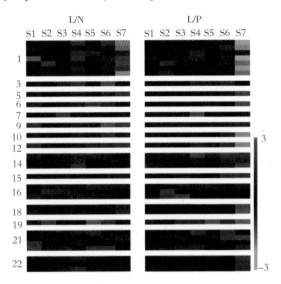

Fig.2 Expression profiles of DGbetween *LYP9* and its parents in carbohydrate biosynthesis pathway.Genes involved in carbohydrate metabolism were identified according to their Enzyme Commission annotation, and those genes that differentially expressed at least once were shown.The log$_2$-transformed ratio between the hybrid and either parent was used (L, *LYP9*; N, *9311*; P, *PA64s*).Each row represents a single gene, and the number indicates a group of isoenzymes in the pathway according to its position in the path and order.Red and green colors denote up-and down-regulated genes, respectively.The genes are listed as follows: (1) ribulose-bisphosphate carboxylase, (3) glyceraldehyde-3-phosphate dehydrogenase, (5) fructose-bisphosphate aldolase, (6) fructose-bisphosphatase, (7) glucose-6-phosphate isomerase, (9) transketolase, (10) sedoheptulose-1, 7-bisphosphatase (SBPase), (12) phosphoribulokinase, (14) ADP-glucose pyrophosphorylase, (15) UDP-glucose pyrophosphorylase, (16) sucrose-P synthase, (18) sucrose synthase, (19) invertase, (21) starch synthase, and (22) branching enzyme.

Nonadditive–Expressed Genes. Concerning the relative level of gene expression among a hybrid-parent triad, we often expect 2 scenarios to come into play.In the first scenario, gene expression in the hybrid exhibits a cumulative mode, contributed by each allele from the respective parents.In the other scenario, the expression deviates from the midparental level.The former scenario is additive, indicating that all eles from both parents may contribute to gene expression in the hybrid, attributable mostly to a *cis*-regulation mechanism.The latter scenario is nonadditive, in which other regulators probably contribute to an altered expression of the corresponding alleles in the hybrid, attributable mostly to *trans*-regulation (3).In comparison with gene expression among the *LYP9* triad, we detected 860 up-regulated and 1,095 down-regulated nonadditive genes (NAG).The number of NAG in each sampling triad ranged from 195 to 497 (Table 3); they composed 0.5% – 1.4% of the total gene set and 29.6% – 53.7% of DG$_{HP}$ identified at 7 tissues.

Table 3　Nonadditive-expressed genes in LYP9

Sample	Number of NAG				Number of NAG in DG$_{HP}$					
	Up	Down	Total	$a\%$	DG$_{HP}$	$b\%$	DG$_{HPU}$	$c\%$	DG$_O$	$d\%$
S1	80	115	195	0.5	144	38.3	108	50.2	36	22.4
S2	97	180	277	0.8	184	34.8	148	45.1	36	17.9
S3	163	147	310	0.8	206	29.6	168	39.6	38	14.0
S4	182	220	402	1.1	261	41.6	222	49.7	39	21.7
S5	140	126	266	0.7	239	39.2	209	52.1	30	14.4
S6	103	177	280	0.8	218	39.6	189	55.4	29	14.3
S7	158	339	497	1.4	426	53.7	264	52.3	162	56.1
Total	860	1095	1846	5.0	1481	46.5	1245	55.1	488	38.1

Note：$a\%$ denotes the percentage of NAG in the total gene set (36,926), $b\%$, $c\%$, and $d\%$ denote the percentage of NAG in total numbers of DG$_{HP}$, DG$_{HPU}$ and DG$_O$, respectively.

DGHP Are Enriched in Known QTLs. We were able to map 2,673 DG$_{HP}$ to 3,128 QTLs classified into 9 categories and 209 traits in the rice genome (www.gramene.org).One important piece of evidence supporting the correlation between the 2 types of data is the fact that the fraction of DG$_{HP}$ in the transcriptome profiles (36, 926 expressed genes) is 8.6% as compared with the fractions of DG$_{HP}$ mapped to QTLs—10.1% and 11.8% in the QTL intervals that harbor less than 50 and 10 genes, respectively (Table S4).Among DG$_{HP}$-related QTLs, many are well characterized, including 1000-*seed weigh* (e.g., AQCY015, CQAS23, AQAI076, and CQAS23), *filled grain number* (e.g., AQCY010, AQCY059, AQAK009, and AQAK011), *grain number* (CQB22, AQDR015, AQDR059, and AQED038), and *grain yield per panicle* (AQDR091, AQDR103, and AQDR104).The potential association between DG$_{HP}$ and QTLs were also suggested within many QTL regions, such as Starch synthase Ⅲ (Os055024_01) to AQCY010 for *filled grain number*, *putative sugar transporter* (Os055048_01) toAQAI076 and AQEY022 for 1000-*seed weight*, and *auxin response factor* (Os016758_01) to CQK15 for *panicle number*. To help portray this DG$_{HP}$-QTL correlation, we aligned DG$_{HP}$ over yield-related QTL regions covering less than 100 genes on rice chromosomes (Fig.3).

Discussion

Complex Regulatory Mechanisms Probably Underlie Gene Expression Changes in Hybrid. Transcriptomes are not only always specific to cell types but also are regulated at different levels, such as transcription and splicing, and through genetic or epigenetic mechanisms.Although in this current report we are unable to show detailed sequence comparisons and validations for different alleles of annotated DG, allelic sequence variation-especially those in the regulatory sequence/element-is undoubtedly one of the causes of gene-expression change in hybrids.We will certainly proceed in identifying these allelic differences of all DG in our dataset.Another class of gene regulators is *trans*-regulators, such as transcription factors (TFs).The dosage effect of such regulatory genes had been proposed to affect

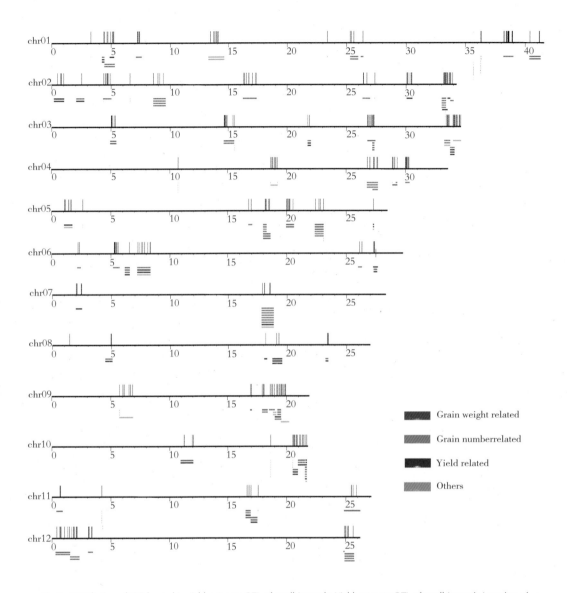

Fig.3 Distribution of DG located in yield-category QTL of small intervals. Yield-category QTL of small intervals (number of genes ≤ 100) that harbor DG_HP were aligned to TIGR's rice pseudochromosome version 5. The long horizontal lines represent 12 rice chromosomes ,the short horizontal lines represent QTL intervals , and the short vertical lines represent DG_HP.

phenotypes in hybrids (22) .We indeed found that 187TFs exhibited differential expression in the hybrid compared with either parent.It is quite a coincidence that a recent study using seedling tissue of 2 hybrid triads, based on the genomic sequence of *93-11* and *nipponbare*, also suggested that altered gene expression caused by interactions between transcription factors and the allelic promoter region in the hybrids was one plausible mechanism underlying heterosis in rice (23) .

Moreover, we noticed that among those differentially expressed TFs, the AP2-EREBP family—potential targets of miRNA (24) —is over represented.Noncoding RNAs are involved in epigenetic regulations, and other epigenetic mechanisms including DNA methylation, acetylation and deacetylation of histones, and chromatin remodeling.It had been reported that the degree of methylation in hybrids is different from that in inbred lines in *Arabidopsis* and rice (25, 26).A recent study reported that epigenetic regulation of a few regulatory genes (CCA1 and LHY in this case) induced cascade changes both in downstream genes (TOC1, GI, etc.) and in physiological pathways, and ultimately induced growth and development, which also indicates the presence of a general mechanism for the growth vigor and increased biomass commonly observed in hybrids (27).In the present survey, we also found that among DG_{HP} there were many epigenesis-related genes, including methyltransferase, hydroxymethyltransferase, serine Oacetyltransferase, histone acetyltransferase, acetyl-CoA acyltransferases, and chromodomain helicase-DNA-binding protein 3.The expression of these genes is being verified experimentally as is their involvement in related biological pathways.

Gene Expression Variations in Hybrid Suggests Correlation to Genetic Mechanisms Responsible for Heterosis. The dominance and overdominance hypotheses (28) were proposed to explain heterosis before the molecular concepts of genetics were formulated, and these hypotheses are not closely allied with molecular principles.We categorized DG between hybrid and parents (DG_{HP}) into 5 basic categories: overdominance (H2P), underdominance (L2P), dominance (CHP and CLP), and midparent (B2P).We found that dominant expression was the most prevalent among DG_{HP} (81.6% - 91.8%).Additive and nonadditive expression represent another possible genetic model for gene expression in hybrids (3).Whether or not a transcript shows nonadditive expression is most likely to be influenced by the contributions of *cis*- and *trans*-acting factors of this gene (29 - 32).In our data, the majority of genes in the hybrid showed additive expression, and the phenomenon suggests that *cis*-acting elements usually play a major role in the control of general gene expression.Nonadditively expressed genes in our entire dataset constituted only 0.5% - 1.4% of the total discovered in each sampled tissue but accounted for 29.6% - 53.7% of DG_{HP}.A similar result was observed in a study of maize heterosis, in which the nonadditive expressed genes were found to contain 2.2% of the total genes and 22% of DG (6).It should be noted that we were unable to detect those genes where the silencing of 1 allele was compensated by overexpression of the other, which might cause underestimation of nonadditive genes in hybrids, as mentioned previously (31).The analysis of nonadditive gene expression indicates that allelic expression in hybrids may not just be a combination of alleles from the 2 parents but is rather regulated by other genes or epistatic mechanisms.Nonadditive gene expression was also considered as midparent heterosis or heterotic expression (10, 23).

　　A study in gene expression in maize endosperm revealed heterochronic expression of 3 allele pairs (33).In the present study, 85% of DG were detected only once in 7 tested tissues.For those DG that appeared more than once, 63% (*LYP9/PA64s*) to 75% (*LYP9/93-11*) differed in the same direction; i.e., either upor down-regulated.This trend indicates that their corresponding regulatory mechanisms may function in the same way in different tissues and under different conditions.In contrast, 25% - 39% of those genes follow a different trend; they differ in the opposite direction, so that a gene in the hybrid may be under a different control mechanism or the regulatory factors may function in a different way under variable conditions.

DG Are Candidates for Genes That Play an Important Role in Heterosis. Microarray-based expression studies allowed us to identify genes that are differentially expressed between a hybrid and its parents, and these DG are often found to be expressed in a biased pattern in comparison with regular transcriptomes. For example, we found that the DG_{HP} involved in the carbohydrate-metabolism pathway had a larger fraction of up-regulated genes than down-regulated genes, similar to the recent studies (23, 27). However, of the genes taking part in oxidative phosphorylation, there were more down-regulated genes identified in the hybrid than in the parental lines. In addition, heading stage is an important period for panicle development and grain-yield formation, and our previous serial analysis of gene expression (SAGE) analysis showed that genes related to protein biosynthesis and peptide transport were up-regulated in the panicle of the hybrid *LYP9* (16). Based on our current data, a similar conclusion was reached in the analysis of gene expression in flag leaves of heading stage and flowering stage. It was interesting to find that sucrosetransport genes are up-regulated in *LYP9* panicles as compared with those in *93-11* panicles, suggesting that the transportation of carbohydrate from the source to the sink in *LYP9* is more efficient than in at least one of the parents.

An altered expression of the maize domestic gene *tb1* was characterized as the cause of observed quantitative phenotypic changes by a fine-mapping approach (34), and a transcription activator was demonstrated to be responsible for the significant plant-height changes in an *Arabidopsis* hybrid (35). Recently, a major quantitative gene in rice, *Ghd7*, isolated by map-base cloning and encoding a CCT domain protein, was considered as a crucial factor for increasing productivity and adaptability of an elite hybrid cultivar, *Shanyou 63*, and some other *indica* varieties (36). In our current study, not only have we found many TFs in our DG collections, but we also mapped a high fraction of DG to the intervals of grain-yield-related QTLs. These results led us to believe that DG between the hybrid and parents may contribute in a significant way to heterosis. We also have constructed databases integrating heterosis-related genes among major crops and experimental plants, identifying altered sequences among differentially expressed alleles (37), and mapping relative DG to QTLs discovered in this study.

Materials and Methods

Rice Whole Genome Oligonucleotide Array. The whole-genome array was developed based on annotated and predicted genes from the genome assembly of *indica* rice *93-11* (11, 12). Oligonucleotides were arrayed onto 2 poly-L-lysine-coated microscope slides as a set with a SpotArray72 microarrayer (Perkin-Elmer) in the microarray laboratory at Beijing Genomic Institute, and the slides were processed according to a standard procedure (38).

Plant Materials and Data Processing. *LYP9* and its parental lines (*93-11* and *PA64s*) were grown in a greenhouse for the seedling samples and in the rice field for all other samples. The plant tissues were collected and stored at −80 ℃. RNA samples were isolated (39), quantitated by using a Nano Drop1000 spectrophotometer (Nanodrop Technologies), and labeled (40, 41). Each sample had at least 3 biological replications to minimize systematic errors. Separate tiff images of Cy5 and Cy3 channels were obtained by ScanArray Lite scanner (Perkin-Elmer), and spot intensities were quantified by using the

Axon GenePix Pro 5.1 image analysis software.

We categorized our raw data with 3 simple criteria.First, features were flagged as "bad" either by using Genepix or by manual investigation, second, a false positive rate ≤ 5% in reference to the controls was found, and third, legitimate features were found in at least 2 of the 3 replicate sets or 3 of the 4 replicate sets.The processed data were normalized based on the mean of all expressed genes.The normalization of the 2-channel data for each array was done by using the intensity-based Loess method with R language.DG were defined by a log-scale ratio between paired samples with a P value <0.01 (Ztest).

Functional Annotation. For each gene identified, we performed detailed functional annotations by using standard tools, such as BLAST (42, 43) and HMMer (44), against public data, including (*i*) the The Institute for Genomic Research (TIGR) Rice Pseudomolecules and Genome Annotation database (release 5.0, http：//rice.plantbiology.msu.edu)；(*ii*) the knowledge-based Oryza Molecular Biological Encyclopedia (http：//cdna01.dna.affrc.go.jp/cDNA/)；(*iii*) the TIGR Rice Gene Index (http：//compbio.dfci.harvard.edu/tgi/)；and (*iv*) the UniProtKB/Swiss-Prot (www.ebi.uniprot.org). We also used the Kyoto Encyclopedia of Genes and Genomes (http：//www.genome.jp/kegg/) and Gene Ontology databases (45) for protein annotation (E value<10^{-7}).The HMMpfam program (http：//hmmer.janelia.org/) wasused to search Pfam hidden Markov models retrieved from Pfam release 18 (46) for structural domains E value (<0.001).

Because some categories are larger (i.e., involve more genes) than others, they tend to show more frequently in any set of genes；thus, it is essential to identify the statistically significant categories in a set of genes.We took the whole set of genes as the default background distribution and used the reported method (47) to decide the significance of DG in each category, with P value cutoff of 0.05 as the significance threshold.

Mapping DG to QTL. Rice QTL data with physical positions on the TIGR release 5genome were acquired from Gramene (www.gramene.org) and 2,685 DG were mapped to 2,729 riceQTL, covering 9QTL categoriesand 211QTL traits.For better demonstration of the relationship between DG and QTL, we classified yieldrelated QTL according to the number of genes in each chromosome region and performed an enrichment test according to the method described in ref.47.

ACKNOWLEDGMENTS. The authors thank Dr.Chengzhi Liang (Cold Spring Harbor Laboratory, Cold Spring Harbor, NY) for his supply of mapping data of rice QTL to Tigr5 pseudogenome, and Xiaojun Tan and Dr.Xiting Yan for technical support of data analysis.This project was funded by Chinese Academy of Sciences Grants KSCX2-SW-306 (to L.Z.), KSCX1-SW-03 (to Z.Z.), and KSCX1-SW-03-01 (to J.Y.), National Natural Science Foundation of China Grants 90208001 and 30550005 (to L.Z.) and 30221004 (to J.Y.and G.L.), National Basic Research Program of China Grants 2004CB720406 (to Z.Z.) and 2006CB101706 (to G.L.), Ministry of Science and Technology Grants 2002AA229021 (to J.Y.) and 2006AA10A101 (to Z.Z.), and Ph.D.Programs Foundation of Ministry of Education of China Grant 20060533064 (to L.Y.).

400

References

1. Bennetzen JL (2000) Comparative sequence analysis of plant nuclear genomes: Microcolinearity and its many exceptions. *Plant Cell* 12: 1021–1029.

2. Fu H, DoonerHK (2002) Intraspecific violation of genetic colinearity and its implications in maize.*Proc Natl Acad Sci USA* 99: 9573–9578.

3. Birchler JA, Auger DL, Riddle NC (2003) In search of the molecular basis of heterosis.*Plant Cell* 15: 2236–2239.

4. Song R, Messing J (2003) Gene expression of a gene family in maize based on noncollinear haplotypes.*Proc Natl Acad Sci USA* 100: 9055–9060.

5. Guo M, et al. (2004) Allelic variation of gene expression in maize hybrids.*Plant Cell* 16: 1707–1716.

6. Swanson-Wagner RA, et al. (2006) All possible modes of gene action are observed in a global comparison of gene expression in amaize F1 hybrid and its inbred parents. *Proc Natl Acad Sci USA* 103: 6805–6810.

7. VuylstekeM, van Eeuwijk F, VanHummelenP, KuiperM, ZabeauM (2005) Genetic analysis of variation in gene expression in Arabidopsis thaliana. *Genetics* 171: 1267–1275.

8. Yao Y, et al. (2005) Identification of differentially expressed genes in leaf and root between wheat hybrid and its parental inbreds using PCR-based cDNA subtraction. *Plant Mol Biol* 58: 367–384.

9. HeG, et al. (2006) Haplotype variation in structure and expression of a gene cluster associated with a quantitative trait locus for improved yield in rice. *Genome Res* 16: 618–626.

10. Huang Y, et al. (2006) Heterosis and polymorphisms of gene expression in an elite rice hybrid as revealed by a microarray analysis of 9198 unique ESTs. *Plant Mol Biol* 62: 579–591.

11. Yu J, et al. (2002) A draft sequence of the rice genome (Oryza sativa L.ssp.indica) . *Science* 296: 79–92.

12. Yu J, et al. (2005) The genomes of Oryza sativa: A history of duplications. *PLoSBiol* 3: e38.

13. Xie Z, et al. (2006) Pedigree analysis of an elite rice hybrid using proteomic approach. *Proteomics* 6: 474–486.

14. Wang W, et al. (2008) Proteomic profiling of rice embryos from a hybrid rice cultivar and its parental lines.*Proteomics* 8: 4808–4821.

15. Zhou Y, et al. (2003) Gene identification and expression analysis of 86, 136 Expressed Sequence Tags (EST) from the rice genome.*Genomics Proteomics Bioinformatics* 1: 26–42.

16. Bao J, et al. (2005) Serial analysis of gene expression study of a hybrid rice strain (*LYP9*) and its parental cultivars.*Plant Physiol* 138: 1216–1231.

17. Song S, Qu H, Chen C, Hu S; Yu J (2007) Differential gene expression in an elite hybrid rice cultivar (Oryza sativa, L) and its parental lines based on SAGE data. *BMC Plant Biol* 7: 49.

18. Ma L, et al. (2005) A microarray analysis of the rice transcriptome and its comparison to Arabidopsis. *Genome Res* 15: 1274–1283.

19. JZ Yi, XiaoW (2000) The production technology of the Liang-You-Pei-Jiu (LYP9) . *Hybrid Rice* 5: 76–77 (in Chinese) .

20. Malkin R, Niyogi K (2000) in *Biochemistry*

and Molecular Biology of Plants, eds Buchanan B, Gruissem W, Jones R (American Society of Plant Biologists, Rockville, MD), pp 610–619.

21. Dennis D, Blakeley S (2000) in *Biochemistry and Molecular Biology of Plants*, eds Buchanan B, Gruissem W, Jones R (American Society of Plant Biologists, Rockville, MD), pp 630–672.

22. Birchler JA, Riddle NC, Auger DL, Veitia RA (2005) Dosage balance in gene regulation: Biological implications.*Trends Genet* 21: 219–226.

23. Zhang H-Y, et al. (2008) A genome-wide transcription analysis reveals a close correlation of promoter INDEL polymorphism and heterotic gene expression in rice hybrids. *Mol Plant* 1: 720–731.

24. Shigyo M, Hasebe M, Ito M (2006) Molecular evolution of the AP2 subfamily.Gene 366: 256–265.

25. Xiong LZ, Xu CG, Saghai Maroof MA, Zhang Q (1999) Patterns of cytosine methylation in an elite rice hybrid and its parental lines, detected by a methylation-sensitive amplification polymorphism technique. *Mol Gen Genet* 261: 439–446.

26. Madlung A, et al. (2002) Remodeling of DNA methylation and phenotypic and transcriptional changes in synthetic Arabidopsis allotetraploids.Plant *Physiol*/129: 733–746.

27. Ni Z, et al. (2008) Altered circadian rhythms regulate growth vigour in hybrids and allopolyploids. *Nature* 457: 327–31.

28. Crow JF (1948) Alternative hypotheses of hybrid vigor.*Genetics* 33: 477–487.

29. Doss S, Schadt EE, Drake TA, Lusis AJ (2005) Cis-acting expression quantitative trait loci in mice. *Genome Res* 15: 681–691.

30. Pastinen T, Hudson TJ (2004) Cis-acting regulatory variation in the human genome. *Science* 306:

647–650.

31. Stupar RM, Springer NM (2006) Cis-transcriptional variation in maize inbred lines B73and Mo17 leads to additive expression patterns in the F1 hybrid. *Genetics* 173: 2199–2210.

32. Ronald J, Brem RB, Whittle J, Kruglyak L (2005) Local regulatory variation in Saccharomyces cerevisiae. *PLoS Genet* 1: e25.

33. Guo M, Rupe MA, Danilevskaya ON, Yang X, Hu Z (2003) Genome-widem RNA profiling reveals heterochronic allelic variation and a new imprinted gene in hybrid maize endosperm. *Plant J* 36: 30–44.

34. Clark RM, Wagler TN, Quijada P, Doebley J (2006) A distant upstream enhancer at the maize domestication gene tb1 has pleiotropic effects on plant and inflorescent architecture. *Nat Genet* 38: 594–597.

35. Su N, Sullivan JA, Deng XW (2005) Modulation of F1 hybrid stature without altering parent plants through trans-activated expression of a mutated rice GAI homologue. *Plant Biotechnol J* 3: 157–164.

36. XueW, et al. (2008) Natural variation in Ghd7 is an important regulator of heading date and yield potential in rice. *Nat Genet* 40: 761–767.

37. Song S, et al. (2008) HRGD: A database for mining potential heterosis-related genes in plants. *Plant Mol Biol* 69: 255–260.

38. Eisen MB, Brown PO (1999) DNA arrays for analysis of gene expression. *Methods Enzymol* 303: 179–205.

39. Bachem CWB, Oomen RJFJ, Visser RGF (1998) Transcript imaging with cDNA-AFLP: A step-by-step protocol. *Plant Mol Biol Reporter*16: 157–173.

40. Ma L, et al. (2002) Genomic evidence for COP1 as a repressor of light-regulated gene expression and development in Arabidopsis. *Plant Cell* 14: 2383–

402

2398.

41. Ma L, et al. (2001) Light control of Arabidopsis development entails coordinated regulation of genome expression and cellular pathways. *Plant Cell* 13: 2589–2607.

42. Altschul SF, Gish W, Miller W, Myers EW, Lipman DJ (1990) Basic local alignment search tool. *J Mol Biol* 215: 403–410.

43. Altschul SF, et al. (1997) Gapped BLAST and PSI-BLAST: A new generation of protein database search programs. *Nucleic Acids Res* 25: 3389–3402.

44. Durbin R, EddySR, KroghA, Mitchison GJ (1998) Biological Sequence Analysis: *Probabilistic Models of Proteins and Nucleic Acids.* (Cambridge Univ Press, Cambridge, U.K.).

45. Ashburner M, et al. (2000) Gene ontology: Tool for the unification of biology.The Gene Ontology Consortium.*Nat Genet* 25: 25–29.

46. Mistry J, et al. (2006) Pfam: Clans, web tools and services. *Nucleic Acids Res* 34: D247–D251.

47. Mao X, Cai T, Olyarchuk JG, Wei L (2005) Automated genome annotation and pathway identification using the KEGG Orthology (KO) as a controlled vocabulary. *Bioinformatics* 21: 3787–3793.

作者: Gang Wei   Yong Tao   Guozhen Liu   Chen Chen   Renyuan Luo   Hongai Xia   Qiang Gan   Haipan Zeng   Zhike Lu   Yuning Han   Xiaobing Li   Guisheng Song   Hongli Zhai   Yonggang Peng   Dayong Li   Honglin Xu   Xiaoli Wei   Mengliang Cao   Huafeng Deng   Yeyun Xin   Xiqin Fu   Longping Yuan[*]   Jun Yu[*]   Zhen Zhu[*]   Lihuang Zhu[*]

注: 本文发表于 *Proceedings of the National Academy of Sciences of the United States of America* 2009 年第 106 卷第 19 期。

发展杂交水稻　保障粮食安全（B）

【摘　要】针对人口不断增加而耕地不断减少的状况，解决粮食短缺问题的唯一途径显然就是提高单位面积产量。水稻是最主要的粮食作物，提高水稻产量对保障粮食安全意义重大。实践证明，利用水稻杂种优势是提高水稻单产最有效的途径之一。从杂交水稻研究的最新进展出发，论述了通过研究和推广超级杂交稻，可以大幅度提高水稻产量，保障我国的粮食安全。

【关键词】杂交水稻；单位面积产量；粮食安全

目前世界人口已超过 60 亿人，到 2030 年将达到 80 亿人，同时，每年转作其他用途的土地面积为 1 000 万～3 500 万 hm^2，其中有一半来自农田。我国的情况与此类似，2020 年人口将增加到 14.5 亿人，且每年要减少数百万亩耕地。面对这种巨大的人口增长压力和严重的耕地减少情况，解决粮食短缺问题的唯一途径显然就是通过科技进步，大幅度提高单位土地面积的粮食产量。

水稻是最主要的粮食作物，世界上一半以上的人口以稻米为主食，在我国更高达 60% 以上，因此，增加水稻产量对保障粮食安全和减少贫困具有极其重要的意义。

国际水稻研究所所长齐格勒估计，目前每公顷稻田可提供 27 人的粮食，到 2050 年必须提供 43 人的口粮才能满足需要。也就是说，水稻的单产要在现有的基础上提高 60%。任务之艰巨，由此可见。理论上，水稻蕴藏着巨大的产量潜力，有许多方法可以提高水稻的产量，例如灌排工程的建设、土壤条件的改善、栽培技术的提高和高产品种的培育等。简言之，夺取水稻高产，必须要有良种、良法和良田的配套。良种是核心，良法是手段，良田是基础，三者缺一不可。

在良种选育方面，我们正在攻关第 3 期超级杂交稻，采取分三步走的策略。第 1 步的目标是每公顷产 12.5　t（即亩产 830　kg 左右），已在2009年实现；第2步的目标是每公顷产13 t（即亩产860 kg左右），计划2012年达到；第3步争取在2015年以前实现每公顷产13.5 t（即亩产900 kg）。技术上采取的是常规手段与分子技术相结合的路线，即形态改良、提高杂种优势水平和利用远缘有利基因，后者又以转C_4基因为重点，以期育成产量潜力为每公顷13.5～15.0 t（即亩产900～1 000 kg）、具有优良的株型、强大的杂种优势和高光效的C_4型超级杂交稻。

在良法方面，有一项新的突破，是最近研制成功的超级稻专用肥（以下简称专用肥），肥效非常好。2009年冬至2010年春在三亚试验的结果是，在等量N肥（210 kg/hm^2）的条件下，专用肥比常规施肥法增产22%，而且其增产效果是综合性的，即株高、穗数、粒数、粒重均有所提高，实在难能可贵。2010年在湖南十多个中稻和一季晚稻超级杂交稻"百亩示范片"安排了常规肥和专用肥大面积的对比试验，同时有正规的辅助性试验。初步观察和测产表明，专用肥的效果显著超过常规施肥。例如，湖南溆浦县对Y两优2号"百亩示范片"施专用肥的测产为15.9 t/hm^2，按八五折计算，单产可达13.5 t/hm^2；桃源县2个组合各1个"百亩示范片"，用专用肥和常规肥各3.33 hm^2（即50亩）进行对比，目前尚处在乳熟期，现将测产的数据列于表1。

表1 超级稻专用肥与常规肥料增产潜力比较

品种名称	肥料种类	株高/cm	穗数/($10^4 \cdot hm^{-2}$)	每穗总粒数	结实率/%	千粒重/g	理论产量/($kg \cdot hm^{-2}$)
Y两优2号	超级稻专用肥	115.2	258.75	244.7	90	27	15 385.8
	常规肥（CK）	103.6	225.30	249.1	90	27	13 637.7
	比CK增减/%	11.2	14.8	−1.8	0	0	12.8
P88S/1128	超级稻专用肥	115.8	198.7	283.4	85	28	13 402.0
	常规肥（CK）	103.7	164.3	296.8	85	28	11 605.0
	比CK增减/%	11.7	20.9	−4.5	0	0	15.5

虽然现在还未有实际产量结果，但表1的数据和溆浦县的测产已足够说明专用肥比常规肥具有显著的增产效果，这的确是良法方面的一项重大进展。超级杂交稻加专用肥，如虎添翼，有望提前实现第3期单产13.5 t/hm^2（即亩产900 kg）的目标。

第3期超级杂交稻研究成功，对保障我国的粮食安全将发挥重要作用。杂交水稻的发展历史表明，凡在育种上有所突破，就能使水稻的产量跃上一个新台阶，20世纪80年代育成的高产三系杂交稻汕优63，大面积单产7.5 t/hm^{2}左右；20世纪90年代末期育成的第1

期两系超级杂交稻两优培九，大面积单产为 8.25 t/hm² 左右；"十一五"期间育成的第 2 期超级杂交稻 Y 两优 1 号、P88S/0293 等组合，大面积单产在 9.0 t/hm² 以上。预计，第 3 期超级杂交稻推广后，单产又能再提高至 750 kg/hm² 以上。从 1996 年我国超级稻育种计划立项开始，每 5 年左右就登上一个新台阶，这个台阶的高度很高，示范田是每公顷 1 500 kg，大面积生产田是每公顷 750 kg，但是我们都跨越了。这些成绩充分显示发展杂交水稻的巨大增产潜力和我国在这一研究领域的国际领先地位。

到 2020 年，为了满足我国 14.5 亿人口的粮食需求，国家发展和改革委员会出台了要增产 500 亿 kg 粮食计划。按水稻占粮食作物的比重为 40% 计算，需增产稻谷 200 亿 kg。从技术上的角度看，这项任务能够完成。近年来国家杂交水稻工程技术研究中心在积极推进"种三产四"丰产工程，取得了较好的效果。所谓"种三产四"是指种"三亩"超级杂交稻，产原有水平"四亩"田所产的粮食。2007 年率先在湖南省启动，20 个示范县，有 18 个县达标；2009 年 32 个示范县，总面积 9.8 万 hm²，有 29 个县达标。安徽芜湖市、河南信阳市也在实施这项工程，效果同样良好。广东、广西、云南、贵州对实施该项工程亦很积极。因此，我强烈建议农业部应将"种三产四"丰产工程纳入"高产创建"计划，作为其中重要的具体内容予以立项，从而就能更有力地推动该项工程在全国稻区推广。2015 年发展到 400 万 hm²，等于增加了 133 万 hm² 稻田，按每公顷产 7 500 kg 计算，可增产粮食 100 亿 kg；2020 年发展到 800 万 hm²（约占杂交水稻种植面积的一半），相当于增加了 267 万 hm² 稻田，可增产 200 亿 kg 稻谷。这意味着，只要努力实施好这项工程，国家下达的到 2020 年增产稻谷的任务就能圆满完成，为保障中国的粮食安全做出应有的贡献。

作者：袁隆平

注：本文收录于《第 1 届中国杂交水稻大会论文集》2010 年。

不同施氮量和密度对超级杂交水稻产量及群体结构的影响

【摘　要】以超级杂交水稻两优 0293 为供试材料，设 3 个施肥水平：低 N、135 kg/hm²、中 N、270 kg/hm²，高 N、405 kg/hm²，不施 N 肥为对照；密度也设 3 个水平：33.3 cm×33.3 cm，26.6 cm×26.6 cm，26.6 cm×20.0 cm。试验采用裂区试验，以肥料为主区，以密度为副区，小区均随机排列，试验在不同栽插密度（行距）条件下，研究不同施肥量对超级稻产量及群体结构的影响，结论为以施 N（270 kg/hm²）、密度（20.0 cm×26.6 cm）处理产量最高，群体结构最好。

【关键词】超级杂交水稻；施 N 量；密度；产量

近年来，随着生产条件的改善、品种生产力的演进以及增施氮肥和改进栽培技术等，水稻单产不断提高，对缓减粮食压力起了重要作用[1-5]。本试验在不同栽插密度（行距）条件下，研究不同施肥量对产量及群体结构的影响，为水稻优质高产高效施肥技术提供理论和实践依据。

1　材料与方法

1.1　试验地点

试验分别于 2006 年、2007 年在湖南杂交水稻研究中心试验田进行。重复性好。

1.2　供试品种

超级杂交中籼稻：两优 0293。

1.3　试验设计

两优 0293 设 3 个施肥水平：N_1（不施 N 肥）为对照；N_2（低 N，135 kg/hm²）；N_3（中 N，270 kg/hm²）；N_4（高 N，405 kg/hm²）。

基蘖肥与穗粒肥用量比为 6∶4；N、P、K 用量比为 1.0∶0.5∶0.8。密度也设 3 个水平：
M_1（33.3 cm×33.3 cm），M_2（26.6 cm×26.6 cm），M_3（26.6 cm×20.0 cm）。试
验采用裂区试验，以肥料为主区，以密度为副区，小区均随机排列。4 次重复。小区拉线划行
作埂，埂面覆盖塑料薄膜，面积 32 m^2，单灌单排。5 月 21 日播种，湿润育秧。6 月 11 日
移栽，每穴 2 苗，秧苗带茎蘖数、叶龄基本一致，田间管理同大田栽培。全生育期采用无污
染浅水灌溉，不晒田。全程进行病虫防治。

1.4　测定项目及方法

叶龄进程、茎蘖动态：移栽后 15 d，每小区定点 10 蔸，每 5 d 观测 1 次群体苗数，直
到抽穗。

叶面积指数（Leaf area index，LAI）和干物质积累量：各小区分别于分蘖高峰期、幼
穗分化期、孕穗期、齐穗期按梅花形取样，每小区取 5 蔸，用长宽系数法测定叶面积，于孕
穗期、抽穗期、齐穗期按梅花形取样，每小区取 5 蔸，然后将植株按茎、叶、穗分别装袋，
于 105 ℃杀青 30 min，经 80 ℃烘干至恒重，考察干物质积累量。

成熟期测定产量结构。割方测定实产：于收获期每小区按对角线取 12 蔸考察产量构成因
素；小区中心 5 m^2 收割脱粒后晒干测产，折算成 13.5% 水分的单位面积产量。

1.5　数据处理

在 DPS 数据处理软件平台[6]上进行数据处理分析。

2　结果与分析

2.1　不同处理对产量与产量构成因素的影响

分析试验结果表明，各施肥处理间的产量差异达极显著水平，不同施氮处理的每公顷
产量分别在 7 t，8 t，9 t，10 t 4 个台阶，说明两优 0293 的产量潜力较大，氮肥运筹对
超级杂交稻产量的影响十分明显，总体上为施氮量越低，产量越低（表 1、表 2）。如施氮水
平低于 N_2（135 kg/hm^2），产量较低，低于 9 t/hm^2，这与别人的结论一致。唐启源等[7]
的试验结果表明施氮水平在 0～130 kg/hm^2，产量在 9 t/hm^2 以下。施氮水平等于 N_2
（135 kg/hm^2），产量大于 9 t/hm^2，但小于 10t/hm^2，施氮水平 N_3 处理（270 kg/hm^2）
的平均产量达 10.07 t/hm^2，个别小区达 12 t/hm^2。施氮量与产量的关系呈单峰曲线，以
270 kg/hm^2 施氮量处理产量最高。在施氮量 135～270 kg/hm^2 的范围内增产效应明显。

密度对产量的影响，理论产量都是 $M_1<M_2<M_3$，实际产量，不施氮时，以中等密度（M_2，26.6 cm×26.6 cm）产量最高。低氮水平 N_2 下，以低密度（M_1，33.3cm×33.3cm）最高。中氮水平 N_3 下，以高密度（M_3，20.0 cm×26.6 cm）产量最高。高氮水平 N_4 下，以中等密度（M_2，26.6 cm×26.6cm）产量最高。综合来看，以 N_3M_3 处理产量最高，实际产量达 10.9 t/hm²。各施氮处理的株高没有明显差异，株高随着施氮水平提高而有所升高，超过一定水平后，随施氮水平进一步提高，株高反而有所回落，但升高降低都不显著。穗长也呈同样的趋势，随施氮水平增加而增加，超过一定限度后，不升反降。

表1　供试土壤特性

pH	有机质 /(g·kg⁻¹)	全氮 /(g·kg⁻¹)	全磷 /(g·kg⁻¹)	全钾 /(g·kg⁻¹)	速效磷 /(mg·kg⁻¹)	速效钾 /(mg·kg⁻¹)	水解性氮 /(mg·kg⁻¹)
5.3	30.5	2.02	0.71	17.0	16.4	119	204.2

表2　两优0293不同处理的产量及产量构成

处理	株高 /cm	穗长 /cm	有效穗 /(10⁴·hm⁻²)	每穗总粒数	结实率 /%	千粒重 /g	理论产量 /(t·hm⁻²)	实际产量 /(t·hm⁻²)	收获指数
N_1M_1	102.6	25.4	189.5	212.9	82.1	24.0	7.95	7.31d	0.541 4
N_1M_2	99.9	23.4	221.2	182.6	87.8	24.2	8.58	7.90d	0.526 1
N_1M_3	105.4	24.4	229.7	184.4	84.9	24.4	8.77	7.65d	0.508 1
平均	102.6	24.4	213.5	193.3	84.9	24.2	8.43	7.65d	0.525 2
N_2M_1	112.4	26.2	216.0	231.3	82.9	23.6	9.77	9.70b	0.504 5
N_2M_2	111.6	25.9	250.6	197.8	83.1	23.9	9.84	9.40b	0.506 2
N_2M_3	107.4	25.0	270.0	196.3	83.0	23.9	10.50	9.30b	0.488 7
平均	110.5	25.7	245.5	208.5	83.0	23.8	10.04	9.50b	0.599 8
N_3M_1	111.9	24.0	222.3	222.3	83.6	23.1	9.50	9.00b	0.548 1
N_3M_2	115.0	25.4	273.0	201.5	83.5	24.0	11.02	10.30a	0.588 7
N_3M_3	114.0	24.0	283.1	184.8	86.7	24.3	11.02	10.90a	0.584 2
平均	113.6	24.4	259.5	202.9	84.6	23.8	10.51	10.07a	0.573 7
N_4M_1	108.5	23.4	221.4	196.5	82.1	23.1	8.25	8.10c	0.387 6
N_4M_2	107.4	24.7	264.6	183.3	83.3	23.4	9.45	9.10b	0.441 4
N_4M_3	113.9	23.5	270.0	182.6	81.2	23.7	9.49	8.80c	0.445 9
平均	109.9	23.9	252.0	187.5	82.2	23.4	9.06	8.80c	0.425 0

　　从产量结构看，各施肥处理间的千粒重差异显著，不施肥处理（N_1）千粒重最大，随着施氮水平提高，千粒重下降。但同一施肥处理不同密度之间差异极小。有效穗数随施氮水平的提高而增加，但超过一定限度，不升反而有所下降，但幅度不大。如果只从施氮量来进一步分析，施氮量 135 kg/hm² 以上和以下的处理间，有效穗数存在较大差异，但施氮量 135 kg/hm² 以上的处理间基本差异很小。同一氮水平下，都是稀密度（M_1）有效穗数远小于中密度（M_2）和高密度（M_3）。但中密度（M_2）和高密度（M_3）之间差异很少。其中以 N_3M_3 处理有效穗数最多，达 283.1 万穗 /hm²。每穗粒数与施氮水平存在显著正相关，从 N_1 到 N_2，随施氮水平上升每穗粒数增加，到了一定水平，再增加施氮水平，N_3 到 N_4，每穗粒数反而下降。中期施穗粒肥，促进了颖花数增多，说明施氮水平对超级杂交稻产量形成的影响，在较低施氮水平下表现为对穗数的影响，但在较高施氮水平下主要表现为对每穗粒数的影响。在同一施氮水平内，都是 M_1 显著大于 M_2、M_3，而 M_2、M_3 间基本没差异。结实率除高氮处理（N_4）外，其余相差较小。同一施氮水平内，不同密度间基本没什么差异。

2.2　不同处理的群体结构特点

2.2.1　不同处理对分蘖发生动态的影响

　　不同时期不同施氮处理的茎蘖动态图表明，各施氮处理的茎蘖动态差异，从分蘖中期后没有太大的变化，说明两优0293的茎蘖数较为稳定。不同处理的分蘖高峰期发生在移栽后35～40 d，高峰期苗数均随施氮水平上升而增加，随密度增加而增加。不同密度的分蘖动态都是稀的（M_1）小于密的（M_2、M_3），不施氮时，M_2 与 M_3 基本没差异，施氮水平提升后，M_2 与 M_3 差异也较小，但都与 M_1 有显著差异，分蘖数远大于 M_1（表3）。本试验条件下，说明 M_1 太稀，不适宜超级稻高产栽植，不能保证高产所需足够的分蘖数和有效穗数。以有效穗数最多、中等、最少3种处理为代表，作综合影响分蘖动态图（图1），进一步分析发现，在 135 kg/hm² 及其以上，M_3（20.0 cm×26.6 cm）的栽植密度时，有效分蘖数均能达到270万 /hm² 以上，而且处理间差异极小。其中以 N_3M_3 处理有效分蘖数最多，达 283.1 万个 /hm²（表3）。

表3　不同处理对分蘖发生动态（每平方米苗数）的影响　　　　　　　　单位：苗

处理	移栽后天数 /d											
---	15	20	25	30	35	40	45	50	55	60	65	90
N_1M_1	61.2	80.1	139.5	200.7	224.6	256.9	221.4	203.9	203.0	198.0	195.0	189.5
N_1M_2	84.7	124.6	198.8	305.2	343.7	326.2	266.0	250.6	246.4	238.0	231.0	221.2

续表

处理	移栽后天数 /d											
	15	20	25	30	35	40	45	50	55	60	65	90
N_1M_3	138.8	172.5	285.9	403.1	458.4	412.5	285.9	268.1	265.0	261.6	257.8	229.7
平均	94.9	125.7	208.1	303.0	342.2	331.9	257.8	240.9	238.1	232.5	227.9	213.5
N_2M_1	54.5	79.2	147.6	220.5	246.2	257.9	234.9	223.7	220.0	217.0	216.7	216.0
N_2M_2	95.2	141.4	231.0	363.3	408.8	394.8	310.1	287.0	282.0	276.0	273.0	250.6
N_2M_3	107.8	165.0	273.8	409.7	490.3	489.4	323.4	299.1	290.1	281.2	277.5	270.0
平均	85.8	128.5	217.5	331.2	381.8	380.7	289.5	269.9	264.0	258.1	255.7	245.5
N_3M_1	65.3	99.0	184.5	298.4	354.6	344.3	294.3	275.5	271.2	269.1	268.7	222.3
N_3M_2	126.0	198.1	323.4	466.9	527.8	493.5	355.6	306.6	300.1	290.2	286.3	273.0
N_3M_3	121.9	189.4	309.4	468.8	531.6	534.4	373.1	319.7	312.1	302.9	301.9	283.1
平均	104.4	162.2	272.4	411.4	471.3	457.4	341.0	300.6	294.5	287.4	285.6	259.5
N_4M_1	67.5	114.3	190.4	279.5	351.9	353.7	275.4	272.3	272.0	271.8	271.2	221.4
N_4M_2	114.1	178.5	291.2	456.4	540.4	518.7	351.4	328.3	318.2	308.3	304.5	264.6
N_4M_3	135.9	227.8	394.7	574.7	646.9	682.5	411.6	366.6	346.1	331.2	329.1	270.0
平均	105.8	173.5	292.1	436.9	513.1	518.3	346.1	322.4	312.1	303.8	301.6	252.0

2.2.2 不同处理条件下的干物质积累与分配特点以及 LAI

植株干物质积累量是水稻产量形成的重要基础。不同施肥处理间的地上部干物质积累差异，随生育进程的推进而逐渐加大，施氮处理对光合产物累积的影响主要表现在中后期。

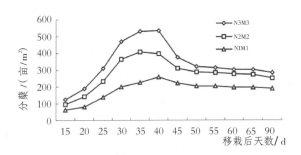

图 1 不同施肥与密度综合影响分蘖动态

干物质在植株各部位的分配比例，各施氮处理间在抽穗以前没有明显不同，但在抽穗以后存在规律性的差异（表 4）。各生育期特别是孕穗期、齐穗期叶片的干物质比例，以施氮量较少、产量较低的处理较小，以 N_3M_3 处理产量最高，叶片的干物质比例最大（图 2）。说明孕

穗期、齐穗期叶片维持一定水平的光合产物比例，对于超级稻防止早衰和夺取高产是有利且必
要的。成熟期残存于秸秆中的干物质比例，与抽穗期相反，以产量较低的处理较高，表明这些
施氮处理的光合产物转运不完全，可能与其早衰、导致运输不畅有关。

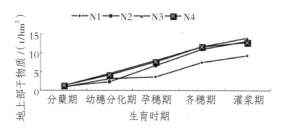

图 2　不同施肥处理的干物质积累动态

表 4　两优 0293 不同时期不同处理下的干物质分配比例　　　　　　　单位：%

处理	孕穗期（8月9日）		抽穗期（8月21日）			成熟期（9月28日）		
	叶	茎鞘	叶	茎鞘	穗	叶	茎鞘	穗
N₁M₁	33.3	66.7	22.3	46.2	31.5	31.8	29.9	38.3
N₁M₂	35.2	64.8	23.6	40.6	35.8	34.2	32.2	33.6
N₁M₃	32.6	67.4	19.7	40.9	39.4	33.2	31.1	35.7
平均	33.7	66.3	21.9	42.6	35.6	33.1	31.1	35.8
N₂M₁	37.1	62.9	28.2	47.4	24.4	32.1	26.8	41.1
N₂M₂	34.8	65.2	24.5	45.2	30.3	38.4	24.9	36.7
N₂M₃	36.5	63.5	26.8	45.8	27.4	31.7	26.6	41.7
平均	36.1	63.9	26.5	46.1	27.4	34.1	26.1	39.8
N₃M₁	42.2	57.8	34.4	48.1	17.5	39.1	28.2	32.7
N₃M₂	36.9	63.1	29.8	49.6	20.6	39.5	37.2	23.3
N₃M₃	42.0	58.0	36.2	49.0	14.8	34.2	30.6	35.2
平均	40.4	59.6	33.5	48.9	17.6	37.6	32.0	30.4
N₄M₁	45.4	54.6	32.9	52.6	14.5	32.6	30.2	37.2
N₄M₂	44.3	55.7	35.0	47.0	18.0	32.8	29.3	37.9
N₄M₃	40.3	59.7	30.5	45.2	24.3	33.6	31.5	34.9
平均	43.3	56.7	32.8	48.3	18.9	33.1	30.3	36.6

叶面积指数是反映群体生长快慢的一个重要指标。由表 5 可知，LAI 的趋势表现为，随

施氮量增加，LAI 逐渐增大（r=0.978 3[**]）；随着生育期推进，LAI 增大，除 N_1（不施肥，对照）以孕穗期的 LAI 最大外，其余处理 N_2、N_3、N_4 均以齐穗期的 LAI 最大，齐穗期后逐渐下降。同一施肥水平下，随着密度增加，LAI 增加，但齐穗期 N_3、N_4 施肥处理有波动（表5）。

表5 不同时期两优 0293 不同处理下的 LAI

处理	分蘖期 （7月6日）	幼穗分化后期 （7月25日）	孕穗期 （8月9日）	齐穗期 （8月27日）
N_1M_1	0.74	2.03	4.33	4.01
N_1M_2	1.05	2.84	5.41	5.05
N_1M_3	1.01	3.19	6.02	5.12
平均	0.93	2.69	5.25	4.73
N_2M_1	0.92	2.48	6.33	6.67
N_2M_2	1.48	3.99	5.90	6.30
N_2M_3	1.40	4.09	6.08	8.70
平均	1.27	3.52	6.10	7.22
N_3M_1	1.04	2.55	5.41	8.40
N_3M_2	1.88	4.98	6.51	8.01
N_3M_3	1.99	5.56	8.40	9.37
平均	1.64	4.36	6.77	8.59
N_4M_1	1.01	2.68	6.95	8.99
N_4M_2	1.42	3.91	7.26	9.49
N_4M_3	2.41	5.10	6.83	8.23
平均	1.61	3.90	7.01	8.90

2.2.3 不同处理对群体结构的影响

不同处理的基本苗随密度增加而增加，随施氮水平上升而增加。最高苗数也是随施氮水平上升而增加，随密度增加而增加。有效穗随着施氮水平的上升而增加，N_3 后稍有下降。在同一施氮水平下，随密度增加而增加，M_2、M_3 比 M_1 有显著增加。成穗率随施氮水平上升而减少，但差异不显著。在同一施氮水平下，随密度增加，成穗率减少，差异显著。但 N_3M_2 稍有特例，可能是误差所致。抽穗期叶面积指数随施氮水平上升而升高，但跟密度无显著相关性。凌启鸿等认为抽穗期单茎茎鞘重对群体质量具有较大意义[8]。在本试验中，随施氮水平上升，抽穗期单茎茎鞘重稍有下降，差异不显著，不施氮最低。单茎叶片重随施氮水平上升而增重，

但 N_3 后稍有下降（表 6）。产量在 8 t/hm^2 以下的 N_1，其单茎叶重相应较低（< 1.0 g），这可能与两优 0293 极好的株叶型有关，同时也表明超级杂交稻超高产对源器官的要求与一般杂交稻组合不同，其抽穗期单茎叶片重应达到一定水平（1.1～1.6 g）。

表 6　不同处理的植株群体结构及抽穗期干物重、粒叶比

处理	基本苗数 /（万株·hm^{-2}）	最高苗数 /（万株·hm^{-2}）	有效穗数 /（万穗·hm^{-2}）	成穗率 /%	抽穗期			粒叶比 /（粒·cm^{-2}）
					LAI	单茎叶片 干重 /g	单茎茎鞘 干重 /g	
N_1M_1	61.2	256.9	189.5	73.8	4.01	0.910	1.887	1.004
N_1M_2	84.7	343.7	221.2	64.4	5.05	0.967	1.664	0.810
N_1M_3	138.8	458.4	229.7	50.1	5.12	0.902	1.874	0.878
平均	94.9	353.0	213.5	52.7	4.73	0.926	1.808	0.897
N_2M_1	54.5	257.9	216.0	83.7	6.67	1.324	2.229	0.750
N_2M_2	95.2	408.8	250.6	61.3	6.30	1.073	1.981	0.791
N_2M_3	107.8	490.3	270.0	55.1	8.70	1.422	2.433	0.592
平均	85.8	385.7	245.5	66.7	7.22	1.273	2.214	0.711
N_3M_1	65.3	354.6	222.3	62.7	8.40	1.628	2.272	0.586
N_3M_2	126.0	527.8	273.0	51.7	8.01	1.259	2.094	0.687
N_3M_3	121.9	534.4	283.1	53.0	9.37	1.419	1.919	1.107
平均	104.4	472.3	259.5	55.8	8.59	1.435	2.095	0.660
N_4M_1	67.5	353.7	221.4	62.6	8.99	1.714	2.742	0.492
N_4M_2	114.1	540.4	264.6	49.0	9.49	1.531	2.053	0.514
N_4M_3	135.9	682.5	270.0	39.6	8.23	0.708	1.050	0.559
平均	105.8	525.5	252.0	50.4	8.90	1.318	1.948	0.522

LAI 和粒叶比的趋势表现为，随施氮量增加，LAI 逐渐增大，虽颖花数也相应增加，但粒叶比呈减少趋势，说明施氮对叶面积的增加作用超过了对颖花的促进作用。其中施氮量最高的 N_4 处理 LAI 最高，而粒叶比最低，表现为群体结构开始失调。因此，如何协调 LAI 和粒叶比的关系，是超级杂交稻施肥需要探讨的重要问题。

2.2.4　不同处理对株叶型的影响

不同处理施氮水平对株高有显著影响，N_1、N_2、N_3、N_4 水平下的株高差异显著。随着施氮水平上升，株高增高。在同一施氮水平下，移栽密度不同，株高也有差异。但不同的施氮水平下，密度影响株高的显著性不同。N_1 水平下，不同密度下的株高差异不显著。N_2 水平下，稀

植（M_1，33.3 cm×33.3 cm）和密植（M_3，20.0 cm×26.6 cm）差异显著，但都与中等密度（M_2，26.6 cm×26.6 cm）差异不显著。N_3 水平下，M_1 株高显著高于 M_2、M_3 的株高，N_4 水平下，M_1 株高高于 M_2，显著高于 M_3 的株高。综合来看，施氮水平最高移栽密度最稀的 $N_4 M_1$ 处理的株高最高（表7）。

表7　不同处理成熟期的株高以及顶部3片叶与垂直穗轴夹角

处理	株高 /cm	倒3叶与穗轴夹角 /（°）	倒2叶与穗轴夹角 /（°）	倒1叶（剑叶）与穗轴夹角 /（°）
$N_1 M_1$	120.6a	46.25a	31.75a	28.75a
$N_1 M_2$	119.2a	39.00a	27.25b	29.75a
$N_1 M_3$	116.9a	40.25a	28.00ab	34.25a
N_1	118.9d	41.83a	29.00a	30.92b
$N_2 M_1$	135.7a	37.25a	26.25b	34.50b
$N_2 M_2$	131.9ab	39.75a	27.50b	45.25a
$N_2 M_3$	131.1b	34.75a	34.75a	37.25ab
N_2	132.9c	37.25bc	29.50a	39.00a
$N_3 M_1$	141.5a	40.25a	28.25b	27.50b
$N_3 M_2$	137.2b	37.25a	28.25b	34.50b
$N_3 M_3$	137.2b	38.50a	33.00a	46.00a
N_3	138.6b	38.67ab	29.83a	36.00ab
$N_4 M_1$	143.6a	36.00a	28.75a	36.75a
$N_4 M_2$	142.1ab	32.75a	30.25a	43.40a
$N_4 M_3$	138.9b	34.25a	32.50a	43.25a
N_4	141.5a	34.33c	30.50a	41.13a

注：相同施 N 水平条件下同列数据后带相同小写字母表示在 0.05 水平上差异不显著。

由上可见，倒3叶与垂直穗轴的夹角，不同施氮水平间有显著差异，但同一施氮水平不同密度处理间，无显著差异。N_1 水平下倒3叶夹角显著大于 N_2、N_4 处理下的夹角，不显著大于 N_3 处理下夹角。说明倒3叶夹角与密度无关，而与施氮水平有关。倒2叶与垂直穗轴的夹角，不同施氮水平间无显著差异，但同一施氮水平不同密度处理间，有的有显著差异。N_1 水平下，M_1 与 M_2 有显著差异；N_2、N_3 水平下，M_3 与 M_1、M_2 有显著差异；N_4 水平下，不同密度间无显著差异。不同密度间，M_3 与 M_1、M_2 有显著差异，M_1 与 M_2 间无显著差异。倒1叶与垂直穗轴的夹角，不同施氮水平间 N_1 与 N_2、N_3、N_4 处理间有显著差异，同一施氮 N_1、

N_4 水平下不同密度处理间，也无显著差异。

3　讨论

本试验施肥水平，纯 N 在 135 kg/hm^2 与 270 kg/hm^2 之间或 270 kg/hm^2 与 405 kg/hm^2 之间是否还有更合适的施 N 量值得进一步深入研究，因为本试验处理与处理之间施氮量差距太大。同时，密度处理，本试验以 20.0 cm×26.6 cm、33.3 cm×33.0 cm、26.6 cm×26.0 cm 为代表，其他密度处理也值得探讨。

综合来看，以 N_3（270 kg/hm^2）M_3（20.0 cm×26.6 cm）处理产量最高，实际产量达 10.9 t/hm^2；千粒重随着施氮水平提高而下降，各施肥处理间千粒重差异显著，但同一施肥处理不同密度之间差异极小；有效穗数以 N_3M_3 处理最多，达 283.1 万穗 /hm^2。施氮水平对超级杂交稻产量形成的影响，在较低施氮水平下表现为对穗数的影响，但在较高施氮水平下主要表现为对每穗粒数的影响。

不同施氮处理间的地上部干物质积累差异，随生育进程的推进而逐渐加大，施氮处理对光合产物累积的影响主要表现在中后期；随施氮量增加，LAI 逐渐增大，同一施氮水平下，随密度增加，LAI 增加，颖花数也相应增加，但粒叶比呈减少趋势。

施氮与密度影响株叶形态，施氮水平最高移栽密度最稀的 N_4M_1 处理的株高最高；倒 3 叶与垂直穗轴的夹角，施氮水平间有显著差异，但同一施氮水平不同密度处理间无显著差异；倒 2 叶与垂直穗轴的夹角，施氮水平间无显著差异，密度处理间的差异显著性与施氮水平有关；倒 1 叶与垂直穗轴的夹角，N_1 处理与 N_2、N_3、N_4 处理间有显著差异，N_1、N_4 水平下不同密度处理间，也无显著差异。

──────────── References ────────────

参考文献

［1］许世觉，姚必仁，王细国，等. 两优培九超高产示范栽培技术 [J]. 杂交水稻，2001，16（2）：33-34.

［2］杨春献，向邦豪，张其茂，等. 两系超级杂交稻 "百亩片" 平均单产 12.26 t/hm^2 的栽培技术 [J].

416

杂交水稻, 2003, 18（2）: 42-44.

［3］吴朝晖. 超级杂交稻新组合 P88S/0293 在海南三亚单产超 12 t/hm² 的栽培技术 [J]. 杂交水稻, 2003, 18（6）: 36-37.

［4］孟卫东, 王效宁, 邢福能, 等. 超级杂交稻新组合 P88S/0293 在海南大面积示范单产超 12 t/hm² 的栽培技术 [J]. 杂交水稻, 2005, 20（1）: 46-49.

［5］石庆华, 程永盛, 潘晓华, 等. 施氮对两系杂交晚稻产量和品质的影响 [A]. 第 7 届全国栽培理论与实践学术研究会交流材料汇编 [C].1999.

［6］唐启义, 冯光明. 实用统计分析及其 DPS 数据处理系统 [M]. 北京: 科学出版社, 2002.

［7］唐启源, 邹应斌, 米湘成, 等. 不同施氮条件下超级杂交稻的产量形成特点与氮肥利用 [J]. 杂交水稻, 2003, 18（1）: 44-48.

［8］凌启鸿. 作物群体质量 [M]. 上海: 上海科技出版社, 2000.

作者: 吴朝晖　袁隆平 *

注: 本文收录于《第 1 届中国杂交水稻大会论文集》2010 年。

水稻亚种间重组自交系及其回交群体产量性状的相关与通径分析

【摘 要】对培矮 64S/9311 的重组自交系 F_7 群体（RILs F_7）及其回交 F_1 群体（RILs BCF_1）这 2 个群体的产量及产量构成因素性状进行研究，分析了各产量和产量构成因子在 2 个群体中的遗传变异性及相关性与通径关系。结果表明，性状发生了丰富变异，有利于进一步作杂种优势分析与 QTL 定位等研究；相关分析得出 RILs F_7 产量构成因素与产量均呈极显著相关，RILs BCF_1 小区产量和产量的构成因子每穗实粒数、结实率存在极显著的正相关，2 个群体产量和产量的构成因子相关程度均表现为：每穗实粒数＞结实率＞千粒重＞每穗总粒数＞单株有效穗数＞穗长；通径分析各性状对单株产量的直接通径系数依次为：每穗实粒数＞千粒重＞单株有效穗数＞穗长＞结实率＞每穗总粒数，每穗实粒数是产量的重要作用因子，其他产量构成因素性状尤其是结实率，在产量贡献方面有一定的牵制作用。

【关键词】亚种间杂交稻；重组自交系与其回交衍生群体；产量性状；遗传变异；相关分析；通径分析

水稻亚种间强大的杂种优势，一直被认为是用来进一步提高水稻产量的有效途径[1-2]。由于籼、粳亚种的亲缘关系比较远，不亲和，导致杂种受精结实不正常，结实率很低。前人已进行了大量籼、粳亚种特性的遗传分化及杂种优势研究[3-6]。本研究应用具一定亲缘关系的重组自交系与具有亚种间杂种特性的重组自交系回交衍生群体组合，来探讨产量与产量构成因子性状的遗传变异、相关与通径分析，目的在于为亚种间杂交稻育种提供科学参考。

1 材料与方法

1.1 材料

所用水稻材料为培矮 64S/9311 的 F_7 重组自交系（RIL）群

体、其 F_6 群体与培矮 64S 配制的回交 F_1 群体和两优培九。重组自交系（RIL）群体来自于 64S/9311 杂交产生的 F_1，自交产生 F_2，再将 F_2 单株选可育株收取单穗，经单粒传法连续多代套袋自交形成的纯合株系。其中，于 F_4 代应用分子标记辅助选择，借助均匀分布在 12 条染色体上的 151 个多态性 SSR 标记进行辅助选择得到 219 个具有亲本遗传背景的、分子标记覆盖水稻全基因组的重组自交系 RILs。重组自交系回交 F_1 群体（以下简称 RILs BCF_1）来自以 RILs F_6 每个株系取 1 株作父本分别与培矮 64S 回交，获得回交 F_1。

1.2 方法

1.2.1 试验设计　将 RILsBCF$_1$ 和对照两优培九按完全随机区组设计，2006 年在湖南杂交水稻研究中心试验田按 3 次重复，每小区 3 行，每行 11 株，单株种植，株行距为 16.7 cm×23.3 cm；5 月 8 日播种；5 月 27 日移栽，秧龄 20 d。将 RIL F_7 群体于 2007 年夏种植于湖南杂交水稻研究中心试验田，每株系 5 行，每行 10 株，单株种植，株行距为 16.7 cm×23.3 cm；5 月 10 日播种，5 月 29 日移栽，秧龄 20 d。以上试验土壤肥力中等，地力均匀，田间管理同一般大田。成熟时每小区选中间行的中间 5 株收取，按标准进行室内风干考种。考查项目包括株高、单株有效穗数、穗长、每穗总粒数、每穗实粒数、结实率、千粒重和小区产量或单株产量等性状。

1.2.2 数据处理　考种数据按株系、性状求小区单株平均数，再计算各个性状平均值、方差、表型相关系数，并作遗传分析、相关分析、通径分析。用 Micro Excel 2003 软件计算各项平均值；采用 DPS 数据处理系统软件 8.01 版本对上述性状进行多元统计分析，包括简单相关分析、应用亲子回归估计遗传力等；采用 DTWIN 软件进行通径分析。以上分析取各项平均值。遗传力估计原理为应用亲子回归，用性状子代观察值平均数与亲代观察值平均数（中亲值）回归直线的斜率表示，由 DPS 系统根据亲代和子代性状值，获得遗传力估计值。

2　结果与分析

2.1 重组自交系 RILs F_7 与回交 RILs BCF_1 群体产量性状遗传力与变异分析

RILs F_7 与 RILs $BCF_1$2 个群体作单季种植的产量性状的遗传与变异情况表现见表 1。考察各个性状极差值，RILs BCF_1 群体单株产量、单株有效穗数、单株实粒数的极差均大于 RILs F_7 的相应性状的极差，除千粒重外，其余性状极差值都远远高于相同性状最小值，每穗实粒数最大，达 49 倍；即使极差值小的穗长也达最小株系的 64%，最小的千粒重达最小株系的14.3%，表明 RILs BCF$_1$ 群体各株系间存在明显的差异，分离广泛。RILs BCF$_1$ 群体因其单株均为 F$_1$，与其相应的亲本群体 RILs F$_7$ 相比，由于杂种优势的存在，单株产量及其构成因子每穗总粒数、千粒重性状表现平均观测值高于 RILs F$_7$ 的平均观测值，试验显示产量的优势最大，产量构成因子中以千粒重的优势最小。比较 RILs F$_7$ 与 RILs BCF$_1$ 群体性状平均值间的差异，RILsBCF$_1$ 的每穗总粒数性状比 RILs F$_7$ 的差异大，平均值高出 36.18 粒，表现出穗大粒多；而对于结实率则出现相反趋势的较大差异，主要因为具有亲本 9311 遗传背景的重组自交系与培矮 64S 回交，表现为亚种间杂交，使结实受到影响。

表1　重组自交系（RILs）与回交 RILs 群体性状遗传与变异情况

变量	EP	PL/cm	SPP	GPP	PSS/%	GW/g	HY$_{plant}$/g
RILs 最大值	13.33	29.45	257.09	200.88	92.87	31.15	41.45
RILs 最小值	4.33	14.25	76.00	24.78	16.16	7.71	2.57
RILs 极差	9.00	15.20	181.09	176.10	76.71	23.44	38.88
RILs 平均值	8.29	20.74	155.51	107.64	69.20	22.80	19.31
RILs 标准差	1.51	1.98	33.29	30.96	13.88	3.13	6.22
RILs BCF$_1$ 最大值	17.90	27.06	244.18	189.99	81.28	24.70	52.98
RILs BCF$_1$ 最小值	7.10	19.70	119.37	8.57	5.14	23.35	3.57
RILs BCF$_1$ 极差	10.80	7.37	124.80	181.42	76.14	1.36	49.40
RILs BCF$_1$ 平均值	12.08	22.70	191.69	94.77	48.92	23.84	24.83
RILs BCF$_1$ 标准差	2.31	1.10	20.91	37.83	18.48	0.22	0.35
遗传力 h^2/%	81.44	89.43	91.09	90.81	89.41	99.68	93.56
变异系数 CV/%	25.60	6.50	14.90	48.70	44.60	0.90	46.20

注：EP. 单株有效穗数；PL. 穗长；SPP. 每穗总粒数；GPP. 每穗实粒数；PSS. 结实率；GW. 千粒重；HY$_{plant}$. 单株产量。后同。

考察 RILs F$_7$ 与 RILs BCF$_1$ 的遗传性，估计各性状的遗传力（从亲代稳定传递给子代的遗传传递能力）以千粒重最高，达 99.68%；单株产量、每穗实粒数、每穗总粒数分别为 93.56%，90.81%，91.09%；穗长、结实率分别为 89.43%，89.41%；单株有效穗数最低，为 81.44%。表明与 RILs F$_7$ 比较，RILs BCF$_1$ 的性状差异总体趋势无大的出入，并且有很好的吻合性，产量及其构成因子中，以千粒重遗传力最大，其次依次为单株产量。

比较 RILs F$_7$ 与 RILs BCF$_1$ 产量各构成因子性状变异系数，从大到小排序可依次为：每穗实粒数（48.7%）＞单株产量（46.2%）＞结实率（44.6%）＞单株有效穗数（25.6%）＞每穗总粒数（14.9%）＞穗长（6.5%）＞千粒重（0.9%），表明 RILsBCF$_1$ 群体在每穗实粒数、单株产量、结实率性状上变异丰富，穗长、千粒重性状差异较小，而单株有效穗数、每穗总粒数的表现介于二者之间。总体来说，除千粒重产生变异仅为 0.9%、穗长产生变异 6.5% 外，其余性状变异都超过 10%，最大为每穗实粒数，其产生 48.72% 的变异；其次为单株产量，产生 46.2% 的变异；结实率第三，产生 44.6% 的变异。

2.2 重组自交系 RILs F7 与回交 RILs BCF1 群体产量及构成因素性状相关分析

RILs F_7 产量性状各构成因素间的相关分析见表 2。产量构成因素中，单株有效穗数与每穗总粒数呈极显著负相关（r = -0.367 3），与每穗实粒数呈显著负相关（r = -0.210 7），株高与单株有效穗数（r = -0.054）、单株有效穗数与穗长（r = -0.187 1）、结实率与每穗总粒数（r = -0.013）均呈负相关关系，但未达显著水平；其余产量构成因素之间均呈正相关，但株高、单株有效穗数、穗长与结实率，以及千粒重与单株有效穗数、每穗总粒数相关系数 r 分别为 0.188 6，0.078 7，0.048 1，相关性均未达显著水平；株高与穗长、每穗实粒数、每穗总粒数相关系数 r 分别为 0.312 1，0.326 4，0.266 1，相关性呈 0.01 水平极显著，穗长与每穗实粒数、每穗总粒数、千粒重相关系数 r 分别为 0.352 1，0.432 1，0.268 4，相关性呈 0.01 水平极显著，每穗实粒数与结实率、每穗总粒数、千粒重相关系数 r 分别为 0.667 4，0.723 3，0.202 8，相关性呈 0.01 水平极显著，结实率与千粒重相关系数 r 为 0.237 5，相关性呈 0.01 水平极显著；7 个产量构成因素与产量均呈极显著相关，其相关程度依次为：每穗实粒数（r = 0.740 6）>结实率（r = 0.620 7）>千粒重（r = 0.460 8）>每穗总粒数（r = 0.419 1）>单株有效穗数（r = 0.342 9）>穗长（r = 0.303 8）>株高（r = 0.276 5）。表明每穗实粒数、结实率对单株产量有较大促进作用，单株有效穗数、穗长、株高也对促进单株产量增加起作用。

表 2　重组自交系（RILs）产量及构成因素性状相关分析

变量	PH	EP	PL	GPP	PSS	SPP	GW	HY_{plant}
PH	1							
EP	-0.054	1						
PL	0.312 1**	-0.187 1	1					
GPP	0.326 4**	-0.210 7*	0.352 1**	1				
PSS	0.188 6	0.078 7	0.048 1	0.667 4**	1			
SPP	0.266 1**	-0.367 3**	0.432 1**	0.723 3**	-0.013	1		

续表

变量	PH	EP	PL	GPP	PSS	SPP	GW	HY_{plant}
GW	0.160 1[*]	0.068 6	0.268 4[**]	0.202 8[**]	0.237 5[**]	0.050 6	1	
HY_{plant}	0.276 5[**]	0.342 9[**]	0.303 8[**]	0.740 6[**]	0.620 7[**]	0.419 1[**]	0.460 8[**]	1

注：*，** 分别表示达 5% 显著和 1% 极显著水平。后同。

RILs BCF$_1$ 产量性状各构成因素间的相关分析见表3。小区产量和产量的构成因子每穗实粒数、结实率存在极显著的正相关，相关程度分别为 0.65，0.64。此外单株有效穗数与千粒重、穗长与每穗总粒数和每穗实粒数相关水平极显著。产量构成因子之间存在不同程度的负相关，如穗长与千粒重和小区产量之间存在负相关，单株有效穗数与每穗总粒数之间显著负相关。产量构成因素与产量的简单相关分析从表中可以看出，各产量构成因素与产量的相关程度为：每穗实粒数（0.648 6）>结实率（0.643 2）>千粒重（0.064 5）>每穗总粒数（0.050 1）>单株有效穗数（0.035 1）>穗长（−0.029 1）。而只有穗长与小区产量成负向趋势。表明每穗实粒数、结实率与小区产量的关系最为密切。

表 3　回交 RILs 群体产量及构成因素性状相关分析

变量	EP	PL	SPP	GPP	SLPP	PSS	GW	HY_{plot}
EP	1							
PL	0.013 4	1						
SPP	−0.111 7	0.491 98[**]	1					
GPP	0.111 2	0.202 5[**]	0.301 7[**]	1				
SLPP	−0.172 9[*]	0.072 9	0.253 7[**]	−0.844 9[**]	1			
PSS	0.152 4[*]	0.076 3	0.021 6	0.951 4[**]	−0.952 5[**]	1		
GW	0.183 5[**]	−0.086 5	−0.002 5	0.070 2	−0.072 5	0.080 5	1	
HY_{plot}	0.035 1	−0.029 1	0.050 1	0.648 6[**]	−0.627 3[**]	0.643 2[**]	0.064 5	1

各产量构成因素间的简单相关分析表明，单株有效穗数与每穗总粒数呈负相关，且与每穗总粒数负向趋势较大（r = −0.111 7）；每穗总粒数与千粒重呈负相关；而结实率与千粒重呈正相关；小区产量、每穗实粒数、每穗总粒数这 3 个性状之间，全部呈显著正相关；其中，小区产量与每穗总粒数、每穗实粒数之间，每穗总粒数与每穗实粒数之间，呈高度相关。涉及其余 5 个性状的相关性则明显降低：千粒重与其他性状之间，结实率与每穗总粒数之间，都不具

显著相关性。

2.3 回交 RILs 群体产量构成因素性状对单株产量的通径分析

有效穗、穗长、每穗总粒数、每穗实粒数、结实率、千粒重、单株产量的通径分析结果（表 4）表明，各性状对单株产量的直接通径系数大小顺序为：每穗实粒数＞千粒重＞单株有效穗数＞穗长＞结实率＞每穗总粒数。通径分析决定系数 $R^2 = 0.476\,042$，剩余通径系数 $Pe = 0.723\,849$。

表 4　回交 RILs 群体产量性状对单株产量的通径分析

作用因子	相关系数	直接作用	EP→HY$_{plot}$	PL→HY$_{plot}$	SPP→HY$_{plot}$	GPP→HY$_{plot}$	SLPP→HY$_{plot}$	PSS→HY$_{plot}$	GW→HY$_{plot}$	间接效应总和
EP	0.035 1	−0.059 7		−0.001 5	0.145 1	0.366 8	−0.286 3	−0.133 4	0.003 9	0.095
PL	−0.029 1	−0.109 2	−0.000 8		−0.639 1	0.668 0	0.120 7	−0.066 7	0.001 9	0.080 0
SPP	0.050 1	−1.299 3	0.006 7	−0.053 7		0.995 3	0.420 0	−0.018 9	0.000 1	1.349 0
GPP	0.648 6**	3.299 0	−0.006 6	−0.022 1	−0.392 0		−1.398 8	0.832 4	0.001 5	2.650 0
SLPP	−0.627 3**	1.655 5	0.010 3	−0.008 0	−0.329 6	−2.787 4		0.833 3	−0.001 6	2.283 0
PSS	0.643 2**	−0.874 9	−0.009 1	−0.008 3	−0.028 0	3.138 7	−1.576 9		0.001 7	1.518 0
GW	0.064 5	0.021 5	−0.010 9	0.009 4	0.003 3	0.231 7	−0.120 1	−0.070 4		0.044 0

分析表明，在产量构成因素中对小区产量有直接贡献的最大的为每穗实粒数（3.299）；产生负向直接效应最大的是每穗总粒数（−1.299 3），其次是结实率（−0.874 9）。每穗实粒数对小区产量产生的直接效应最大，且与小区产量呈显著相关，表明每穗实粒数是产量的重要作用因子，即通过每穗实粒数的选择可以达到提高产量的目的；但每穗实粒数的间接效应总和有较大负向性（−2.650 0），表明通过其他产量构成因素性状尤其是结实率，在产量贡献方面有一定的牵制作用，影响其对产量的贡献，因此在选择每穗实粒数时要兼顾其他产量构成因素性状。结实率对小区产量的直接影响具有较大负向效应，其对小区产量的影响主要是通过其他产量构成因素性状产生的间接效应，其中通过每穗实粒数的间接效应影响较大（3.138 7），使结实率最终与小区产量的相关性呈极显著正相关。每穗总粒数对小区产量的直接影响负向效应大，且与小区产量呈不显著相关关系，说明其不是影响产量的主要因素；但每穗总粒数对小区产量具有间接的影响，它通过每穗实粒数等其他产量构成因素性状产生的间接效应（0.995 3），对产量造成影响。

3 讨论

在 RILs BCF$_1$ 群体中，株系间呈现明显变异，但株系间性状相对稳定，各个性状的分离在一定区间呈连续分布（除千粒重集中在 $23 \sim 24$ g 之间外），表现为数量性状的分布特点。进一步分析各性状在群体中的遗传变异及分布情况，可进一步作为产量性状杂种优势分析、QTL 定位及分子遗传学研究的基础。

对于主要农作物重要农艺性状，前人已进行了方差分析、相关分析和通径分析等大量研究[7-10]，本研究对重组自交系 RILs F$_7$ 各株系产量与产量构成因子性状的相关分析表明，每穗实粒数、结实率对单株产量有较大促进作用，单株有效穗数、穗长、株高也对促进单株产量增加起作用，类似于马铮[11]、梁世胡[12]、谭酬志[13] 的研究结果。分析的情况表明，强调较好的穗粒结构，综合考虑早代单株选择的适当株高、适宜穗长的株型，可作为较好的亲本材料。RILs BCF$_1$ 群体各株系小区产量与产量的构成因子每穗实粒数、结实率存在极显著的正相关，与对 RILs F$_7$ 的分析相同，表明产量形成过程中，"库"大对形成较高的产量潜力有利；单株有效穗数与千粒重、穗长与每穗总粒数和每穗实粒数呈极显著相关，与超级杂交稻品种选育的技术路线，即在育种中选择中大穗的材料获得高产，有相近的遗传学意义；穗长与千粒重和小区产量、单株有效穗数与每穗总粒数等产量构成因子之间存在不同程度的负相关，且达显著水平，说明在产量潜力一定的前提下，水稻植株具有自行调节的能力。因此，并不能一味地追求大穗或多穗，应该有效地协调各产量构成因素，才能培育出高产的水稻品种。

本研究的结果认为每穗实粒数是对产量起积极作用的主要因素，结实率与每穗总粒数对产量的作用，均较大程度通过每穗实粒数间接影响，与马铮、梁世胡、粟学俊[14] 通径分析认为每穗实粒数对单株产量的直接作用大的观点相同；但千粒重对产量的影响不大，有效穗数、穗长不论直接还是间接的效应，均不构成对产量的主要影响作用，这与梁世胡、粟学俊分析认为千粒重对单株产量的直接作用大的观点不同，这是因为用以研究的 2 个群体均具有相同的遗传背景，各株系之间关系密切，而株系内性状相对稳定，千粒重变异程度最小。

粟学俊研究认为通径分析只能反映参试材料本身的产量构成特点，对整个杂交水稻并无共性，有些类型的杂交组合属于多穗型，有些属于大穗型，有些则属于大粒型，每一类型的杂交稻都有其独特的优势。本研究用通径分析的多元决定系数 $R^2 = 0.476\,042$，也说明了参与讨论的产量构成因素对产量的影响效应仅为 47.6%。各产量构成因素形成相互间制约，甚至一个产量因素的提高会削弱其他因素对产量的作用，这是高产育种复杂之所在。

424

References

参考文献

[1] 肖金华, 袁隆平. 水稻籼粳亚种间杂种一代优势及其与亲本关系的研究 [J]. 杂交水稻, 1988, 3 (1): 5-9.

[2] 袁隆平. 杂交水稻育种的战略设想 [J]. 杂交水稻, 1987, 2(1): 1-3.

[3] 牟同敏, 卢兴桂, 贺道耀. 水稻亚种间杂种 F_1 主要农艺性状的相关、通径和优势分析 [J]. 湖北农业科学, 1990, (11): 6-9.

[4] 覃惜阴, 陈彩虹, 黄英美. 水稻两系亚种间杂种优势及相关分析 [J]. 广西农业科学, 1995, (4): 145-149.

[5] 程新奇, 严钦泉, 周清明, 等. 水稻籼粳中间型 RIL 系主要农艺性状杂种优势分析 [J]. 海南大学学报(自然科学版), 2008. (2): 161-165.

[6] 杨振玉, 刘万友. 籼粳亚种 F_1 的分类及其与杂种优势关系的研究 [J]. 中国水稻科学, 1991, (4): 151-156.

[7] Xiao J, Li L, Ynan L, et al. Genetic diversity and its relationship to hybrid performance and heterosis in rice as revealed by PCR-based markers[J]. Theor Appl Genet, 1996, 92: 637-643.

[8] 程融, 孙明, 李成荃. 杂交粳稻品质性状的遗传研究 IV. 杂交粳稻品质与产量性状间的典范相关分析 [J]. 杂交水稻, 1995, 10(3): 28-30.

[9] 杨惠杰, 杨仁崔, 李义珍, 等. 水稻超高产品种的产量潜力及产量构成因素分析 [J]. 福建农业学报, 2000, (3): 1-8.

[10] 蒋利和. 杂交水稻主要性状的相关性及配合力研究 [D]. 南宁: 广西大学, 2002.

[11] 马铮, 霍二伟, 卢兆成, 等. 杂交水稻主要性状对产量的影响 [J]. 山东农业科学, 2006, (3): 21-23.

[12] 梁世胡, 李传国, 伍应运, 等. 杂交水稻产量构成因素的通径分析 [J]. 广东农业科学, 1999, (6): 4-6.

[13] 谭酬志, 邹小云, 熊春梅, 等. 籼型三系杂交水稻产量性状的相关与通径分析 [J]. 安徽农业科学, 2008, 36(29): 12 629-12 631.

[14] 粟学俊, 陈彩虹, 褟绮琳, 等. 杂交水稻产量构成分析与育种策略 [J]. 广西农业科学, 2002, (6): 283-285.

作者: 辛业芸　袁隆平

注: 本文收录于《第 1 届中国杂交水稻大会论文集》2010 年。

图书在版编目（CIP）数据

袁隆平全集 / 柏连阳主编. -- 长沙 ： 湖南科学技术出版社，2024. 5.

ISBN 978-7-5710-2995-1

Ⅰ. S511.035.1-53

中国国家版本馆 CIP 数据核字第 2024RK9743 号

YUAN LONGPING QUANJI DI-QI JUAN

袁隆平全集 第七卷

主　　编：柏连阳

执行主编：袁定阳　辛业芸

出 版 人：潘晓山

总 策 划：胡艳红

责任编辑：张蓓羽　任　妮　欧阳建文　胡艳红

责任校对：王　贝　唐艳辉

责任印制：陈有娥

出版发行：湖南科学技术出版社

社　　址：长沙市芙蓉中路一段 416 号泊富国际金融中心

网　　址：http://www.hnstp.com

湖南科学技术出版社天猫旗舰店网址：

　　　　　http://hnkjcbs.tmall.com

邮购联系：本社直销科 0731-84375808

印　　刷：长沙超峰印刷有限公司

　　　　　（印装质量问题请直接与本厂联系）

厂　　址：湖南省宁乡市金州新区泉洲北路 100 号

邮　　编：410600

版　　次：2024 年 5 月第 1 版

印　　次：2024 年 5 月第 1 次印刷

开　　本：889mm×1194mm　1/16

印　　张：27.75

字　　数：572 千字

书　　号：ISBN 978-7-5710-2995-1

定　　价：3800.00 元（全 12 卷）